Bumblebees

Their Behaviour and Ecology

Bumblebees

Their Behaviour and Ecology

DAVE GOULSON

School of Biological Sciences, University of Southampton

OXFORD
UNIVERSITY PRESS

Great Clarendon Street, Oxford OX2 6DP

Oxford University Press is a department of the University of Oxford.
It furthers the University's objective of excellence in research, scholarship,
and education by publishing worldwide in

Oxford New York

Auckland Bangkok Buenos Aires Cape Town Chennai
Dar es Salaam Delhi Hong Kong Istanbul Karachi Kolkata
Kuala Lumpur Madrid Melbourne Mexico City Mumbai Nairobi
São Paulo Shanghai Taipei Tokyo Toronto

Oxford is a registered trade mark of Oxford University Press
in the UK and in certain other countries

Published in the United States
by Oxford University Press Inc., New York

© Oxford University Press, 2003

The moral rights of the author have been asserted
Database right Oxford University Press (maker)

First published 2003

All rights reserved. No part of this publication may be reproduced,
stored in a retrieval system, or transmitted, in any form or by any means,
without the prior permission in writing of Oxford University Press,
or as expressly permitted by law, or under terms agreed with the appropriate
reprographics rights organization. Enquiries concerning reproduction
outside the scope of the above should be sent to the Rights Department,
Oxford University Press, at the address above

You must not circulate this book in any other binding or cover
and you must impose this same condition on any acquirer

British Library Cataloguing in Publication Data
Data available

Library of Congress Cataloging in Publication Data
Goulson, Dave.
Bumblebees : their behaviour and ecology / Dave Goulson.
Includes bibliographical references and index.
1. Bumblees–Behaviour. 2. Bumblebees–Ecology. I. Title.
QL568.A6 G69 2003   595.79′9–dc21   20021932051

ISBN 0 19 852606 7 (hbk)
ISBN 0 19 852607 5 (pbk)

10 9 8 7 6 5 4 3 2 1

Typeset by Newgen Imaging Systems (P) Ltd., Chennai, India
Printed in Great Britain
on acid-free paper by Biddles Ltd, Guildford & King's Lynn

Preface

With their large size and colourful furry coat, bumblebees are among the most familiar and recognizable of insects. They enjoy an enviable popularity compared to most insect fauna, for the buzz of foraging bumblebees is intimately associated in our minds with warm summer days and flower-filled meadows. They are widely recognized as being beneficial through their role in pollination, and bumblebees are most reluctant to mar their reputation by stinging; most species only do so when very hard pressed. Despite their familiarity, there is a great deal that we do not know about bumblebees. Many species are very hard to distinguish, rendering fieldwork difficult and discouraging amateur interest. Their nests are exceedingly hard to locate, so that those of some species have never been found. Fundamental aspects of the behaviour of many species, such as mating, have never been seen. Bumblebees are undergoing a widespread decline, but this has not yet caught the attention of the general public to the same extent as, for example, the plight of rare butterflies or birds. But bumblebees are probably of far greater ecological and economic importance than these groups because the pollination of crops and the survival of many wildflowers depends upon them. The intention in writing this book was in part to try to draw attention to the importance of conserving dwindling bumblebee populations. However, that was not my only motivation. Bumblebees have always been popular subjects for scientific study, but research has burgeoned in recent years. Many new discoveries have been made with regard to their ecology and social behaviour, but this information is widely dispersed in the literature. The last twenty years has also seen the commercialization of bumblebee breeding for pollination, and the invasion of new parts of the globe by bumblebee species, with potentially far-reaching consequences. Here I attempt to summarize the current state of knowledge of these fascinating and charismatic organisms, and identify some of the many gaps that remain in the hope of stimulating further research.

A plea for forgiveness is necessary at this point for I am sure that I have made numerous mistakes when attempting to synthesize and explain the work of others.

Acknowledgements

I am indebted to the work of others who long ago laid the foundations for the study of bumblebees. In particular *The Humble-bee* by Sladen (1912, reprinted in 1989), *Bumblebees* by Free and Butler (1958) and *Bumblebees* by Alford (1975) are invaluable reference works. Prys-Jones and Corbet (1991) provide an excellent and accessible introduction to the subject which helped to stimulate my interest in bumblebees. I am also grateful to Juliet Osborne, James Cresswell, Paul Williams, Mike Edwards, Ben Darvill and many others for invaluable discussions. Lastly I would like to thank my wife, Lara, for her unfailing support and encouragement.

Contents

1. Introduction — 1
   - 1.1 Evolution and Phylogeny — 2
   - 1.2 The Life Cycle — 4

2. Thermoregulation — 11
   - 2.1 Warming Up — 11
   - 2.2 Controlling Heat Loss — 13
   - 2.3 Thermoregulation of the Nest — 15

3. Social organization and conflict — 19
   - 3.1 Caste Determination — 19
   - 3.2 Division of Labour — 22
   - 3.3 Sex Determination — 30
   - 3.4 Control of Reproduction and Queen–Worker Conflicts — 31
     - 3.4.1 Timing of reproduction — 33
     - 3.4.2 Matricide — 35
   - 3.5 Sex-ratios in *Bombus* — 36
   - 3.6 Sex-ratios in *Psithyrus* — 39

4. Finding a mate — 41
   - 4.1 Territoriality — 41
   - 4.2 Nest Surveillance — 42
   - 4.3 Scent-marking and Patrolling — 42
   - 4.4 Inbreeding Avoidance — 45
   - 4.5 Evolution of Male Mate-location Behaviour — 46
   - 4.6 Queen-produced Sex Attractants — 46
   - 4.7 Monogamy versus Polyandry — 47

5. Natural enemies — 53
   - 5.1 True Predators — 53
   - 5.2 Parasitoids — 55
     - 5.2.1 Conopidae (Diptera) — 55
     - 5.2.2 Sarcophagidae (Diptera) — 57
     - 5.2.3 Braconidae (Hymenoptera) — 57
     - 5.2.4 Mutilidae (Hymenoptera) — 58
   - 5.3 Parasites and Commensals — 58
     - 5.3.1 Viruses — 58
     - 5.3.2 Prokaryotes (bacteria and others) — 58

5.3.3 Fungi	59
5.3.4 Protozoa	59
5.3.5 Nematodes	61
5.3.6 Mites (Acarina)	63
5.3.7 Other commensals	64
5.4 The Immune System of Bumblebees	64
5.5 Social Parasitism	65
5.5.1 Nest usurpation	65
5.5.2 Cuckoo bees (*Psithyrus*)	66

6. Foraging economics   69

7. Foraging range   73

7.1 Studies with Marked Bees	74
7.2 Homing Experiments	74
7.3 Harmonic Radar	77
7.4 Modelling Maximum Foraging Range	78
7.5 So Why do not Bumblebees Forage Close to their Nests?	79

8. Exploitation of patchy resources   83

8.1 The Ideal Free Distribution	84
8.1.1 Search patterns within patches	84
8.1.2 Non-random choice of patches	87
8.2 The Marginal Value Theorem	88

9. Choice of flower species   95

9.1 Flower Constancy	96
9.1.1 Explanations for flower constancy	97
9.1.2 Can flowers be cryptic?	102
9.2 Infidelity in Flower Choice	105

10. Intraspecific floral choices   107

10.1 Direct Detection of Rewards	107
10.2 Flower Size	108
10.3 Flower Age	108
10.4 Flower Sex	109
10.5 Flower Symmetry	110
10.6 Other Factors	110

11. Communication during foraging   113

12. Competition in bumblebee communities   123

13. Bumblebees as pollinators — 129
    13.1 Pollination of Crops — 129
        13.1.1 Honeybees versus bumblebees — 130
        13.1.2 Approaches to enhancing bumblebee pollination — 132
    13.2 Pollination of Wild Flowers — 137
        13.2.1 Nectar robbing — 138

14. Conservation — 143
    14.1 Causes of Declining Bumblebee Numbers — 144
        14.1.1 Declines in floral diversity — 144
        14.1.2 Loss of nest and hibernation sites — 146
        14.1.3 Pesticides — 146
        14.1.4 Effects of habitat fragmentation — 148
    14.2 Population Structure — 149
    14.3 Why are Some Bumblebee Species Still Abundant? — 151
    14.4 Consequences of Declining Bumblebee Numbers — 152
    14.5 Conservation Strategies — 153

15. Bumblebees abroad; effects of introduced bees on native ecosystems — 161
    15.1 Competition with Native Organisms for Floral Resources — 163
        15.1.1 Effects on foraging of native organisms — 164
        15.1.2 Evidence for population-level changes in native organisms — 166
    15.2 Competition for Nest Sites — 169
    15.3 Transmission of Parasites or Pathogens to Native Organisms — 169
    15.4 Effects on Pollination of Native Flora — 170
    15.5 Pollination of Exotic Weeds — 172
    15.6 Conclusions — 174

References — 177

Index — 229

1
Introduction

Bees (Superfamily Apoidea) belong to the large and exceedingly successful insect order Hymenoptera, which also includes wasps, sawflies, and ants. There are currently approximately 25 000 known species of bee, belonging to over 4000 genera, and undoubtedly many more remain to be discovered. All bees are phytophagous, feeding primarily on nectar and pollen throughout their lives. While many other insects feed on nectar or pollen as adults, very few do so throughout their development. This is simply because pollen and nectar, although nutritious, are sparsely distributed in the environment, and immature insects cannot fly from flower to flower to collect them (they do not have wings). In bees, the adult females gather the food for their offspring, so that the offspring themselves do not need to be mobile. In fact, the larval stage is maggot-like and generally rather feeble, being defenseless and capable of only very limited movement; they are entirely dependant on the food reserves provided for them. To facilitate the gathering of floral resources the mouthparts of adult bees are modified into a proboscis for sucking nectar, and in many species the hind legs of females are modified for carrying pollen (Michener 1974).

As in the wasps (from which bees evolved), bee social behaviour spans a broad spectrum from solitary species to those that live in vast colonies containing tens of thousands of individuals. The social species are more familiar, and it is not widely appreciated that by far the majority of bee species are solitary. In terms of nest architecture and behaviour they are similar to many solitary wasps (the obvious difference being that wasps generally provision their nests with animal prey). Some bee species within the Halictidae and Anthophoridae exhibit primitively social behaviour, living in small colonies in which the females may switch between roles as workers or queens. Approximately 1000 bee species are classed as eusocial (having a non-reproductive caste), although the distinction between primitively social species and eusocial species is sometimes blurred. The most advanced eusocial bees are all within the Apidae, notably *Apis* and the tropical stingless bees (Meliponinae).

Bumblebees (which also belong to the Apidae) are often described as primitively eusocial, since their social organization is said to be simpler than that of the honeybee. Unlike the Meliponinae and *Apis*, most bumblebee species have an annual cycle, with queens single-handedly founding nests. However, some tropical species of bumblebee initiate new colonies by swarming, in a very similar way to honeybees (Garófalo 1974). Temperate species exhibit nest homeostasis (Alford 1975), and have recently been found to recruit nestmates to food sources (Dornhaus and Chittka 1999), attributes normally associated with advanced sociality. Thus the tag of 'primitively eusocial' is probably misleading.

Bumblebees are all fairly large compared with the majority of bee species (or indeed most other insects), and most are covered in dense fur. Due to this combination of size and insulation bumblebees are capable of endothermy, and they are well adapted for activity in cool conditions (Heinrich 1993). It is thus not surprising that bumblebee are largely confined to temperate, alpine and arctic zones. They are found throughout Europe, North America, and Asia (Plate 1). They become scarce in warmer climates such as the Mediterranean, although atypical species are found in the lowland tropics of south east Asia and Central and South America. The mountain chains running through North and South America have allowed these primarily northern temperature organisms to cross the equator, and moderate species diversity is to be found in the Andes. In the Himalayas, they are generally only found at altitudes above about 1000 m rising to 5600 m (Williams 1985a). Species richness peaks in the mountains to the east of Tibet and in the mountains of central Asia (Williams 1994). In Europe, species richness tends to peak in flower-rich meadows in the upper forest and subalpine zones (Rasmont 1988; Williams 1991).

1.1 Evolution and Phylogeny

It is widely accepted that the bees probably first appeared in the early cretaceous approximately 130 million years ago, in association with the rise of the angiosperms (Milliron 1971; Michener 1979; Michener and Grimaldi 1988). Bees evolved from predatory wasps belonging to the Sphecoidea, and indeed primitive bees can be difficult to distinguish from Sphecoid wasps. The earliest known fossil bee is of *Trigona prisca* (Meliponinae), found in amber dating from 74 to 94 mya (Michener and Grimaldi 1988). However, this is an advanced eusocial species so it is reasonable to suppose that a great deal of bee evolution occurred in the 50 million years from the beginning of the Cretaceous to the time when this fossil lived (Michener and Grimaldi 1988).

The earliest fossils attributed to *Bombus* date from the Oligocene (38–26 mya), but we do not know when the group arose (Zeuner and Manning 1976). Inevitably, the fossil record for bumblebees is exceedingly sparse, for such large insects are unlikely to be caught in amber. It seems likely that bumblebees arose in Asia, since this is still the area of greatest bumblebee diversity. Bumblebees probably dispersed westwards from Asia through Europe to North America and finally to South America (Williams 1985a).

A full review of the phylogenetic relationships of bumblebees is beyond the scope of this book. In any case the accepted phylogeny is constantly changing as information becomes available on more species, and new techniques are developed and applied to bumblebees (particularly molecular methods). I provide a brief overview since some taxonomic groupings are based on behavioural and ecological attributes, and we would certainly expect more closely related species to behave more similarly.

The world bumblebee fauna consists of approximately 250 known species, and it is reasonable to assume that the majority of species have now been discovered (unlike most other invertebrate taxonomic groups) (Williams 1985a, 1994, 1998; Pedersen 1996). Recent classifications place all of these species in a single genus *Bombus* (meaning 'booming'). The majority of these species are known as 'true' bumblebees, and have a social worker caste which is more or less sterile (they cannot mate but can laying unfertilized eggs that develop into males). The remaining 45 or so species are known as cuckoo bumblebees, and belong to the subgenus *Psithyrus* (meaning 'murmuring'). These are inquilines that live within the nests of the true bumblebees (they are often described as parasites but strictly speaking this is not accurate, since they do not feed upon their hosts, but only on the food gathered by their hosts). The consensus of most recent authors is that cuckoo bees (subgenus *Psithyrus*) have a monophyletic ancestry (Plowright and Stephen 1973; Pekkarinen et al. 1979; Ito 1985; Williams 1985a, 1994; Pamilo et al. 1987). There is considerable disagreement as to which *Bombus* species are most closely related to *Psithyrus* (discussed in Koulianos and Schmid-Hempel 2000).

Various subdivisions of the genus *Bombus* have been attempted, many of which have subsequently been discarded. Bumblebee taxonomy is notoriously tricky because as a group they are morphologically 'monotonous' (Michener 1990). Early classifications depended heavily on coat colour patterns (Dalla Torre 1880, 1882), but these are now generally regarded as being of limited value, particularly since most species exhibit considerable colour variation both within and between populations, and also because there often seems to be convergent evolution of coat colour driven my Müllerian mimicry (Plowright and Owen 1980; Williams 1991). Such is the confusion in bumblebee nomenclature that there are on average 11 synonyms for each currently recognised species, with *B. lucorum* having over 130.

Classifications based on male genitalia (Krüger 1917; Skorikov 1922) have proved to be more useful. Krüger (1920) split the genus *Bombus* into two sections, Odontobombus and Anodontobombus, based upon the presence or absence of a spine on the middle basitarsus of females. This division is useful since it now appears to broadly correspond to behavioural differences. Sladen (1899) and Sakagami (1976) recognized two groups of *Bombus*, the pocket makers and pollen storers, based on the method used to feed larvae, and these correspond roughly to the Odontobombus and Anodontobombus, respectively. However, data on rearing behaviour is lacking for the majority of *Bombus* species. Recent molecular studies suggest that both the Odontobombus and the Anodontobombus are paraphyletic groups (Pekkarinen et al. 1979; Pamilo et al. 1987; Pedersen 1996; Koulianos and Schmid-Hempel 2000), so clearly this subdivision of the genus *Bombus* requires revision.

Further, the genus *Bombus* is traditionally divided into approximately 50 subgenera, but some of these are also of dubious validity. For example, the subgenus *Pyrobombus*, which contains many of the better-studied North American and

European species, is now thought to be paraphyletic, based on both morphological and molecular studies (Plowright and Stephen 1973; Pederson 1996; Koulianos 1999). There are numerous discrepancies between classifications based upon morphology, behaviour and molecular data, and these will only be adequately resolved when each of these data are obtained for a much larger number of species than are currently available.

1.2 The Life Cycle

Detailed descriptions of the life cycle of bumblebees have been given elsewhere (notably in Alford 1975), and are repeated in brief here. In general, *Bombus* species have an annual life cycle. Queens emerge from hibernation in late winter or spring, and at this time of year they can often be seen searching for suitable nest sites. The timing of emergence differs markedly between species; some, such as *B. terrestris*, emerge in February or March while others, such as *B. sylvarum*, emerge as late as May or June (Alford 1975; Prys-Jones 1982). Most temperate species emerge gradually over several months, but arctic and subarctic species such as *B. frigidus* tend to emerge synchronously, within 24 h of the first appearance of willow catkins (Vogt et al. 1994). Presumably this is an adaptation to the very short season in these regions, in which late emerging queens would not have time to rear a colony.

The sites chosen for nesting also vary between species, both in terms of the habitat type in which they are located (Richards 1978; Svensson et al. 2000), and in their position (Alford 1975). Some species always nest underground using pre-existing holes, very often the disused burrows of rodents (e.g. *B. lucorum*, *B. terrestris*). Other species such as those belonging to the subgenus *Thoracobombus* nest on or just above the surface of the ground within tussocks of grass or other dense vegetation, and again tend to use abandoned summer nests of small mammals. A few species such as *B. pratorum* are opportunistic, employing a variety of nest sites both above and below ground, including old birds' nests, squirrels' dreys, and artificial cavities. All species require a supply of moss, hair, dry grass or feathers from which they form the nest. These materials are arranged into a ball within which is a central chamber with a single entrance.

The queen provisions the nest with pollen, and moulds it into a lump within which she lays her eggs. Generally between 8 and 16 eggs are laid in this first batch. The pollen lump is covered on the outside with a layer of wax (secreted from the ventral abdominal surface of the queen) mixed with pollen. The queen also forms a wax pot by the entrance to the nest, in which she stores nectar. She incubates her brood by sitting in a groove on top of the pollen lump, maintaining close contact between the lump and her ventral surface (Figure 1.1). Queens generate a great deal of heat during this period, maintaining an internal temperature of 37–39 °C, which enables them to maintain a brood temperature of about 30–32 °C (Heinrich 1972a,b). The eggs hatch within about 4 days, and the young larvae consume the pollen. At this early stage they live together within a cavity inside the pollen,

Figure 1.1. Queen of B. lapidarius incubating the brood clump in her newly founded nest. Incubation is energically expensive. The nectar pot is placed just in front of the queen so that she can replenish her energy reserves without losing contact with the brood clump.

known as the brood clump. As well as incubating the brood the queen has to forage regularly to provide a sufficient supply of pollen and replenish her nectar reserves. It seems probable that this is one of the most delicate stages of the bumblebee life cycle, when a shortage of forage or inclement weather could cause the young queen and her colony to perish.

Bumblebees can be divided into two groups according to the way that the larvae are fed. In the so-called 'pocket makers' (Section Odontobombus Krüger), fresh pollen is forced into one or two pockets on the underside of the growing brood clump, forming a cushion beneath the larvae on which they graze. The larvae continue to feed collectively. In the later stages of larval development, the queen pierces holes in the wax cap over the clump and regurgitates a mix of pollen and nectar onto the larvae. In the 'pollen-storers' (Section Anodontobombus Krüger), the brood clump breaks up and the larvae build loose individual cells from wax and silk within which they live until they pupate. They are fed individually for most of their development on regurgitated pollen and nectar.

The larvae have four instars. After approximately 10–14 days of development they spin a strong silk cocoon and pupate. It takes a further 14 days or so for the pupae to hatch, so that the total development time is about 4–5 weeks, depending on temperature and food supply (Alford 1975). The queen continues to incubate the growing larvae and pupae, but those near to the center of the brood clump are kept warmer than those on the periphery. As a result they grow larger and emerge slightly before larvae that develop on the outside. When the first batch of larvae pupate the queen will generally collect more pollen and lay further batches of eggs. When the pupae hatch, the adults must bite their way out of the cocoon, often aided by the queen. Newly eclosed bumblebees are white in appearance; they develop their characteristic coloration after about 24 h. The first batch of offspring are almost invariably workers.

Within a few days of their emergence the queen ceases to forage, presumably because this is a hazardous occupation. This duty of taken over by some of the new workers, while others help her tend to the developing broods.

From this point onwards nest growth accelerates; the nest can increase in weight by tenfold within 3–4 weeks (Goulson et al. 2002a) (Figure 1.2). Several more batches of workers are usually reared, although the size to which the nest grows varies greatly between species. Estimates of worker longevity also vary between species and between studies, from 13.2 days for B. terricola to 41.3 days for B. morio (Chapter 5). Foragers have a shorter life expectancy than nest bees (Chapter 3). Surplus pollen and nectar may be stored in the empty cocoons from which workers have emerged. The temperature of the nest is regulated (Chapter 2); considerable heat can be generated by the workers if necessary, and they keep the brood warm by pressing their bodies against it. They may also ventilate the nest by fanning their wings near the entrance. Prior to emergence of the workers, Cumber (1949a) reported temperatures of 20–25 °C in the nest cavity, increasing to 30–35 °C at the height of nest development. Temperature fluctuations are also greater during early stages of colony development, with variation by no more than about 2.5 °C once many workers are present (Hasselrot 1960).

The failure rate of colonies is high. For example of 80 nests of B. pascuorum in southern England followed by Cumber (1953) only 23 produced any new queens (a further 9 produced only males). Similarly, of 36 B. lucorum nests placed out in the field by Müller and Schmid-Hempel (1992b), only 5 produced queens. Colonies may die out for many reasons; for example because of high rates of parasitism, or they may be destroyed by predators (e.g. badgers) or agricultural practices (e.g. mowing for hay). Availability of a succession of suitable flowers is also vital; Bowers (1985a) found that colonies frequently died out if founded in particular subalpine meadows with a low availability of flowers.

If the nest attains sufficient size, at some time between April and August, depending on the species, the nest switches to the rearing of males and new queens. Some species such as B. polaris that live in the arctic where the season is very short rear only one batch of workers before commencing production of reproductives (Richards 1931). In contrast colonies of B. terrestris can grow to contain up to 350 workers (Goulson et al. 2002a). The duration of nest growth and the size that it attains is not just determined by climate. Within any one region a range of different strategies can be found. In Europe, B. pratorum and B. hortorum nests last for about 14 weeks from founding, compared to about 25 weeks for the sympatric B. pascuorum (Goodwin 1995). In general, no more workers are reared once the colony switches to producing reproductives. The main factor that triggers the switch is thought to be the density of workers in the nest, or perhaps more specifically the ratio of workers to larvae, although it is probably under the control of the queen (Chapter 3). Developing queens require more food over a longer period than worker larvae, so they can only be produced if sufficient food is available, and if there are sufficient workers to feed the larvae. Nests are founded over a prolonged period in spring, but the production of new queens and drones appears to be approximately

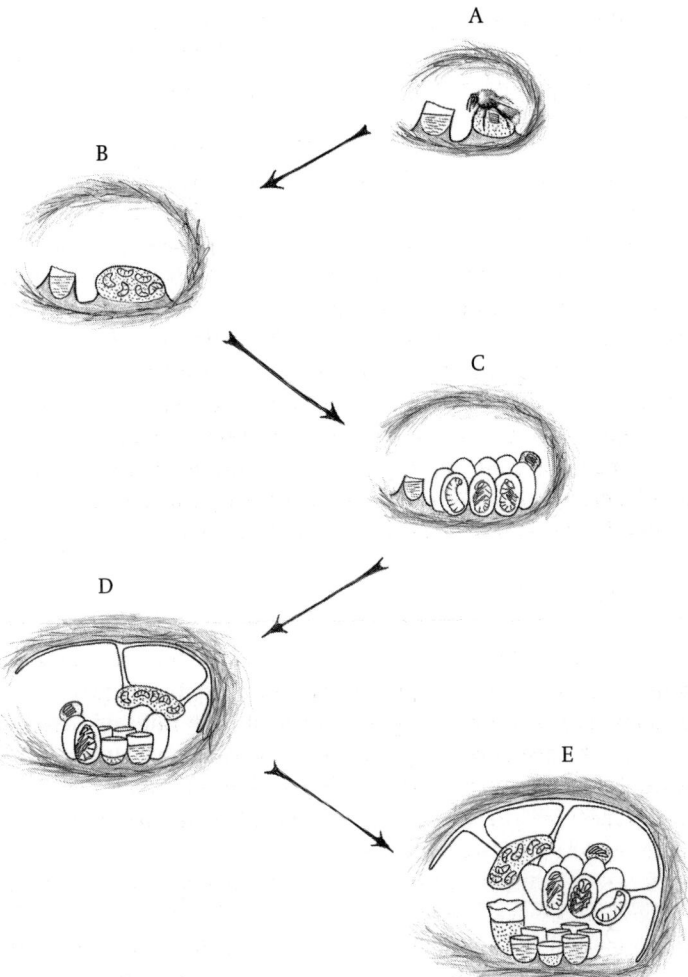

Figure 1.2. Nest development of a generalized *Bombus* species. (A) The queen founds a nest within a ball of dry grass, moss, and animal hair. She constructs a single nectar pot, and lays her first batch of eggs within a brood clump of pollen mixed with nectar and surrounded by a layer of wax. (B) The eggs hatch and the larvae consume the brood clump. The queen alternates incubating the brood with foraging for further nectar (to fuel incubation) and pollen (for the growing larvae). (C) As they near pupation the larvae spin individual silken cells, and cease to feed. Those near the center of the brood tend to pupate first. Once her first batch of larvae cease to feed, the queen will lay another batch of eggs in a brood clump constructed on top of the pupal cells (top right). (D) The first workers emerge. They take over foraging, and also aid the queen in caring for further batches of brood. Old pupal cells are recycled as further nectar pots. A wax cover is often constructed over the nest. (E) The nest grows rapidly as the work force expands. Surplus pollen may be stored in specially constructed tall cells (left). After a variable number of worker broods have been reared, the nest switches to production of new queens and/or males.

synchronized (which means that late-founded nests have shorter durations) (Pomeroy and Plowright 1982; Müller and Schmid-Hempel 1992a).

In Hymenoptera, the males are haploid and females are diploid, so males are produced from unfertilized eggs. This means that the queen can control the sex of her offspring. Workers may also lay eggs, but because they have not mated any eggs that they lay must be male. At the point when the colony switches to rearing of reproductives, some workers often lay eggs, but it is not known what proportion of males are fathered by workers. Owen and Plowright (1982) estimated that 19% of males were the offspring of workers in *B. melanopygus*, but a recent study of *B. hypnorum* (a potentially atypical polyandrous species) found that all males were produced by the queen (Paxton et al. 2001). The number of males and queens reared by a colony varies greatly, and is largely determined by nest size; small nests may rear no reproductives. Moderate-sized nests often rear only males, whilst only the largest nests produce both males and queens (Schmid-Hempel 1998).

The young queens leave the nest to forage, returning at intervals and at night, but they do not provision the nest. They consume large quantities of pollen and nectar, and build up substantial fat reserves. Males play no part in the life of the colony; after a few days they leave, never to return. Once they have left the nest, the males occupy themselves with feeding on flowers (often rather sluggishly), and with searching for a mate (Chapter 4). The mate-location behaviour is unusual. In most *Bombus* species, males deposit pheromone in a number of places in the early morning, choosing leaves, prominent stones, fence posts, or tree trunks. They then patrol these sites on a regular flight circuit during the day (Sladen 1912). Often a succession of males will adopt more or less the same route, so that a continuous stream of males can be observed at any one point. The pheromone is produced by the labial gland, and consists of a complex mixture of organic compounds, mainly fatty acid derivatives and terpene alcohols and esters (Kullenberg et al. 1973). Each bumblebee species employs a different blend, and the scents of some species are readily detectable by the human nose (Sladen 1912). Different species also patrol at different heights, for example *B. lapidarius* tends to patrol circuits at tree-top level, while *B. hortorum* patrols within a meter of the ground (Bringer 1973; Svensson 1979). Presumably species-specific pheromones and distinct patrolling heights facilitate young queens in identifying a mate of the correct species. However, mating is rather rarely observed in the wild in bumblebees, and young queens have never been observed to be attracted to the pheromone-marked circuits of males (Alford 1975). Further studies are required to examine exactly where bumblebee courtship and mating usually takes place in natural situations.

Direct observation and dissection of queens suggests that in most bumblebee species they mate only once (Röseler 1973; Sakagami 1976; VanHonk and Hogeweg 1981). This has been confirmed by molecular studies of a range of European bumblebee species which demonstrated that the offspring of a single queen are usually full siblings (Estoup et al. 1995; Schmid-Hempel and Schmid-Hempel 2000). However, queens of some species including *B. hypnorum* and *B. huntii* do mate up to three times (Hobbs 1967a; Röseler 1973; Estoup et al. 1995). After mating, young queens may continue feeding for a while but before long they begin to search for suitable hibernation sites. As with nest sites, preferences vary between species, but generally queens in the UK are said to prefer north-facing banks with loose soil (Alford 1975). In contrast, subarctic species probably prefer south-facing sites where snow melts first, so that they are stimulated to emerge from hibernation as soon as conditions are favourable (Vogt et al. 1994).

Once they have found a suitable site, the queen rapidly digs down a few centimeters (again, the preferred depth varies between species) and forms a small oval chamber in which she will remain until the following spring. They survive during this long period of inactivity on substantial fat reserves that fill their abdominal cavity; queens that have not laid down sufficient reserves will perish (e.g. in *B. terrestris* the critical weight is about 0.6 g, Beekman et al. 1998). This period of dormancy may begin as early as June in some species, and so it is misleading to refer to it as hibernation.

Once the males and young queens have departed, the nest rapidly degenerates. The remaining workers are old and become lethargic. The foundress is usually worn out and expires. Parasites and commensals consume what remains of the comb.

It has long been suspected that some species, such as *B. jonellus*, *B. pratorum*, and *B. frigidus*, may sometimes have more than one generation per year (Alfken 1913; Hobbs 1967a; Douglas 1973; Alford 1975). Their colonies typically come to an end rather early, in about May, yet sometimes fresh workers are seen foraging late in the summer. Whether these are the result of new queens taking over their mother's nest, or founding new nests of their own has not been established, but Alford (1975) deems the former to be more likely.

There appear to have been some changes in the life cycle of *B. terrestris* in recent years. In New Zealand, where the species is not native, nests can persist through the winter (Cumber 1949b), presumably because the climate is milder that in England (the origin of the New Zealand population). In North Africa and Corsica, this species is active mainly in the winter (Ferton 1901; Sladen 1912), demonstrating that it possesses considerable phenological flexibility. In 1990, workers of *B. terrestris* were found in January and February in Devon (south west England) (Robertson 1991). More recently, *B. terrestris* appears to have become be more-or-less continuously brooded in the Southampton area (southern England); I have observed queens founding nests in December, and workers are seen all winter during warmer weather. Authoritative works on bumblebees such as Sladen (1912) and Alford (1975) make no reference to this, suggesting that it is probably a recent phenomenon. There are few or no native flowers available at this time of year; all visits are to exotic garden plants. It is presumably no coincidence that these observations are confined to the south of England, where the winters are mild. This switch to continuous generations may have been favoured by changes in the climate, and by the availability of exotic flowers providing nectar and pollen through the winter.

The small number of bumblebee species that live within the lowland tropics of South East Asia and South America have atypical life histories. There is no annual cycle, and nests can reach a very large size and contain several thousand workers (Michener and Laberge 1954; Michener and Amir 1977; Brian 1983). As many as 2500 new males and queens can be produced by a single nest of *B. incarum* in Brazil (Dias 1958). In the Brazilian species *B. atratus*, new queens supercede the foundress, and new colonies may be initiated by swarming in the same way as honeybees (Garófalo 1974).

Cuckoo bumblebees (subgenus *Psithyrus*) have annual life cycles similar to those of typical temperate bumblebee species, except that instead of founding their own nest and rearing workers, they steal a nest from a 'true' bumblebee (Chapter 5). *Psithyrus* females emerge later from hibernation, and search for young nests of other *Bombus* species (strictly speaking female *Psithyrus* are not queens because there is no worker caste). Once located, they enter the nest, kill the queen, and take over her role. The bumblebee workers continue to forage and tend to the brood. The *Psithyrus* female lays eggs that develop into either new breeding females or males. Mate location behaviour and hibernation are similar to other *Bombus* species.

2
Thermoregulation

As recently as the 1960s it was widely believed that insects were all essentially ectothermic, so that their body temperature remained close to ambient temperature unless they used external heat sources (generally solar radiation) to heat themselves. Thanks largely to studies of North American moths and bumblebees carried out by Bernd Heinrich in the 1970s, this is now known to be very far from the truth (see particularly Heinrich 1979b). Although many insects, particularly the small species, are unavoidably ectothermic due to their large surface area to volume ratio, larger flying insects such as sphingid moths and bumblebees can generate considerable quantities of metabolic heat, and use this to maintain stable body temperatures many degrees above the ambient temperature. Indeed, they would be entirely unable to fly without this ability. Much of what follows is based on the work of Heinrich. Readers wishing to know more are directed to his excellent book 'Bumblebee Economics' (Heinrich 1979b), and to two more recent general texts on insect thermoregulation, 'The Hot-blooded Insects' (Heinrich 1993) and 'The Thermal Warriors' (Heinrich 1996).

2.1 Warming Up

At rest, bumblebees generally have an internal temperature close to ambient. In the temperate regions where most species live, ambient temperatures in the spring and summer generally fall within the range 5–25 °C. However, to generate the power needed for flight, bumblebees need to raise the temperature of their flight muscles to above 30 °C (sphingid moths operate at even higher temperatures around 47 °C) (Heinrich 1971). To do so, they generate heat through shivering the flight muscles, and probably also through substrate cycling in the flight muscles (Newsholme et al. 1972). In bumblebees, the upward and downward strokes of the wings are each driven by two sets of powerful muscles that in flight contract alternately. During warm-up, they contract at the same time, generating heat but little or no movement (Heinrich 1979b). As they warm, so the speed of contractions can increase, generating yet more heat. Balancing this, heat loss increases as the temperature difference between the thorax and the surrounding air (the temperature excess) increases. Flight can only occur when the muscle temperature reaches the minimum required for flight. The minimum varies greatly between species; some moths that fly in the winter can fly (albeit very weakly) with a thorax temperature of 0 °C (Heinrich and Mommsen 1985). In bumblebees the minimum is about 30 °C, although the optimum thorax temperature is probably closer to 40 °C (Heinrich 1972a,c,d, 1975a, 1993). It is possible that different bumblebee species, which vary in size, hairiness, and the climate to which they are adapted, have different minimum body temperatures at which flight can occur, but most species have not been investigated.

There is an alternative school of thought with regard to the source of heat generated during warm-up in bumblebees. Newsholme et al. (1972) argued that muscle-shivering is not necessary, and that bumblebees are able to burn sugars to generate heat in the flight muscles through substrate cycling. They demonstrated that a key enzyme in this process, fructose bisphosphatase, has unusually high activity in the flight muscles of bumblebees (Newsholme et al. 1972; Prys-Jones and Corbet 1991). In non-flying bumblebees the rate of substrate cycling is inversely related to ambient temperature, enabling the bees to maintain a stable internal temperature when inactive (Clark et al. 1973; Clark 1976). The amount of this enzyme that is present varies greatly between species, and levels appear to correlate with foraging behaviour: bumblebee species such as *B. lapidarius* with high enzyme activity tend to forage on large inflorescences (Newsholme et al. 1972; Prys-Jones 1986). It is proposed that while feeding on an inflorescence these species save energy by allowing their body temperature to drop. However, once the flower is depleted (or if they are attacked by a predator), they need to generate heat rapidly to take off, and they do so through substrate cycling. In contrast, species such as *B. hortorum* tend to feed on solitary flowers, and so when foraging they are almost continuously in flight. Since flight generates heat, they have less need for a rapid warm-up mechanism, and thus have lower enzyme levels.

While this argument is plausible and rather neat, Heinrich (1979b) pointed out that thermogenesis in the proven absence of muscle shivering had never convincingly been demonstrated in any insect, while warm-up in bumblebees always seems to be associated with flight muscle action potentials. Shivering in bumblebees is not externally visible, so it is actually quite hard to prove that they are not doing it. Surholt et al. (1990) attempted to do precisely this, by using a highly sensitive vibration monitoring system to detect muscle contraction in bumblebees during warm-up. They were unable to detect consistent shivering, although some usually occurred at the start of warm-up. In subsequent experiments (Surholt et al. 1991), they apparently demonstrated that the rates of substrate cycling were sufficient to account for observed levels of heat production in bumblebees. However, at about the same time, Esch et al. (1991) were performing an delicate experiment in which they mounted a tiny mirror onto the scutellum of *B. impatiens* onto which they shone a light. The reflected light was picked up using a photovoltaic cell partially obscured so that only a downward-pointing triangle of the cell surface was exposed. The tiniest movements of the scutellum (and the mirror) resulted in movement of the position of the light beam on the cell. Any upward movement would result in the light beam falling on a broader portion of the exposed triangle of the cell, generating more voltage. Conversely, downward movement produced less voltage. Using this set-up they demonstrated shivering during all stages of thermogenesis, as evinced by movement of the scutellum. Whether this experiment has finally laid to rest the substrate cycling hypothesis remains to be seen, for there is still the intriguing cross-species correlation between foraging behaviour and enzyme levels to explain.

Whatever the mechanism of thermogenesis, it is certainly true that bumblebees do generate considerable internal heat. However, there is a limit to the heat that they can generate, and thus there must be a lower limit to the ambient temperature at which they can fly. This limit is determined by the temperature excess that a bee can maintain, which in turn depends on the rate at which it can generate heat and the rate at which heat is lost. Heat balance of any organism can be described by the equation:

$$\frac{dH}{dt} = M - C(T_b - T_a)$$

The change in heat per unit time depends on the amount of heat that is produced (M), and the amount that is lost. The latter depends on the conductance of the body (C) and the temperature difference between the body temperature (T_b) and the ambient temperature (T_a). The amount of heat that can be generated is broadly determined by the muscle mass, which is linearly related to the mass of the bee. The conductance is strongly dependent on the surface area of the bee, and on the degree of insulation. Larger bees have a lower surface area to volume ratio, and thus we would expect them to be able to maintain a higher temperature excess (Stone and Willmer 1989). Bumblebees such as *B. polaris*, which are unusually large and well insulated, are capable of maintaining a temperature excess of 30 °C or more, and so can forage at ambient temperatures close to freezing (Vogt and Heinrich 1994). Similarly, queens of *B. vosnesenskii* and *B. edwardsii* can sustain continuous flight in ambient temperatures ranging from 2 to 35 °C. However, workers are considerably smaller and are unable to maintain an adequate body temperature for flight below 10 °C (Heinrich 1975a). Bumblebee workers vary considerably in size, and in general it is the larger workers that do most of the foraging (Goulson et al. 2002b). One likely explanation for this alloethism is that larger foragers can operate at lower ambient temperatures. They can thus begin foraging earlier in the day, and on cold days. They are also less likely to become grounded when out foraging should the temperature drop.

2.2 Controlling Heat Loss

Most endothermic vertebrates tend to maintain roughly even temperatures throughout their body, although the extremities may be a little cooler. In those insects that thermoregulate, body temperatures are generally very uneven. Large

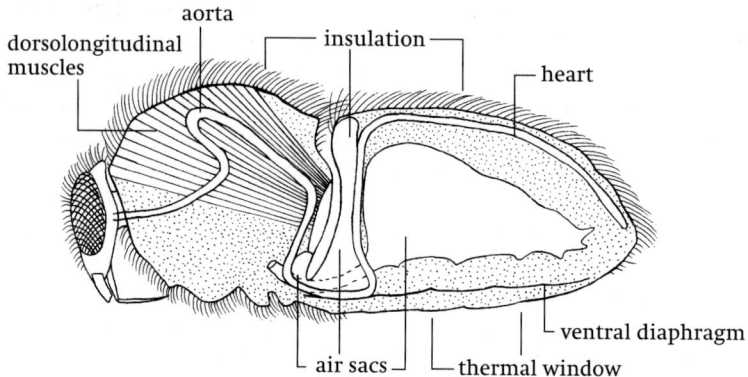

Figure 2.1. Diagrammatic longitudinal section of a bumblebee, showing features involved in thermoregulation (redrawn from Heinrich 1979a). Heat is produced by the flight muscles in the thorax. The thorax is well insulated on the outside with a dense furry coat, and heat loss to the abdomen is minimized by the narrow petiole, thought to act as a heat exchanger, and by insulating air sacs in the abdomen.

flying insects often maintain an elevated and stable thorax temperature, but the rest of the body may be substantially cooler. In flying bumblebees the abdomen is typically 10–15 °C cooler than the thorax (Heinrich 1979b). Heat loss from the thorax to the abdomen is reduced by the narrow waist (the petiole) separating the two, and by an insulating air sac in the anterior section of the abdomen where it contacts the thorax (Figure 2.1). However, the bumblebee heart pumps haemolymph forward from the abdomen to the thorax, from where it flows backwards through the body tissues to the abdomen. Without this flow of fluid to carry carbohydrates to the muscles, flight would not be possible for long. Yet, haemolymph circulation should lead to rapid heat transfer between the thorax and abdomen. Heinrich (1979b) suggested that the petiole acts as a countercurrent heat exchanger. Cool haemolymph in the heart flowing forwards from the abdomen, and in the petiole is forced into intimate contact with the warm haemolymph flowing backwards from the thorax. Inevitably heat will be transferred between the two, so that rather little heat is lost to the abdomen.

Just as there must be a minimum temperature at which insects can fly, so there is also a maximum. In bumblebees, the maximum thoracic temperature that they can tolerate is about 42–44 °C (Heinrich and Heinrich 1983a,b). Here, large size can act against an individual. Flight necessarily generates heat, so that a temperature excess is unavoidable. The larger the insect, the more heat is generated, and the less surface area (proportionally) is available through which to lose it. Thus queens and large foragers are liable to overheating at high ambient temperatures (Heinrich 1975a, 1979b). This presumably explains at least in part why most bumblebee species are found in cool climates.

At moderately high ambient temperatures, large insects such as bumblebees and dragonflies can avoid overheating by shunting heat from the thorax to the abdomen, which increases the surface area from which heat can be dissipated (Heinrich 1976c). If, as Heinrich (1979b) argues, the petiole acts as a countercurrent heat exchanger, how can this be achieved? The size of the aperture between the thorax and abdomen is controlled by the ventral diaphragm; when it contracts, the aperture widens. When the thoracic temperature approaches 44 °C (the approximate lethal limit), several marked physiological changes take place (Heinrich 1979b). Heart beat amplitude increases and the frequency halves, while the frequency of contraction of the diaphragm increases and steadies to match that of the heart. The abdomen also begins to pump at the same frequency (about 350 beats per minute). This leads to alternating pulses of haemolymph between the thorax and abdomen. As the abdomen expands, the diaphragm contracts, drawing a pulse of hot haemolymph from the thorax into the abdomen. As the abdomen contracts, and the heart beats, a pulse of cool liquid flows forwards into the thorax. During each pulse, little or no liquid flows in the opposite direction, so the heat exchange system ceases to operate.

At very low ambient temperatures, shunting heat from the thorax to the abdomen may serve a quite different purpose to avoidance of overheating. *B. polaris* is the northernmost social insect in the world, reproducing well within the Arctic circle. It is a large, unusually hairy bumblebee that is able to exist in regions where, even in the height of summer, ambient temperatures rarely exceed 5 °C (Vogt and Heinrich 1994; Heinrich 1996). As we have seen, all bumblebees have to maintain a high thoracic temperature to remain active. However, Vogt and Heinrich (1994) demonstrated that, unlike other bumblebees that inhabit temperate regions, queens of *B. polaris* also maintain a stable and elevated abdominal temperature (>30 °C). They found that this enables them to develop eggs within their ovaries quickly, something that is presumably important in the short arctic summer. Workers and males of this species have no eggs to develop, and their abdomens are substantially cooler.

2.3 Thermoregulation of the Nest

Depending on the latitude at which they live, bumblebee queens have approximately 2–7 months to found a nest, rear a force of perhaps several hundred workers, and then produce the next generation of reproductives. To compress this cycle into such a short space of time, the immature stages must be incubated to hasten their development. Heating of the abdomen prior to egg laying may be confined to species that inhabit cold climates, but heating of the abdomen to incubate the brood is found in all bumblebees that have been examined. Once the first batch of eggs has been laid the queen spends a considerable amount of time incubating them. She builds the brood clump with a groove on the dorsal surface in which she sits, allowing for close contact between the brood and the ventral surface of her abdomen and thorax (Heinrich 1974). While incubating she produces heat in her thorax, and distributes this to the abdomen by pulsing contractions of the abdomen (Heinrich 1979b). Heinrich (1974) found that *B. vosnesenskii* queens can maintain a brood temperature up to 25 °C above ambient temperature even in the absence of insulation. The amount of heat transferred to the brood is controlled by adjusting the rate of heat transfer from the thorax to the abdomen; in this way a stable brood temperature can be maintained under fluctuating ambient conditions.

Incubation is indoubtedly costly. Silvola (1984) estimated that a *B. terrestris* queen uses about 600 mg of sugar per day at temperatures typical for central Europe, and that to obtain this she may visit up to 6000 flowers. Of course in her absence the brood will rapidly cool, so availability of plentiful, rewarding flowers near to her nest is vital.

Incubation of the brood is aided by the nest site and construction. Queens of some species choose south-facing banks in which to nest, and build their nest above the soil surface where it is exposed to solar warming. Others nest underground, using the insulation provided by abandoned rodents nests. Whether nesting above or below ground, the queen uses the materials that are available to construct an insulated ball within which the brood is reared. As the nest grows this may be supplemented with a wax cap which traps warm air. Once workers are available, they too will incubate the brood. The more workers that are available, the more stable the nest temperature (Seeley and Heinrich 1981). In established nests, the temperature is remarkably stable at around 30 ± 1 °C. Active incubation may become unnecessary as a colony grows, since the activity of many bees can produce sufficient heat to warm the nest. Indeed, large colonies may overheat, at which point some workers switch to fanning the brood with their wings (Vogt 1986). At these times part of the wax cap may also be removed from the nest. Workers may also fan the nest in response to rising CO_2 levels (Weidenmüller *et al.* 2002).

The thermoregulatory capacity of established bumblebee nests is impressive. I once attempted to kill a commercial colony of *B. terrestris* by placing it in its entirety in a domestic freezer at -30 °C. After 24 h, I returned to find the colony alive and buzzing loudly; the workers had gathered into a tight clump over the brood and were presumably shivering at maximum capacity. The queen was hidden in their centre. Subsequent experience has shown that briefly anaesthetising the nest with CO_2 before placing it in the freezer is much more effective.

Although no workers appear to specialize entirely in nest thermoregulation, this task is adopted more readily by some bees than others. It seems that individual bees differ in the threshold at which they respond to either declining or rising temperatures. As nest temperature increases, bees with the lowest threshold for fanning behaviour begin to do so (O'Donnell and Foster 2001). If the nest continues to get warmer, bees with higher thresholds switch to fanning as well. Lifetime effort on fanning and incubation are positively correlated, so that bees that do one tend also to engage in the other. Bees with high thesholds may rarely if ever engage in thermoregulatory behaviour under natural conditions; these individuals presumably specialize in other tasks such as foraging. By containing a range of individuals with varying thresholds, the colony responds appropriately to thermal challenge, allocating effort to thermoregulation as required. What is not known is how the threshold for each bee is determined, and how a range of thresholds can be present among a group of very closely related workers.

Foragers of social insects such as bumblebees and honeybees have an advantage over solitary species with regard to warming up, for they can exploit the warm environment of the nest. Internal heat production is slow at low temperatures, so that it may take a long time for an bee to become warm enough to fly (and at very low temperatures they may be entirely unable to do so). Warming up is a costly activity, for during warm-up energy is being expended without any rewards being accrued. Thus the shorter the duration the better. Bumblebee nests are insulated and maintained at a temperature close to 30 °C through metabolic heat production, so that foragers have little trouble attaining flight temperature in the cool temperatures of early morning. In contrast, solitary species may be unable to forage until much later in the day.

3
Social organization and conflict

With their fat and furry appearance, their slow, meandering flight amongst flowers, and their docile behaviour, it is easy to dismiss bumblebees as charming but dim. Examination of a nest might confirm this opinion; it is, in appearance, a ramshackle affair compared to that of the honeybee. The pupal cells, honey pots, and larvae are haphazardly arranged, and the nest is often over-run with parasites and commensals. For these reasons, and because of the difficulties involved in finding bumblebee nests, researchers were slow to investigate the bumblebee social system. Bumblebee workers were considered to be generalists, each carrying out all tasks rather than dividing up the work in the efficient way that, for example, ants or honeybees do. Similarly, although the honeybee waggle dance has been known for many years, it was erroneously assumed that bumblebees did not communicate about sources of forage (Dornhaus and Chittka 1999). However, in recent years interest in the social life of the bumblebee has undergone a renaissance. Perhaps, in part because some species can now be bred in the laboratory (or the nests bought from commercial suppliers), in the last 10 years, bumblebees have been used for studies of diverse topics including queen–worker conflict, caste determination, polyandry and parasite resistance, and alloethism. This work is revealing that, despite their bumbling appearance, the social life of the bumblebee is every bit as complex as that of other eusocial insects.

3.1 Caste Determination

Bumblebees exhibit marked variation in size (Plate 3). Queens are the largest caste and, in pollen-storers such as *Bombus terrestris* (see Chapter 1) the size distribution of females is strongly bimodal, with little overlap between the size range of queens and that of other workers (Figure 3.1). However, size is not a reliable indicator of caste since, in some species, particularly the pocket-making species, there is considerable overlap (Plowright and Jay 1968). Structurally, queen and worker bumblebees are identical in all other aspects of their external morphology. The most striking difference between queens and workers is in the size of their fat deposits; workers have very little fat, particularly in their abdomen, leaving plenty of room for the honey stomach. In contrast, in young queens the abdomen is largely full of fat. This leads to queens being heavier for their size than workers (Richards 1946; Cumber 1949a).

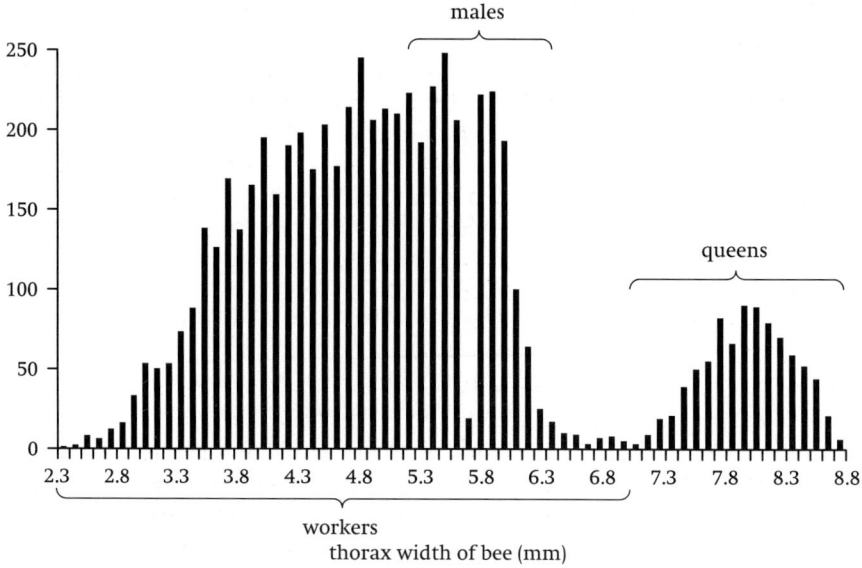

Figure 3.1. Frequency histogram of numbers of the different castes of *B. terrestris*, grouped by thorax width. Based on all the bees in 28 fully developed nests (*n* = 6371) (D.G. unpublished data).

So what determines whether a female bee becomes a worker or a queen? All eggs are capable of developing into either, regardless of when they are laid during colony development. Thus, even the very first batch of eggs laid in a nest can be induced to develop into queens (Sladen 1912; Free 1955c). In honeybees, prospective queens are fed royal jelly which is nutritionally richer than the food given to larvae destined to become workers (Haydak 1943; Brian 1965b). Some authors have suggested that in bumblebees there may be similar differences in the types of food fed to queens versus workers. Lindhard (1912) proposed that the diet of future queens was supplemented with masticated eggs, although this was never substantiated. After their initial period of feeding on pollen within the brood clump, larvae are fed on a mixture of pollen and nectar, combined with proteins secreted by the adult bees (Pereboom 2000). These proteins are probably mainly invertase and amylase produced in the hypopharyngeal gland (Palm 1949; Pereboom 2000). This mixture is regurgitated on to the larvae as a droplet. Ribeiro (1994, 1999) suggested that future queens receive additional glandular secretions, but these have not been identified and this remains speculative. In terms of the total protein, pollen and carbohydrate in the food mixture, larvae of all castes receive the same proportions (Pereboom 2000). In fact, nurse bees often feed queen, worker and male larvae in rapid succession using the same crop content (Katayama 1973, 1975). It thus seems unlikely that there can be qualitative differences in the food received by larvae of different castes.

There are differences in the way that sexual broods are fed in pocket-making species. Worker larvae are fed for most of their development on pollen deposited in pollen pockets. In comparison, male larvae and those destined to become queens are fed on regurgitated food from an earlier age (Alford 1975). Some authors have suggested that caste determination is simply a matter of how much food the larvae receive (Röseler and Röseler 1974; Alford 1975; Ribiero et al. 1999). Increasing the frequency of feeding makes larvae more likely to develop into queens in *B. pascuorum* (Reuter 1998), but not in *B. terrestris* (Pereboom 1997). Feeding rate is presumably dependent upon the ratio of workers to larvae, and this is strongly correlated with queen production in *B. terricola, B. perplexus,* and *B. ternarius* (Plowright and Jay 1968). However, measurement of growth rate of future queens versus workers revealed no difference in *B. terricola* (Plowright and Pendrel 1977) and, contrary to expectation, queens of *B. terrestris* developed more slowly during their early instars than workers of the same age (Ribiero 1994). This is clearly not what we would expect if future queens were fed more than future workers. Larvae that are to become queens are fed more frequently (Röseler and Röseler 1974; Alford 1975; Ribiero et al. 1999) but, as Pereboom (2000) points out, the period of rapid feeding in *B. terrestris* is after the point at which worker larvae have ceased to feed (i.e. caste has already been determined). It seems probable that some of these apparent anomalies are due to differences between species in the mechanisms involved in caste determination.

It is now generally accepted that, in *B. terrestris* at least (but perhaps not in pocket-makers such as *B. pascuorum*), caste is determined during two critical phases in the development of the larvae. The queen appears to excrete a pheromone to which larvae are sensitive at an age of about 2–5 days; if it is present they enter an irreversible pathway toward development as workers (Röseler 1970, 1991; Cnaani et al. 1997, 2000). If this pheromone is not present the larvae become queens, but only if they also receive sufficient food in their final instar (Röseler 1991; Cnaani and Hefetz 1996; Pereboom 1997, 2000; Ribeiro et al. 1999). The pheromone has not yet been identified, but the evidence for its existence is convincing. It seems that the pheromone is not airborne, but is transmitted either directly to the larvae by contact with the queen, or via workers. Röseler (1970) found that larvae separated from the queen by a fine mesh developed into queens, but if workers were regularly moved from the queen's side to the side the larvae were on, then the larvae developed as workers. Although the identity of the pheromone is not known, it seems that it probably acts by suppressing production of juvenile hormone, and low levels of juvenile hormone lead to larvae moulting earlier and at a smaller size. Topical applications of juvenile hormone to first or second instar larvae of *B. terrestris* results in them developing into queens (Bortolotti et al. 2001), and natural levels of juvenile hormone and ecdysteroids are higher in larvae destined to become queens than in larvae destined to be workers (Cnaani et al. 1997, 2000; Hartfelder et al. 2000).

Pheromone signals of this sort are probably not enforceable (Seeley 1985*a*; Keller and Nonacs 1993). If it were in the best interests of the larvae to develop as queens we would expect them to do so (Bourke and Ratnieks 1999). Perhaps, attempting to develop into a queen during the early stages of colony development is a poor strategy for a larva to adopt, for if insufficient workers are available to feed her then the prospective queen would be small, and small queens are likely to die

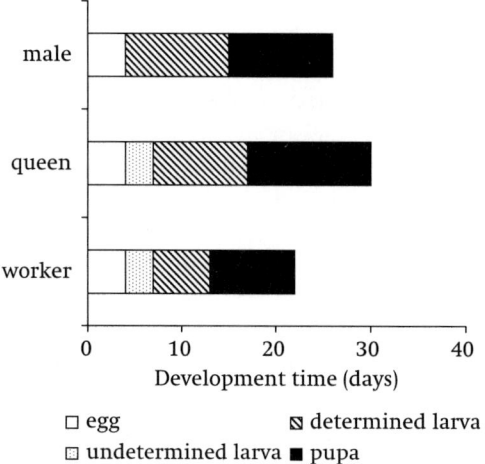

Figure 3.2. Development times of workers, males and new queens of *B. terrestris* (times taken from Duchateau and Velthuis 1988; see Shykoff and Muller 1995). At hatching, diploid larvae have the potential to become either workers or queens, but after about 3 days their pathway becomes determined.

during hibernation (Beekman *et al.* 1998). It seems likely that the pheromone signal from the queen is the best indication that the larvae have as to their optimal course of development.

Whatever the details of the mechanisms involved in caste determination, once caste has been determined, larvae that are destined to become queens enter a different developmental pathway and continue to feed for longer than those that become workers, thus attaining a much greater size (Cnaani *et al.* 1997). They also have a longer pupal development (Frison 1928, 1929; Röseler 1970) (Figure 3.2).

3.2 Division of Labour

Within the worker caste there is great variation in size, even within single bumblebee nests. For example, thorax widths of workers of *B. terrestris* range from 2.3 to 6.7 mm and body mass varies eightfold from 0.05 to 0.40 g (Figure 3.3). *B. terrestris* belongs to the Section Anodontobombus (also known as pollen storers); worker size variation is even greater amongst the Odontobombus (pocket making) species (Pouvreau 1989). Size variation of such magnitude is exceedingly rare in other insects, and is not found in other social bees; for example, workers of *Apis mellifera* are markedly uniform in size, particularly within single colonies. So why do bumblebee workers vary so greatly in size?

In the very first batch of workers reared by a queen, the only source of warmth is provided by the queen herself who incubates the brood. Larvae situated closest

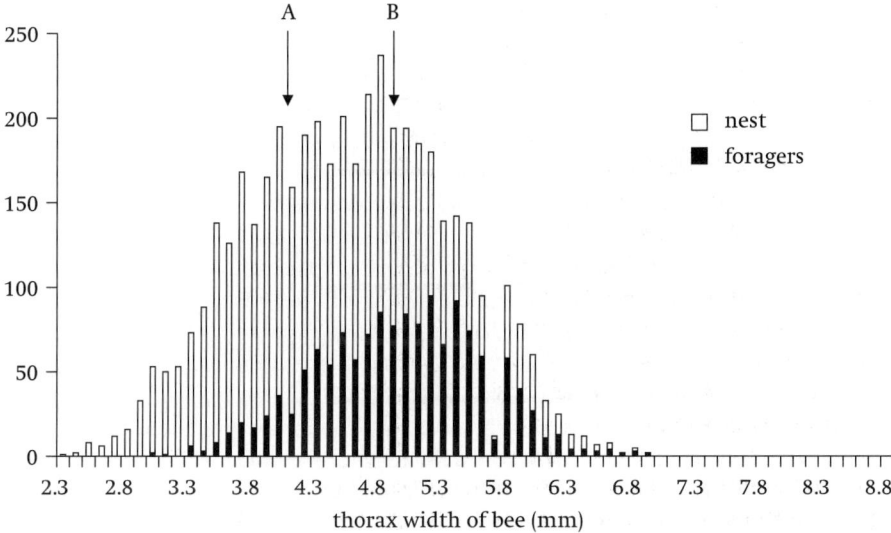

Figure 3.3. Frequency histogram of numbers of worker bees grouped by thorax width. Frequencies for bees caught in the nest ($n = 3077$) are stacked on top of those for foragers ($n = 1417$). The mean sizes of each group are indicated by arrows. From Goulson et al. 2002b.

to the incubation groove in which the queen sits tend to grow larger than those further away (Alford 1975). For subsequent broods the nest temperature is likely to be more even, since it is regulated by a number of workers. However, in pocket-making bumblebee species (Odontobombus), groups of larvae live and feed within a wax covered chamber on pockets of pollen provided by workers. It is likely that the position of larvae within the group affects how much food they receive (they may actually compete for food), so that some grow larger than others (Sladen 1912; Cumber 1949a). This would explain why workers of Odontobombus species vary more in size than do Anodontobombus species (Alford 1975; Pouvreau 1989). In Anodontobombus species, such as *B. terrestris*, larvae spend most of their development in individual silk cells, and are fed directly on droplets of nectar and pollen mixes, regurgitated by the adults directly on to the body of each larva (Alford 1975). The size attained by larvae is directly proportional to the amount of food they are given (Plowright and Jay 1968; Sutcliffe and Plowright 1988, 1990). Thus in this group of bumblebees the size of new workers is under the direct control of the bees rearing them (Ribeiro 1994). Yet, as we have seen, even pollen-storer workers exhibit an eight-fold variation in mass. It seems implausible that this is the result of sloppy parenting skills, the accidental neglect of some larvae at the expense of others. Given that larvae are reared in a controlled environment by a team of specialized workers, it seems far more likely that this size variation has an adaptive function, that colonies benefit from rearing workers of a range of sizes.

So what might this benefit be? The most obvious comparable instance of size variation in social insects occurs in some ant and termite species. Here, size is related to behaviour, with individuals of particular sizes specializing in particular tasks; a phenomenon known as alloethism. For example, in leaf-cutter ants of the genus *Atta*, the largest workers are soldiers, specializing in nest defence against mammals; medium-sized workers forage for food, while the smallest workers tend the fungus garden and initiate alarm responses along trails near the nest (Hughes *et al.* 2001).

Polyethism, the division of tasks among workers, is thought to be the key feature underlying the phenomenal ecological success of the eusocial insects (Wilson 1990). There is disagreement in the literature as to whether bumblebee workers exhibit polyethism. The traditional view is that they exhibit little specialization; they do not exhibit the clear age-based polyethism characteristic of honeybees (*Apis mellifera*), and workers regularly switch between foraging and performing tasks within the nest (Free 1955*a*; van Doorn and Heringa 1986; Cameron and Robinson 1990). However, this view is questionable; there is abundant evidence that bumblebee workers do exhibit polyethism. Young adults only perform within-nest tasks and are more likely to become foragers as they become older (Pouvreau 1989; O'Donnell *et al.* 2000; Silva-Matos and Garófalo 2000). Wax in bumblebees is secreted on the underside of the abdomen, beginning on the second day after adult emergence but declining after the first week (Röseler 1967*a*). Since wax is only required within the nest, young workers are predisposed towards nest maintenance tasks. In terms of age-related polyethism the only difference between honeybees and bumblebees is that, in bumblebees, the age at which individuals switch to foraging is variable and some workers never become foragers. Young foragers generally collect nectar and tend to switch to collecting pollen as they age (Free 1955*a*). Bumblebees probably do exhibit more behavioural plasticity than honeybees. Individuals can switch between tasks in response to colony requirements; for example, nest bees will switch to foraging if the foragers are experimentally removed or if nectar reserves are artificially removed (Kugler 1943; Free 1955*a*; Pendrel and Plowright 1981; Cartar 1989). Similarly, when nectar reserves are removed, foragers switch from pollen to nectar collection, and vice versa (Free 1955*a*; Cartar 1989; Plowright and Silverman 2000). Individual bees differ in the threshold level of resources within the colony to which they respond (van Doorn 1987; Cartar 1992*a*) (similarly, individuals differ in the threshold temperature at which they begin incubating or fanning the brood, O'Donnell and Foster 2001). Specialized foragers bring most food to the nest, while the majority of within-nest tasks are carried out by specialists (O'Donnell *et al.* 2000). Specialists are probably more efficient at their tasks; workers that are primarily foragers occasionally do within-nest tasks, but they do so much less quickly than specialized nest bees (Sakagami and Zucchi 1965; O'Donnell and Jeanne 1992; Cartar 1992*a*).

In addition to foraging and brood maintenance there is at least one other task that workers perform. Large nests of *B. lucorum* and *B. terrestris* generally have one, or occasionally two, guard bees that sit within the nest entrance and scrutinize foragers as they enter the nest (Free 1958). Using marked bees in *B. terrestris* colonies we have found that the same individual carries out this task for many days (D.G. pers. obs.). Thus bumblebee workers clearly do exhibit a range of behavioural specialisations.

Do these differences in behaviour relate to size (i.e. is there alloethism within bumblebee colonies)? It was long ago noticed that foragers of a range of bumblebee species appear to be larger, on average, than bees that remain in the nest (Colville 1890; Sladen 1912; Meidell 1934; Richards 1946; Cumber 1949a; Brian 1952; Free 1955a). Most recently, in samples of 4794 *B. terrestris* workers from 28 nests we have found that nest bees are consistently smaller than foragers (mean thorax widths 4.34 ± 0.01 and 4.93 ± 0.02 (mm $\pm$ SE), respectively) (Figure. 3.3) (Goulson et al. 2002b). All workers remain within the nest and engage in care of the brood for the first few days of life (much like honeybees) (Pouvreau 1989). It seems that the difference in average size between foragers and nest bees comes about because large workers tend to switch from within-nest tasks to foraging at an earlier age, while the very smallest workers never switch to foraging (Pouvreau 1989). Thus bumblebees do exhibit alloethism.

Why then do larger workers tend to be foragers? In leaf-cutter ants the explanation for alloethism is partially obvious; the large soldiers with their huge jaws are far better equipped to inflict damage on an attacking predator such as an armadillo than their smaller siblings (although the full explanation for alloethism in leaf-cutter ants is far more complicated than this, see Hughes *et al.* 2001 for a discussion). For bumblebees, it is not obvious why larger bees might be better suited to foraging.

Rather than trying to explain why foragers are large, Free and Butler (1959) suggested an explanation as to why nest workers should be small; they argued that they would be better able to manoeuvre within the cramped confines of the nest. In a test of this hypothesis, Cnaani and Hefetz (1994) experimentally manipulated the size of nest workers of *B. terrestris* and demonstrated that larvae reached a larger size when tended by large workers, compared to when they were tended by an equal number of small workers. However, this does not fully refute the hypothesis, for a fairer comparison would have been between larvae tended by an equal biomass consisting of either a few large workers or many small ones. One cannot help but suspect that, in this situation, many hands may well make light of the work and produce larger offspring. This experiment remains to be done, but even if it did find that small bees are advantageous within the nest, it would still not explain why large worker bees are reared at all. We need to demonstrate a positive advantage for large size in foragers.

A number of possible explanations for the larger size of foragers have been proposed. Free and Butler (1959) suggested that large workers could carry more forage. This is intuitively obvious and subsequent experiments have confirmed it to be true. If foraging *B. terrestris* are captured as they return to the nest, and the mass of forage measured, there appears to be a more-or-less linear relationship between the thorax width of the forager and the mass of forage they are carrying (Goulson *et al.*

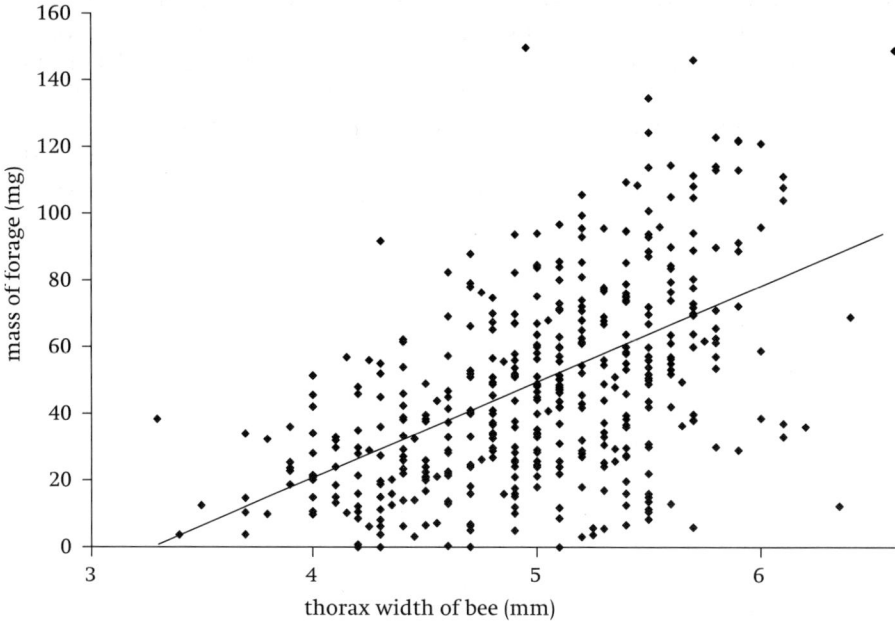

Figure 3.4. Relationship between mass of foraging *B. terrestris* workers and the mass of forage (pollen and nectar combined) that they bring back to the nest in a single foraging trip (from Goulson *et al.* 2002b).

2002b) (Figure 3.4). The mass of forage, calculated as a proportion of the body mass of the bee, did not vary with body size, with bees carrying on average a mass of pollen and/or nectar equivalent to 23.1% of their unladen body mass (remarkably, the heaviest load carried amounted to 77.1% of the unladen body mass of the bee).

Although larger foragers can carry more food, this does not explain why foragers are large. Presumably the cost to the colony of rearing a worker is approximately proportional to its size. For example, for every worker of 250 mg the colony could have reared two workers of 125 mg. The single large bee or the two small bees would each be expected to bring back about 58 mg of forage per trip, but it seems likely that the single large bee would take much longer to do so (for it would have to single-handedly visit twice as many flowers as each of the small bees). There is no evidence that larger bees are more rapid foragers, although few studies have examined foraging efficiency with respect to size. As noted earlier, Stout (2000) found that smaller workers of *B. terrestris* and *B. lapidarius* were better able to trigger the complex flowers of *C. scoparius* than were large workers and were also faster at handling flowers that had previously been triggered. Morse (1978b) found no differences in the foraging speeds of large and small workers of *B. vagans* foraging on *Vicia cracca*. To test whether large workers do bring back more food per unit time than small workers, we arranged *B. terrestris* nests so that bees leaving or entering the nest walked over the pan of a balance, enabling us to record both the duration of each foraging trip and the net mass gained during foraging (Goulson et al. 2002b). The data demonstrated that larger bees are more efficient foragers when collecting nectar, but not pollen (Figure 3.5). However, whether the greater efficiency of larger bees is sufficient to offset their greater rearing costs remains to be determined.

Why might larger bees be able to gather nectar more quickly? Pouvreau (1989) suggested that larger workers are at an advantage in foraging because they have

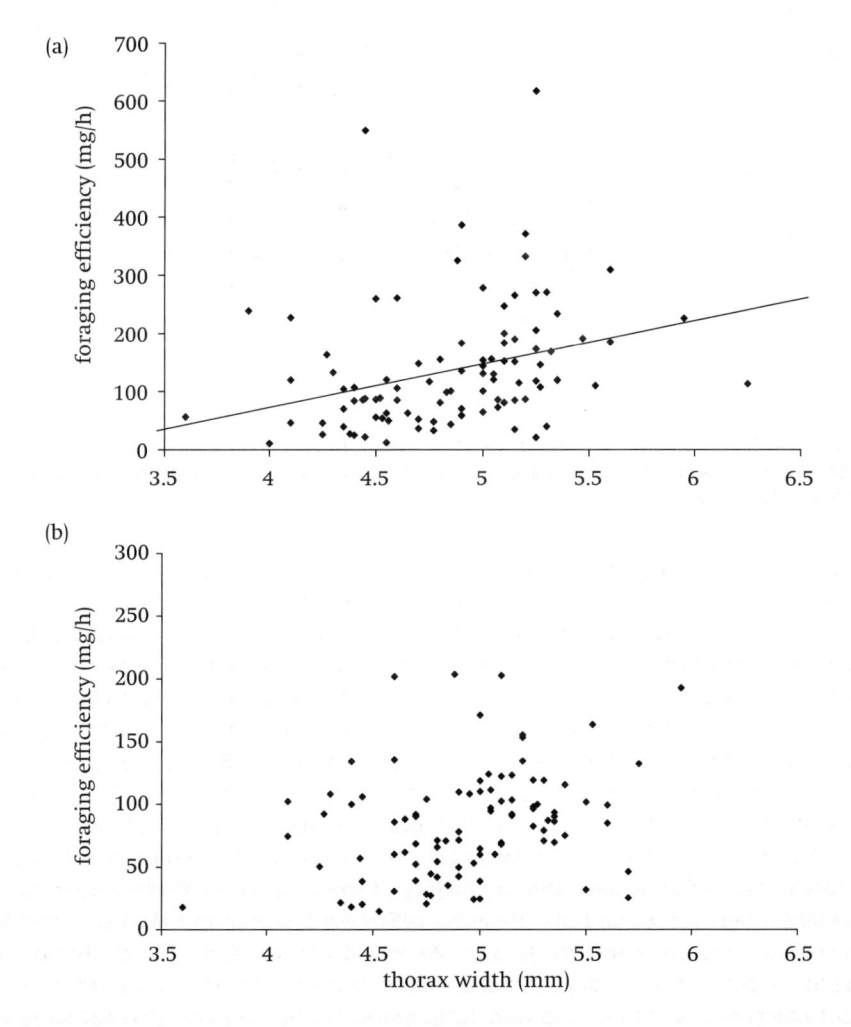

Figure 3.5. Relationship between foraging efficiency (mass of forage gathered per time spent foraging) and thorax width. Based on 98 *B. terrestris* foragers from 3 nests (from Goulson *et al.* 2002b). Trips were recorded as pollen gathering if pollen was visible in the pollen baskets of returning bees. (a) Nectar gathering trips. (b) Pollen gathering trips (in which nectar may also have been gathered).

longer tongues and so are able to feed on deeper flowers (the relationship between overall size and tongue length is proportional (Medler 1962a,b; Pekkarinen 1979; Goulson et al. 2002b)). However, having a long tongue is not necessarily an advantage. Bees with short tongues can forage more quickly on shallow flowers (Plowright and Plowright 1997). In fact the most common bumblebee species in the UK are all relatively short-tongued species, where most long-tongued species are on the brink of extinction (Edwards 1999). If having a long tongue provided an automatic advantage, we would expect foragers to have evolved longer tongues, not a larger size.

Another possibility is that larger bees may have greater visual acuity, and so be better able to find flowers. Visual acuity is likely to greatly affect search times (Spaethe et al. 2001). In *Cataglyphis* ants, the larger castes have been demonstrated to have greater visual acuity (Zollikofer et al. 1995), and the visual acuity of bumblebees is greater than that of honeybees (which are smaller) (Macuda et al. 2001). To my knowledge, intraspecific variation in acuity of bumblebees has not been examined.

Morse (1978a) suggested that large workers may be able to forage over greater distances. Large bumblebee species tend to go on foraging trips of longer duration than smaller species and thus may cover larger distances (Free 1955b). However, there have been no studies of the distance or duration of foraging trips in relation to size variation within species and, in general, very little is known of the foraging range of bumblebees (see Chapter 7). Similarly, we do not know how flight speed relates to size; if larger bees fly faster then this clearly would provide some advantage in foraging. The relative flight speeds of foragers in relation to their size have never to my knowledge been examined. Cresswell et al. (2000) calculated the upper limit of the foraging range of bumblebees to be about 10 km, this being the maximum distance from which a bee foraging for nectar could return with a net profit. The limit is imposed by the rate at which energy is burned on the flight back to the nest; this must be less than the total amount of energy that can be contained in the honey stomach. The energetic cost of foraging is approximately proportional to weight (Heinrich 1979b), and it has been shown that the amount of nectar that can be carried is proportional to weight. If the assumptions of Cresswell's model are correct then the maximum foraging distance of foragers should be independent of body size.

Yet another possibility relates to predation. Foraging is a dangerous task that probably increases worker mortality in social insects (van Doorn 1987; O'Donnell and Jeanne 1995; O'Donnell et al. 2000). Silva-Matos and Garófalo (2000) found that worker mortality in the tropical bumblebee *B. atratus* was strongly correlated with frequency of foraging. Similarly, longevity of workers of *B. diversus* is longer in queenless colonies in which little foraging occurs, compared to queenright colonies (Katayama 1996). Estimates of worker longevity vary between species and studies, from 13.2 days for *B. terricola* (Rodd et al. 1980) to 41.3 days for *B. morio* (see also Brian 1952, 1965a; Garófalo 1976; Goldblatt and Fell 1987; Katayama 1996). However, mark recapture studies of *B. vagans* and *B. terricola* visiting patches of flowers suggest that foragers rarely live for more than a week or two (Morse 1986b). Garófalo (1976) estimated the mean longevity of *B. morio* nest bees to be 72.6 days, compared to 36.4 days for specialist foragers. Foraging appears to reduce the ability of *B. terrestris* workers to encapsulate foreign bodies (suggesting that they have less resistance to parasitoids) (König and Schmid-Hempel 1995). It seems likely that larger bees are less prone to predation, particularly by spiders, than small bees. Conversely the conopid fly *Sicus ferrugineus*, which attacks bees while they are foraging on flowers, preferentially parasitises large workers (Schmid-Hempel and Schmid-Hempel 1996a). If, overall, large bees do have a longer life expectancy as foragers, then sending larger bees out to forage may be the safest option for the colony. No data is available on the longevity of foragers in relation to size; this would be an interesting and relatively straightforward area for study.

Perhaps the most promising candidate explanation for alloethism in bumblebees relates to thermoregulation. Free and Butler (1959) pointed out that larger workers would be better able to forage in adverse weather. All bees are limited to foraging within a particular temperature range and, in general, the lower limit of this range shifts downwards as body size increases (Stone and Willmer 1989). For example, queens of *B. vosnesenskii* and *B. edwardsii* can sustain continuous flight in ambient temperatures ranging from 2 to 35 °C, but workers are unable to maintain an adequate body temperature for flight below 10 °C (Heinrich 1975a). It seems that all bumblebees, from the smallest workers to the largest queenss, have to maintain their thoracic temperature within the range 31–42 °C to be able to fly (interestingly, males often feed on massed flowers and allow their temperature to fall below this range) (Heinrich and Heinrich 1983a,b). Thus, larger foragers are presumably able to become active at lower ambient temperatures than small foragers but conversely, they are more prone to overheating in warm weather (Heinrich 1975a, 1979b). Indeed, these are arguments used to explain why bumblebees are superior pollinators to honeybees in cool climates and why the distribution of bumblebees is largely confined to temperate regions. The nest itself is maintained at a more-or-less constant temperature, so individual-level thermoregulation is not an issue for bees working within the nest.

If thermoregulation were the explanation for alloethism in bumblebees, one might predict that the queen should rear a few large foragers early in the season but that workers reared in summer should be smaller. Several studies have examined changes in worker size during the season, with variable results. Knee and Medler (1965) found an increase in worker size for three American species late in the season. Plowright and Jay (1968) found an increase in worker size as the season progressed in some species but not in others. Röseler (1970) describes an initial decline in the mean size followed by a general increase in *B. terrestris*. It seems that foragers are not, in general, larger early in the season. Of course the sizes produced may not be the optimum with regard to thermoregulation, particularly if the colony is constrained by a shortage of pollen. This is particularly likely to be the case in early spring when the queen has to singlehandedly gather food for her offspring.

Between them, it is to be hoped that one of these hypotheses explains why larger workers are able to gather nectar more quickly than their smaller siblings. What none of them address is why there is so much size variation within foragers (Figure 3.3). If it is in some way advantageous for foragers to be large and for nest bees to be small, why is there not a bimodal size distribution? It may be that having foragers of a range of sizes allows them each to specialize in flower types appropriate to their morphology and so improves overall foraging efficiency of the colony while minimizing intra-colony competition. Different size classes do tend to visit different flower species (Cumber 1949a; Heinrich 1976a; Morse 1978b; Inouye 1980a; Barrow and Pickard 1984; Johnson 1986). For example, Cumber (1949a) found that large workers of *B. pascuorum* tended to visit *Lamium album*, which has a deep corolla, while the smaller workers visited *Lamiastrum galeobdolon* which has a substantially shallower corolla. Overall, the mean corolla depths of the different flower species visited varies in accordance with the tongue lengths of different sized workers (Prys-Jones 1982). Interestingly, Johnson (1986) found that it was only the large foragers of *B. ternarius* and *B. pennsylvanicus* that selected deeper flowers. The smaller foragers visited both deep and shallow flowers. Johnson (1986) suggests that this may be because the small bees were primarily nest bees that had been forced to forage due to a food shortage in their colonies, and thus they were inexperienced.

Studies have also found that there are differences in the mean size of foragers engaged in gathering pollen versus those that gather nectar, but they do not agree in the direction of this difference. Some studies have found that it is the larger foragers that tend to collect pollen, while the smaller foragers collect nectar (Brian 1952; Free 1955a; Miyamoto 1957; Pouvreau 1989), while Goulson et al. (2002b) found the reverse. It may be that the size of bees specializing in each of these tasks depends on which flowers are locally available. The structure of particular flowers suits bees of a particular size; for example, small foragers of *B. terrestris* and *B. lapidarius* are better able to trigger flowers of *Cytisus scoparius* (Stout 2000). Since these flowers only provide pollen, we might expect pollen gatherers to tend to be small in nests situated near to patches of flowering *C. scoparius*. Thus, the relative sizes of pollen and nectar gatherers may vary between nests and at different times of the year according to the flowers that are locally available. Having foragers of a range of sizes enables both resources to be gathered efficiently.

Differently sized foragers are also likely to differ in their optimal ambient temperature range for activity. We might also predict that smaller workers would be better suited to foraging in warm weather and in the middle of the day, and larger workers in poor weather and in the early morning and evening. It would be informative to examine whether the propensity of workers of particular sizes to engage in foraging is affected by ambient temperature.

It would be possible to test whether having foragers of a range of sizes is beneficial to bumblebee colonies by artificially varying the size distribution of workers in experimental nests. Young workers can readily be moved between colonies, so that it would be possible to create colonies with only large or small workers, or with a range of sizes, and then measure nest growth and foraging efficiency.

3.3 Sex Determination

The sex of Hymenoptera is determined in an unusual way, using a system known rather dauntingly as parthenogenetic arrhenotoky (Crozier and Pamilo 1996). Put simply, fertilized eggs develop into diploid females, while unfertilized eggs develop into haploid males. Actually, it is slightly more complex than this. In many Hymenoptera (including bumblebees), individuals are male if they are homozygous at one or more sex-determining loci (Paxton et al. 2000). Heterozygotes at this locus are females. Since haploids are inevitably homozygous at all loci, all haploids are males. But it is quite possible for a diploid to be homozygous at this particular loci; such individuals develop as sterile males.

In *B. terrestris*, it appears that only one locus is involved (Duchateau et al. 1994). This is also the case in honeybees and many other bee species that have been examined (Mackensen 1951; Woyke 1963, 1979; de Camargo 1979; Kukuk and May 1990). In fact, use of a single sex determining locus seems to be the norm in most Hymenoptera (Cook and Crozier 1995). The fewer loci involved, the more likely it is that diploid males will occur. Diploid males appear to have very low or zero fitness in bumblebees (Duchateau and Mariën 1995). In honeybees, diploid male larvae are consumed by workers, but in bumblebees they are reared to adulthood (Plowright and Pallet 1979; Duchateau et al. 1994). Their production is thus particularly undesirable since it places a burden on the colony for no gain. To counter this, species that depend on just one locus have large numbers of alleles, so that the probability of homozygotes occurring is remote. In *B. terrestris*, there are thought to be at least 46 alleles at the sex-determining locus (Duchateau et al. 1994). Diploid males have not been recorded in field situations. However, it has been suggested that they may be a problem for commercial rearers of bumblebees, where inbreeding in the captive stock is likely (Duchateau et al. 1994).

3.4 Control of Reproduction and Queen–Worker Conflicts

Social Hymenoptera with an annual life cycle generally produce new reproductives at the end of the colony cycle (Wilson 1971; Michener 1974), and bumblebees are no exception. There is thought to be a trade-off between maximum colony growth and reproduction, constrained by impending ecological (often seasonal) changes that will soon make conditions unsuitable for either. Limiting reproduction to the end of the cycle ensures that the largest possible workforce is present to rear reproductives (Oster and Wilson 1978). In bumblebees, colonies of many species disband long before the end of the flowering season, and it is probably mounting pressure from parasitoids that curtails their development (Schmid-Hempel et al. 1990).

The early, pre-reproductive phase of colony development in bumblebees is generally harmonious, but in the later reproductive phase violent, even fatal, conflicts may occur between members of the colony. Some workers within a colony become more aggressive than others, both towards intruders and to their siblings, and these individuals tend to show a greater degree of ovarian development (Free 1958). These bees are generally nest bees rather than foragers. Late in the development of the colony, such workers will sometimes construct egg cells and lay their own (unfertilized and thus male) eggs. The foundress queen will retaliate by eating these eggs and then laying her own in the egg cells (Free et al. 1969). In turn, the workers may eat the queen's eggs, often doing so as she is laying them (Honk et al. 1981; Duchateau and Velthuis 1989; Bloch and Hefetz 1999). Egg-eating by workers has frequently been observed in a range of bumblebee species including *B. lapidarius*, *B. terrestris*, *B. lucorum* and *B. fervidus*, so is probably a widespread phenomenon (Plath 1923a; Free 1955c; Röseler 1967b). Eggs are generally only eaten within the first 24 h after being laid (Huber 1802). Perhaps after this time it becomes impossible for either the queen or workers to distinguish between their own eggs and those of others. Workers may also throw larvae out of the nest at this time, although the identity of these larvae has not been established (Röseler 1967b; Pomeroy 1979). Egg-eating leads to fights within the nest and, on occasion, the queen may be killed by her own workers (Van Honk et al. 1981; Van Doorn and Heringa 1986; briefly reviewed in Bourke 1994).

So what causes this conflict? The answer lies at least in part in the unusual patterns of relatedness found within Hymenoptera due to their haplodiploid sex-determination. In the vast majority of bumblebee species, the queen mates only once (Estoup et al. 1995; Schmid-Hempel and Schmid-Hempel 2000) and colonies are founded by single queens. The sons of the colony queen carry 50% of their mother's genes, whereas the sons of workers carry only 25% of the queen's genes. Thus we expect the queen to favour rearing her own sons rather than allowing her daughters to lay their own eggs. However, from the point of view of the workers, their own sons carry more of their genes (50%) than do their brothers (25%). Even their nephews (the sons of other workers) are more closely related to them (sharing 37.5% of their genes) than are their brothers (incidentally, patterns of relatedness become far more complicated in species where queens mate more than once, or as in some ant species where nests may be founded by more than one queen). The interests of the queen and of the workers are opposed; each would prefer to rear their own sons (Hamilton 1964; Trivers and Hare 1976).

Almost all studies of colony development and conflicts in bumblebees have been of *B. terrestris*. In this species the onset of conflicts within the colony (known as the competition point) appears to be closely correlated with the time when the colony commences rearing new reproductives (Van der Blom 1986; Van Doorn and Heringa 1986; Duchateau and Velthuis 1988). Up to this point, the foundress queen appears to produce a pheromone that induces diploid larvae to develop as workers rather than queens (Röseler 1970, 1991). If she dies or is removed, workers will often rear new queens and lay their own eggs earlier than would otherwise occur. Similarly, suppression of the queen by an invading *Psithyrus* can induce workers to lay eggs. Müller and Schmid-Hempel (1992b) monitored nests of *B. lucorum* for attack by *Psithyrus* and removed any *Psithyrus* queens within 3 days of their arrival. Nests that had been briefly attacked produced significantly more males, suggesting that even this very brief suppression of queen dominance can lead to significant worker reproduction.

Why does colony harmony break down at the competition point? What prevents workers from laying eggs earlier? Duchateau and Velthuis (1988) and Röseler (1991) hypothesize that worker aggression steadily increases until eventually the queen loses her dominance and ceases production of the pheromone. They argue that it is the pheromone that inhibits worker reproduction. It has been experimentally demonstrated that the queen does cease pheromone production at this time. Young female larvae placed with a queen taken from a colony before the competition point become workers, whereas if they are placed with a queen from a colony which has passed the competition point they become queens (Cnaani et al. 2000).

As Bourke and Ratnieks (2001) point out, there is a flaw in the argument put forward by Duchateau and Velthuis (1988) and Röseler (1991). Suppression of worker reproduction by a pheromone is unenforceable; selection would favour workers that ignored this signal and laid eggs anyway (if it were in their interests to do so). Other authors have also disputed the suggestion that worker reproduction is inhibited by a queen pheromone and have provided some experimental evidence against this idea (Keller and Nonacs 1993; Bloch et al. 1996; Bloch 1999; Bloch and Hefetz 1999). It seems unnecessary to argue that the switch to rearing reproductives should be forced upon the queen, since it is in her interests to do so at some point. Bourke and Ratnieks (1999, 2001) suggest a subtly different interpretation. They hypothesize that the queen ceases pheromone production of her own volition (rather than because she is being oppressed by the workers). Workers do not lay eggs before this time because it is not in their interests to do so; worker reproduction early in the colony cycle would slow colony growth (because males do not work) and reduce production of full sisters (who are more closely related to workers than their sons).

Interestingly, there is evidence that ovarian development in young workers is prevented by the presence of the queen and, also, by the presence of dominant workers in the absence of a queen (Röseler and Röseler 1977; Röseler et al. 1981; Bloch et al. 1996; Bloch and Hefetz 1999). Perhaps attempting to develop a reproductive capacity is a dangerous strategy for a young worker since it will place her in conflict with older (and often larger) colony members. Queens and dominant workers seem able to recognize, and are more aggressive towards, workers with developed ovaries (Van Doorn and Heringa 1986; Duchateau 1989; Röseler and Van Honk 1990).

3.4.1 *Timing of reproduction*

B. terrestris colonies differ greatly in the timing of the reproductive phase. Some colonies switch from rearing workers to rearing reproductives at a relatively early stage in colony development, about ten days after the emergence of the first adult workers (Duchateau and Velthuis 1988). These colonies produce mostly males. Other colonies switch later, about 24 days after emergence of the first workers, and these tend to produce mainly new queens. In both colony types, the competition point occurs about 31 days after emergence of the first workers. Thus in early-switching colonies, the competition point does not occur until about 21 days after the queen commences laying male eggs. In contrast, in late-switching colonies, the competition point occurs about 7 days after the first eggs are laid that are destined to become new queens (Duchateau and Velthuis 1988) (it must be noted that there is considerable variation in the timing of these events).

Bourke and Ratnieks (2001) suggest that the timing of the switch is under the control of the queen. In early-switching colonies, she commences laying male eggs while presumably continuing to release the pheromone that prevents female offspring from developing into queens. They argue that workers are unable to detect that they are rearing males until the male larvae are about 10 days old (15 days after the eggs were laid). Since it is probably not in the interests of workers to lay their own eggs before the switching point and they are not able to detect that the switching point has occurred for 15 days, we would expect a substantial delay between the switching point and the onset of the competition point in early-switching colonies. Interestingly, Bloch (1999) demonstrated that the onset of worker reproduction and of intra-colony conflict could be accelerated simply by replacing the queen's diploid eggs with haploid eggs. Presumably this leads to workers detecting male brood at an earlier stage in colony development.

Once the workers detect the presence of male larvae, they could throw them out of the nest and attempt to replace them with their own offspring. However, it seems that they rarely do so. Bourke and Ratnieks (2001) argue that it is not in their interests to do so. Bumblebees exhibit marked protandry (Beekman and Van Stratum 1998) and it seems likely that early-emerging males enjoy greater mating success than do late-emerging males (Bulmer 1983) (although there is no experimental evidence available on mating frequency in relation to emergence time). Since the sons of workers would be at least 15 days younger than those of the queen, their expected mating success may be much lower. Also, the queen's sons are nearing the completion of their development by the time they are detected; if they are detected 15 days after they were laid, they are only about 3 days from pupation (timing of development taken from Duchateau and Velthuis 1988). If the workers lay their own eggs at this point, they will not have hatched until after the oldest of the queen's sons have pupated. Thus they will not be in competition for food and so, for the workers, there is little or nothing to be gained from destroying the first male larvae that they detect.

Consider now late-switching colonies. Here the switch is presumably determined by the queen ceasing to produce the pheromone that prevents female larvae from developing into queens (Cnaani *et al.* 2000; Bourke and Ratnieks 2001). Experimentally increasing the number of workers in the nest can bring forward the production of new queens (Bloch 1999), suggesting that the queen's decision as to when to cease pheromone production is flexible and dependent on availability of a sufficiently large work force to rear new queens (which require more food than workers). Larvae are sensitive to the queen's pheromone at about 2–5 days old (7–9 days after their eggs were laid). The workers commence laying their own eggs about 7 days after the first eggs that are destined to become queens are laid (Duchateau and Velthuis 1988). As Bourke and Ratnieks (2001) point out, this corresponds precisely with the presumed time at which the queen ceases pheromone production.

Neat though these explanations for the timing of onset of the competition point are, there are some anomalies that require further investigation:

(a) Bloch (1999) found that in a few early-switching colonies, the competition point *preceeded* the switching point. Bourke and Ratnieks (2001) put this down to worker error.
(b) As Bourke and Ratnieks (2001) concede, their hypothesis falls 6 days short of explaining the 21-day lag generally observed between the switching point and the competition point in early-switching colonies. These 6 days cannot be explained by the time needed for ovary development in workers, for Duchateau and Velthuis (1989) demonstrated that the ovaries of some workers are fully developed before the switching point (and in late-switching colonies egg laying by workers occurs very promptly).
(c) It is not clear why workers should lay their own eggs in late-switching colonies. Their sons are likely to compete for food with the developing queens (remember that the workers are more closely related to the new queens than they are to their own sons). Also, given the strong male-bias in the ratio of males to new queens found in many bumblebee populations (Beekman and Van Stratum 1998) and the probable early-male advantage discussed above, the expected reproductive success of worker-produced males in late-switching colonies is very low. Bourke and Ratnieks (2001) suggest that competition between worker-produced males and future queens is minimal; an experimental test of this supposition would be valuable. Even if this proves to be so, it then begs the question why does the queen not lay more diploid eggs, and so increase production of new queens? If the colony has sufficient resources to rear worker-laid males, then it could presumably rear more future queens instead.

Overall, Bourke and Ratnieks' hypothesis fits the available data reasonably well and is certainly the closest we have yet come to a full explanation for the reproductive strategies adopted by bumblebee nests. It would be very useful to obtain data on other species, since almost all studies to date have focussed exclusively on B. terrestris. Identification of the queen pheromone would be invaluable, for it would enable experimental manipulations to test various aspects of the hypothesis (unfortunately, analysis of queen exocrine secretions suggests that there are at least 500 candidate compounds, Hefetz et al. 1996). Also, at present, there are few data on the proportion of bumblebee males that are produced by workers and as to how skewed parentage of worker-produced males is towards the more dominant individual workers. Such data could be obtained easily using established microsatellite markers (Estoup et al. 1995, 1996) and would provide an insight into the degree of control that queens and workers each have over colony reproduction.

3.4.2 Matricide

Matricide has frequently been observed in the later stages of colony development in a range of bumblebee species (reviewed in Bourke 1994). It appears to be the result of a gradual process whereby conflict between the queen and workers steadily increases, resulting in a loss in queen condition and, sometimes, ultimately leading to her death. Why is this in the interests of the workers? Bourke (1994) considers the conflicting pressures on workers with regard to matricide in detail, although at this time it had not become apparent that *B. terrestris* colonies adopt one of two alternative reproductive strategies. In colonies specializing in male production, one would expect competition to be most fierce. If workers kill the queen they may well increase their own chances of reproducing and, since their sons are more valuable to them than brothers, matricide may be their best strategy. However, in early-switching colonies the queen generally lays some diploid eggs (which develop into queens) towards the end of colony development, so that matricide still has a cost to workers in terms of lost sisters. In late-switching colonies matricide would seem to be a poor strategy for workers, since it will prevent production of a (potentially large) number of fertile sisters. Unfortunately no data are available on the frequency of matricide in early-switching versus late-switching colonies.

The optimum strategy for workers may well depend upon the condition of the queen, as well as her decision to switch early or late. If her reproductive potential has been reduced through injury or parasitization, this may favour matricide. Since injury may occur during queen–worker conflict over male production, there may be positive feedback; conflict reduces the queen's condition which, in turn, pushes the optimum worker strategy towards further conflict leading to matricide. If the queen's condition becomes sufficiently low it may actually benefit her to die, rather than to continue fighting with her daughters (who in her absence will produce more of her grandsons) (Bourke 1994).

3.5 Sex-ratios in *Bombus*

The population sex-ratio in bumblebees varies greatly, but tends to exhibit a strong male-bias (reviewed in Bourke 1997; see also Beekman and Van Stratum 1998). A male bias is unexpected in social Hymenoptera. Kin-selection theory predicts that in a colony founded by a monogamous queen, workers should favour a 3 : 1 investment (females to males), while the queen should favour equal investment in both sexes (Trivers and Hare 1976). Neither the queen nor the workers should benefit from a male-biased sex-ratio.

So why are bumblebee sex-ratios often male biased? Possible explanations have been considered in depth by Bourke (1997) and Beekman and Van Stratum (1998). The answer must be linked to the frequency with which colonies adopt an early-switching or late-switching strategy, since the former produce mostly males and the later mostly new queens. Bourke and Ratnieks (2001) argue that queens adopt an early-switching, male-producing strategy with a probability of 0.5 (observation of natural nests suggests that about half are early-switching and half are late-switching, Duchateau and Velthuis 1988). If the queen chooses to adopt a strategy of producing males then the workers have no choice but to comply, since they cannot lay their own diploid eggs (Bulmer 1981). If half of all colonies specialize in male production, it is in the interests of both the queens and workers in remaining colonies to specialize in queen production (Bourke and Ratnieks 2001). A very similar system is thought to operate in the ant *Pheidole desertorum* (Helms 1999). This argument is plausible but appears to contradict that made earlier that, in late-switching colonies, the workers commence rearing their own male offspring as soon as they detect the switching point. Also, it does not in itself explain why bumblebee sex-ratios are often male biased. A number of compatible explanations have been proposed (see Bourke 1997), of which the most plausible are outlined below:

(a) Perhaps the simplest explanation for male biased sex ratios is that males are cheaper to produce (Beekman and Van Stratum 1998). Fisher (1930) predicted that most organisms should invest equally in sons and daughters, not that most organisms should produce equal numbers of sons and daughters. In bumblebees, males are markedly smaller than queens and so are cheaper to rear. For example, in *B. terrestris* the weight ratio of males to queens is 1 : 2.1 (Owen *et al.* 1980; Duchateau and Velthuis 1988). When the calorific value is taken into account, the differential is even greater, at about 1 : 3.3 for mature new queens (Beekman and Van Stratum 1998). All else being equal, we would expect colonies that specialize in male production to rear about three times as many males than a colony specializing in queen production can rear queens. However, both Bourke (1997) and Beekman and Van Stratum (1998) conclude that this alone is not sufficient to account for observed population sex-ratios.

(b) Worker reproduction can, in theory, lead to male biased sex ratios, particularly if one worker can monopolize reproduction. But Bourke (1997) argues convincingly that this is highly unlikely to lead to male-bias of the magnitude that is frequently found. We have little information as to what proportion of males produced by colonies are the offspring of workers; Owen and Plowright (1982) detected 19% of males to be worker-laid in *B. melanopygus*, but a recent study of *B. hypnorum* (a potentially atypical polyandrous species) found that all males were produced by the queen (Paxton *et al.* 2001). Worker reproduction in colonies in which the queen has died prematurely (for example due to infection by parasites) may contribute to male biased sex ratios for, in this position, workers have no choice but to rear sons (Owen *et al.* 1980). However, rates of orphaning that have been recorded are not adequate to explain observed male biases (Bourke 1997).

(c) Bulmer (1981, 1983) demonstrated that male biased sex ratios can arise because of protandry. Protandry (the emergence of males before new queens) has been recorded in a range of bumblebee species, and appears to be the norm (Hobbs 1964b; Pomeroy and Plowright 1982; Shelly *et al.* 1991; Foster 1992; Müller *et al.* 1992; reviewed in Bourke 1997). It is thought to occur because early emerging males probably enjoy a higher mating success than do late emerging males, since the former have more opportunities to encounter virgin queens during their lifetime. Bulmer's models predict that a male-biased sex-ratio can be optimal under certain conditions (particularly if male production is under control of the queen), and Bourke (1997) concluded that this provides the most convincing explanation for male-biased sex-ratios in bumblebees.

(d) Beekman and Van Stratum (1998) develop Bulmer's (1983) model further by demonstrating that, under conditions of low resource availability, the best strategy for a colony is to produce males, because male size is probably not closely linked to their reproductive success, whereas in new queens an adequate size is crucial for surviving hibernation. Thus a stressed colony should produce males, while a colony with plentiful resources can afford to produce new queens. Certainly, studies of *B. melanopygus, B. terricola,* and *B. lucorum* have found that it is large bumblebee colonies that produce mostly queens, while small colonies tend to produce mainly males (Owen et al. 1980; Owen and Plowright 1982; Müller et al. 1992) (Figure. 3.6). In contrast, other studies of *B. lucorum, B. terrestris* and *B. hypnorum* found no such relationship (Müller et al. 1992; Paxton et al. 2001). Artificially induced stress on colonies of *B. terrestris* did not result in a switch to male production, but actually resulted in production of fewer males and had no effect on the number of new queens produced (Müller and Schmid-Hempel 1992a). In early-switching colonies, worker production is curtailed by the queen. Whether she does so as a result of perceived resource availability, or for other reasons, is not known. Clearly the success of early versus late-switching depends on what other colonies are doing. If early-switching is triggered unconditionally by low resource availability then, in poor years, we would expect the vast majority of colonies to produce mainly males, which would lead to very low reproductive success.

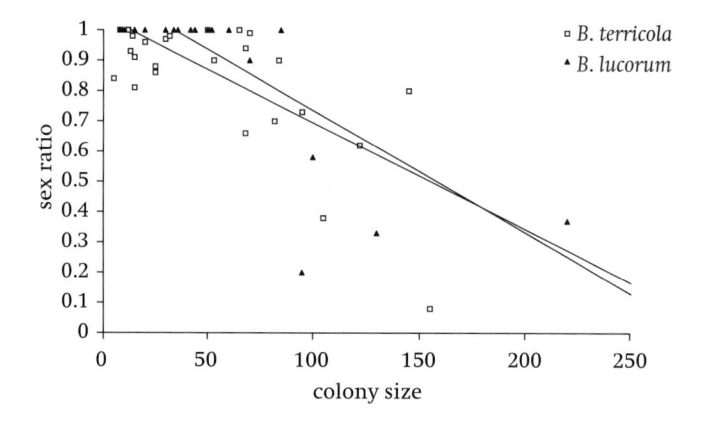

Figure 3.6. Relationships between the sex ratio of offspring produced by colonies and colony size (numbers of workers) for two bumblebee species, *B. terricola* (data from Owen and Plowright 1982) and *B. lucorum* (data from Müller et al. 1992). Sex-ratio is defined here as the number of males divided by the total number of sexuals produced.

Much of the work to date on bumblebee sex-ratios has been theoretical; empirical studies are conspicuously rare. The data that are available suggest that bumblebee sex-ratios are highly variable and this is not adequately explained by any of the models that have been presented. In particular, the factors that determine which reproductive strategy a foundress queen adopts require investigation.

3.6 Sex-ratios in *Psithyrus*

In contrast to the remainder of the genus *Bombus*, sex-ratios of the subgenus *Psithyrus* appear to be female biased, although the data available are rather sparse and based on laboratory studies (Fisher 1992; see also Bourke 1997). Since *Psithyrus* have no worker caste, we would predict equal investment in sons and daughters, leading to a numerical bias towards male offspring. *Psithyrus* also appear to be protandrous, which should further increase male bias (Bulmer 1983). Fisher (1992) suggested that female bias may result from local mate competition among males (Hamilton 1967). However, given that *Psithyrus* have very similar mate location behaviours to other *Bombus* species (see Chapter 4), it is not clear why *Psithyrus* should be any more prone to local mate competition (Bourke 1997). An alternative explanation is that the sex-ratio in *Psithyrus* is influenced by the behaviour of the host workers and queen. Both may reproduce after invasion by a *Psithyrus* female (Fisher 1987, 1992). Interestingly, the total ratio of new sexuals (*Psithyrus* and host combined) may be similar to the ratio produced by the unparasitised host (Fisher 1992). Because of reduced queen dominance, the sex-ratio of reproductives produced by the hosts is heavily male biased and it has been suggested that the female bias shown by *Psithyrus* represents the 'balance' left if workers rear an overall sex-ratio appropriate to the unparasitized colony (Fisher 1992). However, this suggests that the *Psithyrus* female is unable to exert much influence over colony reproduction (for in this explanation the female bias in *Psithyrus* offspring is maladaptive), which seems unlikely given her dominance. Also, the female bias in *Psithyrus* offspring occurs even when the hosts do not reproduce (Fisher 1992). Far more research into the interactions between *Psithyrus* and their hosts is needed if we are to unravel the underlying strategies of *Psithyrus* reproduction.

4
Finding a mate

Mating in bumblebees is rarely observed in the wild. However, the pre-mating behaviour of the males is often conspicuous, and has been the subject of many studies. At least three distinct mate location strategies have been recorded in different bumblebee species.

4.1 Territoriality

Males of some *Bombus* species station themselves by a prominent landmark (either perched or hovering) and await a female. It appears that they search visually for queens, and they have unusually large compound eyes to facilitate this. Despite their large eyes the males are clearly not able to distinguish queens from other organisms at a distance, since they will readily chase after any small flying organism, including birds. At least eight species bumblebee are known to use this system of mate location; two from Europe (*B. confusus* and *B. mendax*) (Saunders 1909; Krüger 1951; Schremmer 1972), three from North America (*B. nevadensis, B. griseocollis*, and *B. rufocinctus*) (Frison 1917; Hobbs 1965a; Alcock and Alcock 1983; O'Neill et al. 1991), and three from Asia (Williams 1991). O'Neill et al. (1991) describe the behaviour of the North American species in detail. Males stake out territories, usually centered on a prominent perch such as a fence post or tree. They dart out from their perch at passing insects, and if they encounter a queen they attempt to grasp her in midair with their legs; if successful the couple often continue in flight (and are then lost from view), but sometimes they fall to the ground. Observations of couples that fall to the ground suggest that copulation rarely ensues; more frequently the queen escapes by crawling away through dense vegetation. Males are faithful to their territories for long periods, up to 26 days. However, they fight fiercely over territories, and resident males are frequently ousted. Fights can lead to severe injuries (Williams 1991). The territories are not based on resources that might be expected to attract females (flowers), often containing no flowers. Different species tend to choose different positions in which to perch; for example, *B. nevadensis* prefers open sunny areas, and *B. griseicollis* favours shaded spots near large trees.

Territorial species also deposit scent-marks on vegetation within their territory, usually doing so early in the day (Alcock and Alcock 1983; O'Neill et al. 1991). The compounds deposited originate in the labial gland (Kindl et al. 1999). In *B. confusus* they have been identified, and consist primarily of geranylcitronellol and (Z)-9-octadecenyl acetate (Hovorka *et al.* 1998). These marks presumably serve to either attract females and/or repel other males, but this has not been examined.

The distinctive large compound eyes of males of all species that are known to be territorial may be indicative of this type of mate location behaviour; for example, *B. regeli*, *B. niveatus*, *B. morrisoni*, and *B. crotchii* also have large eyes (Krüger 1951; O'Neill *et al.* 1991), but their mating system has not yet been examined.

4.2 Nest Surveillance

Males of at least five *Bombus* species (*B. subterraneus*, *B. californicus*, *B. sonorus*, *B. fervidus*, *B. latreillelus*, and *B. ruderarius*) have been seen to stake out the entrance to nests from which young queens are about to emerge (Smith 1858; Tuck 1897; Krüger 1951; Lloyd 1981; Free 1987; Villalobos and Shelly 1987; Foster 1992). Many males may be observed outside each nest, and they seem readily able to distinguish between queens and workers, suggesting that queens have a distinctive odour (see below). When a queen emerges the males may fight furiously (Smith 1858). They may even pursue queens in to the nest and mate with them there (Tuck 1897; Krüger 1951). Where several nests are available within a small area, males may regularly fly between them (Svensson 1980).

4.3 Scent-marking and Patrolling

The males of many species of *Bombus* (including the subgenus *Psithyrus*) patrol regular circuits, in a manner similar to the trap-lining behaviour of foragers. This appears to be by far the most common mate-location mechanism in bumblebees. The behaviour was described in 1851 by Newman, and subsequently by Darwin (1886). Further details were elucidated by Sladen (1912). On these circuits the males pause at short intervals by particular objects, which may be tree trunks, fence posts, leaves, or protruding rocks. They hover by each object for a few seconds before proceeding on to the next. Each object has been marked with a pheromone, placed there by the male in the early morning, and replenished after rain (Alford 1975). Kullenberg *et al.* (1973) observed that the pheromone is smeared onto the chosen object and others nearby using the mandibles, aided also by the proboscis and the underside of the body. Awram (1970) suggested that the beard found on the mandibles of males of many species of bumblebees may be adapted as a brush for this purpose.

The pheromones differ between species, and are often detectable by the human nose (Frank 1941). They are secreted primarily from the labial glands (Kullenberg *et al.* 1973). The constituents of these pheromones have been examined in great detail in a sequence of studies conducted in Sweden, focusing predominantly on Scandinavian species, and most recently by a group working in the Czech Republic.

These have revealed that they consist largely of blends of fatty acid derivatives and terpene alcohols and esters (Stein 1963; Bergström *et al.* 1968; Calam 1969; Kullenberg *et al.* 1970; Svensson and Bergström 1977, 1979; Bergström *et al.* 1981, 1996; Cederberg *et al.* 1984; Descoins *et al.* 1984; Svensson *et al.* 1984; Lanne *et al.* 1987; Appelgren *et al.* 1991; Bergman *et al.* 1996; Bergman and Bergström 1997; Urbanová *et al.* 2001; Valterová *et al.* 2001). The blend is generally species-specific, usually with one or two major components. For example *B. terrestris* uses primarily 2,3-dihydro-6-*trans*-faresol with smaller amounts of geranylcitronellol, while the closely related *B. lucorum* uses a markedly different blend based on ethyl dodecanoate and ethyl tetradecenoate (Bergström *et al.* 1981). Although most research has focused on identifying compounds present in the labial glands, analysis of scent-marks on leaves, and of volatiles in the air around marked leaves, confirm that the same compounds are deposited (Bergman and Bergström 1997).

These compounds may have value in taxonomic studies. Differences have been found between light and dark forms of *B. lucorum* (Bergström *et al.* 1981), perhaps indicating the presence of cryptic species (interestingly, allozyme data from Scandinavia also suggest that *B. lucorum* may contain two species, Pamilo *et al.* 1984). Similarly, labial gland pheromone analysis indicated the presence of two species within samples of *B. lapponicus* from Scandinavia, one of which was subsequently identified as *B. monticola* (Svensson and Bergström 1977; Svensson 1979).

The effects of male scent-marking compounds on female behaviour are poorly known. On rare occasions females have been recorded as being attracted to scent-marked objects (Free 1971; Svensson 1979, 1980), and thus they presumably encounter a male and mate. Very few bioassays of the effects of male scent-marks on queen behaviour have been carried out, and we do not really know how they work (Free 1987). In the carpenter bee *Xylocopa varipuncta*, similar compounds act as long-range attractants to females (Minckley *et al.* 1991), and this seems their most likely function. But in addition to attracting queens, do they have arrestant or aphrodisiac effects? This would appear to provide a relatively straightforward opportunity for further study.

Presumably the species-specific nature of the male pheromone blend facilitates females in locating males of the correct species. It seems that this may be further ensured by the height of the circuit marked by the males. Bringer (1973) and Svensson (1980) found that males of each species tend to mark and circuit objects at particular heights; thus for example *B. hortorum* and *B. (P.) sylvestris* tend to remain within 1 m of the ground. In contrast *B. lapidarius*, *B. terrestris*, and *B. lucorum* may follow routes at tree-top level, up to 17 m (Haas 1949; Awram 1970). However, as Prys-Jones and Corbet (1991) point out, these studies do not take into account the different habitat preferences of the species which may constrain the height of features that are available. There are records of *B. lapidarius* and *B. lucorum* patrolling near ground level in habitats without trees (Krüger 1951). On Salisbury Plain (UK), an extensive area of grassland with few trees or shrubs, I have seen male *B. lapidarius* scent-marking grass stems no more than 70 cm tall. Different species of bumblebee also tend to visit different sorts of landscape feature; for example *B. hortorum* chooses dark hollows, *B. lapidarius* the highest points that are available, and *B. terrestris* visits a range of points along shrubs and trees (Fussell and Corbet 1992b). On a larger scale, different species occupy different habitats. Thus in combination the habitat, height and location of focal points and the pheromone blend all serve to prevent interspecific hybridization. In a study of *Bombus* communities in northern Scandinavia, Bergström *et al.* (1981) found that species that had similar pheromone blends always differed in the habitat they occupied and/or in the height at which males patrolled. Comparable results have also been described for the subgenus *Psithyrus* (Cederberg *et al.* 1984).

The routes followed by males are said to be more-or-less linear, and are generally followed in the same direction (Alford 1975). In *B. hortorum*, Frank (1941) found one particular route to be 300 m long and consisting of 28 marked points, varying from 30 cm to 33 m apart. An individual male was observed to repeat this circuit 20 times within an hour. In between circuits, or when the weather becomes unfavourable, males tend to feed and rest on flowers. It seems that an ability to detect the scent-marks is essential to males during the development of a patrolling circuit, but once they have learned the route they continue to follow it using visual cues. Removal of antennae from inexperienced males prevents them from patrolling on a regular route, but once the route is established bees remain able to follow it even if their antennae are removed (Awram 1970; Free 1987). Indeed, these bees remarked the route on every circuit, suggesting that their inability to detect the scent stimulated them to replenish it.

When patrolling, the males always pause at more-or-less precisely the same points in space (sometimes known as the focal point). Yet, the scent is applied over a number of objects up to 3 m from this point, suggesting that it acts primarily as a long-range attractant (Free 1987). When several males are visiting the same scent mark, they usually pause at slightly different focal points. Presumably these are chosen arbitrarily, and visual cues used to locate them.

The same scent-marking points are often used in successive years (Svensson 1979), but whether this is because some scent remains or simply because the sites are particularly suitable in some way is not known. Interestingly, the features marked by males are rarely if ever flowers. In contrast, males of many species of solitary bee commonly scent-mark flowers to attract females (Kullenberg 1956; Haas 1960). It has been suggested that the move to scent-marking other objects evolved in bumblebees to minimize confusion over potential partners (Awram 1970). Flowers attract workers, which are generally very similar to queens in appearance, so that a male patrolling marked flowers may waste much time attempting to mate with workers.

An intriguing feature of the male behaviour is that often a number of males will adopt similar or overlapping routes, so that a stream of individuals can be observed passing by (Alford 1975). This has been observed particularly frequently in *B. hortorum* (Darwin 1886). This is in marked contrast to the territorial behaviour of other *Bombus* species (described above). Indeed, the pheromones deposited by patrolling species appear to be attractive to other males (Kullenberg 1956); for example, if leaves scent-marked by males of *B. (P.) bohemicus* are moved to new locations they attract further males, which deposit more scent in the vicinity (Kullenberg 1973). Similarly, if crushed heads of males are smeared on to leaves, further males of the same species can be attracted in substantial numbers (Free 1987). In fact attraction of males to male scent-marks has been recorded far more frequently than attraction of queens.

One might intuitively expect that males would do better by actively avoiding each other and establishing their own distinct routes. However, a system whereby many males are attracted to the same places may benefit females by providing an opportunity for mate choice; since females of most species only mate once (while males can mate many times) it is particularly important for a female to choose a high quality male. Unfortunately since female attraction to male circuits, and subsequent mating, have so scarcely been recorded (Free 1971; Svensson 1979), we can only speculate as to exactly what happens and how it may have evolved. Perhaps a single male cannot deposit enough pheromone to attract females; by only responding to multiple pheromone marks, females might ensure themselves a choice of males. Alternatively, a female may only mate once she has had the opportunity to evaluate several potential partners, so a male circuiting on his own would not obtain a mate.

It is odd that mating has been so rarely observed in bumblebees. Free (1971) suggests that pairs may immediately leave the patrolling route, to avoid encounters with further males, but this has not been observed. Another as yet unexplained phenomenon is that mate location behaviour is very rarely seen in some common bumblebee species. For example *B. pascuorum* and *B. lucorum* are both abundant in much of Europe, but have only very rarely been seen patrolling (and have never been seen to use other mate-location mechanisms) (Awram 1970; Fussell and Corbet 1992b). Similarly, *B. (P.) vestalis* is generally the most common *Psithyrus* species in southern UK, and males can be exceedingly abundant, yet, we have no records of its mating behaviour. One cannot help but suspect that there may be other mating systems used by bumblebees that are not easily observed (perhaps they take place in the canopy of trees). There is clearly need for further research on this fascinating but poorly understood subject.

4.4 Inbreeding Avoidance

Foster (1992) demonstrated that *B. frigidus* (and possibly also *B. bifarius*) preferentially mated with non-siblings when confined in flight cages with a choice of potential partners. Inbreeding frequently results in offspring of reduced fitness (Shields 1982; Bateson 1983; Partridge 1983) (and in bumblebees is likely to lead to production of sterile diploid males). Experimentally inbred colonies of *B. atratus* showed reduced growth compared to outbred colonies (Plowright and Pallet 1979). There are thus clear benefits to be gained by avoiding mating with siblings. However, Foster (1992) found that in contrast to *B. frigidus*, both *B. californicus* and

B. rufocinctus readily mated with siblings. He suggests that this variation between species may correspond to their mating system. Males of *B. californicus* and *B. rufocinctus* both use nest surveillance to find mates, and probably do not survey their own nest in this way. Thus they are unlikely to encounter sisters, and have no need for an inbreeding avoidance mechanism. In contrast, *B. frigidus* and *B. bifarius* are patrollers, and may frequently encounter sibling queens. In these circumstances a means of detecting and avoiding siblings is beneficial.

4.5 Evolution of Male Mate-location Behaviour

Surprisingly, the distribution of different mate-location behaviours does not appear to correspond to phylogeny. Even allowing for some errors in the phylogeny, it seems certain that some of these behaviours must have evolved more than once. For example, when compared against the phylogeny of Williams (1985a), bumblebees that are territorial fall into five separate subgenera. Similarly, the patrolling species fall within eight subgenera (O'Neill *et al.* 1991). These two behaviours clearly have much in common. The scent-marking compounds produced by both groups appear to be similar; for example one of the main components of the scent-mark of the territorial *B. confusus* is geranylcitronellol, which is also a major component of the scent-marks of the patrolling species *B. hypnorum* and *B. lapponicus* (Bergström *et al.* 1981; Hovorka *et al.* 1998). In both groups the compounds are secreted by the labial glands, and are applied mainly in the morning (Kullenberg *et al.* 1973; Kindl *et al.* 1999). However, it is not obvious which mating behaviours are primitive and which are derived, or how species could readily switch between a territorial system with fierce male–male interactions to a patrolling system where males actively seek out sites being used by other males. It is easier to imagine how nest surveillance could evolve into patrolling behaviour (or vice versa), since species which use nest surveillance have been observed to patrol regular circuits between nests. Both systems include tolerance of other males, at least until a queen is available for mating. Mating systems are extremely variable within the Apoidea, and so provide little information as to what the mating system of the ancestral bumblebee might have been.

4.6 Queen-produced Sex Attractants

It seems that queens also produce pheromones which stimulate mating attempts by males. If virgin queens are tethered close to a focal point they usually attract males which attempt to mate with them, but if they are tethered a few meters away they are ignored by males (Free 1971). This suggests that males are only capable of detecting queens over short distances, and probably do so visually (Awram 1970; Free 1971). However, after the initial approach it seems that queen odour is necessary to stimulate the male to attempt copulation, for they will not attempt to copulate with workers, other males, or old, mated queens (Free 1971). By experimentally removing body parts, Free (1971) demonstrated that the source of the pheromone was probably the head (males rarely attempted to mate with decapitated queens!). Van Honk *et al.* (1978) subsequently deduced that the pheromone is probably produced by the mandibular gland. Males rarely attempt to mate with queens in which the mandibular gland has been destroyed, but can be encouraged to do so by smearing these queens with mandibular gland secretions. However, the compounds involved have never been identified.

4.7 Monogamy versus Polyandry

Eusocial behaviour, where some individuals never attempt to reproduce, but devote their energies to helping others to reproduce, is very rare in nature. Outside of the Hymenoptera, it is found in a handful of arthropods (notably the termites) and one obscure mammal. Yet, in the Hymenoptera it is common, and is thought to have evolved independently on a number of occasions (Hölldobler and Wilson 1990). Eusociality appears to be contrary to the Darwinian view of natural selection, and in fact its occurrence was of great concern to Darwin; he was never able to fully reconcile it with his belief that natural selection operated at the individual level. Hamilton (1964) was the first to provide a convincing explanation for eusociality in evolutionary terms. He argued that by helping relatives to breed, an organism was passing on its own genes indirectly (a behaviour now known as kin selection). Related individuals may thus be united by the common interest of passing on their shared genes. A crucial part of Hamilton's theory was that the effectiveness of helping relatives to reproduce as a means of passing on genes depends upon the degree of relatedness; all else being equal, helping a close relative is a better strategy than helping a distant one.

An interesting consequence of haplodiploid sex determination (see Chapter 3) is that full sisters are more closely related to each other than is generally the case; on average, they can expect to have 75% of their genes in common, whereas in most organisms full sisters share 50% of their genes. This predisposes them to cooperate, and is thought to be one major reason why Hymenoptera have repeatedly evolved eusociality, whilst it remains rare in other organisms. However, thus far we have only considered the situation where sisters share both parents. While the majority of social insects have female mating frequencies close to one, in some social Hymenoptera the queens mate many times: the so-called 'supermaters' (Boomsma and Ratnieks 1996). For example, queens of *Apis* spp. may mate with 50 or more males during their nuptial flight (Seeley 1985a; Moritz *et al.* 1995; Oldroyd *et al.* 1995). This can greatly reduce the relatedness of their offspring; half sisters share only 25% of their genes.

What are the pros and cons of single versus multiple mating? Mating with a number of males is presumably costly to the queen in terms of time and exposure to predators (Moritz 1985; Crozier and Pamilo 1996). On the other hand, multiple mating provides the queen with a substantial reserve of sperm, which may be particularly important if she is long lived (Cole 1983). It may set up the opportunity for sperm competition and thus result in better genes for her offspring. Multiple mating may also render the colony less vulnerable to pathogens and parasites (Sherman et al. 1988). Social behaviour predisposes organisms such as bees or ants to epizootics of such organisms, for they live at high densities and have frequent contact with siblings. Parasites and pathogens have long been suspected to act with positive frequency dependence, so that rare host genotypes are less likely to be infected (Haldane 1949; Hamilton 1980). Thus genetic variability within a colony of a social organism, created by multiple mating of the foundress, is likely to reduce the impact of parasites (Tooby 1982; Hamilton 1987; Sherman et al. 1988). Experimental tests with bumblebees support this hypothesis; infections of the protozoan parasite *C. bombi* spread more slowly among unrelated workers than among related workers (Shykoff and Schmid-Hempel 1991a,b). Similarly, under field conditions, colonies of *B. terrestris* with artificially enhanced genetic variability have fewer parasites (Liersch and Schmid-Hempel 1998; Baer and Schmid-Hempel 2001).

Possibly the biggest constraints on mating behaviour are imposed by the social structure of the colony. Single mating promotes cooperation between workers, but may lead to conflicts between the queen and her workers over the sex-ratio of the offspring that are reared (Crozier and Page 1985; Ratnieks and Boomsma 1995). Conversely, multiple mating reduces queen–worker conflict but reduces the incentive of daughters to cooperate with each other, leading to conflicts between daughters over resource allocation and reproductive opportunities (Crozier 1979; Boomsma and Grafen 1991; Pamilo 1991; Sundström 1994).

In contrast to honeybees, it seems that queens of most bumblebee species are monogamous in natural situations (although in a few species, notably *B. hypnorum*, queens do mate more than once) (Röseler 1973; Estoup et al. 1995; Crozier and Pamilo 1996; Schmid-Hempel and Schmid-Hempel 2000). Thus the workers are generally full siblings and are (on average) 75% related to each. This should render bumblebee colonies particularly susceptible to epizootics of parasites and pathogens. Unlike honeybees, bumblebee colonies are also likely to be more prone to queen–worker conflict over offspring sex-ratios, but less prone to worker–worker conflicts. So why do bumblebees and honeybees differ so markedly in their mating behaviour?

The number of times a queen bumblebee mates may be constrained by the selfish interests of males. In honeybees, mating is extremely rapid (taking only a few seconds, Winston 1987) and takes place in flight. In contrast, bumblebees mate while resting on the ground or sometimes high up in vegetation (Figure 4.1). They have occasionally been observed to fly (propelled by the efforts of the queen while the male hangs limply), but are very clumsy. Lie-Pettersen (1901) records beating numerous pairs of copulating *B. terrestris* and *B. pascuorum* from the foliage of deciduous trees. Copulation is in general prolonged, lasting from 10 to 80 mins in those species that have been studied (Alford 1975; van Honk et al. 1978; Foster 1992; Duvoisin et al. 1999). Mean duration appears to be about 36–44 min. During this period the pair are presumably vulnerable to attack.

Figure 4.1. Mating *B. terrestris*. Pairs remain in copula for up to 80 min, but despite this mating pairs are rarely observed in the wild.

Why does copulation take so long in bumblebees? Duvoisin *et al.* (1999) demonstrated that most of the sperm is successfully transferred within the first 2 min of copulation in *B. terrestris*. During the remainder of the time, the male transfers a gelatinous plug to the female genital tract, which completely fills the bursa copulatrix. The plug consists of a mix of palmitic, linoleic, oleic, and stearic acids, and a cyclic peptide, cycloprolylproline, a compound not known from any other insect species (Duvoisin *et al.* 1999; Baer *et al.* 2000). The plug persists within the queen for up to 3 days, and appears to partially block sperm transfer if she mates again during this time (although plugs placed in queens artificially are not very effective at blocking sperm, Sauter *et al.* 2001). Duvoisin *et al.* (1999) conclude that the prolonged copulation is probably imposed on the queen by the male. It serves to prevent her from remating until the male's sperm have reached the spermatheca (which takes 30–80 min). It also allows transfer of the plug which hampers further mating.

In other insects, peptides in accessory secretions of males serve to reduce receptivity in females (Chen *et al.* 1988). Baer *et al.* (2001) conducted experiments in which they transferred components of the sperm plug to queen *B. terrestris* and examined their willingness to remate. Their working hypothesis was that cycloprolylproline was the most likely active component of the plug. Contrary to expectation, only linoleic acid inhibited further mating, and this compound did so effectively for at least one week. However, Baer *et al.* (2001) did not test whether the other compounds played some other role such as blocking sperm. It would seem to be a simple step for females to evolve to ignore the linoleic acid signal if it were in their interests to do so. Perhaps, it is not worthwhile for a bumblebee queen to attempt to mate a second time, since it will commit her to another lengthy copulation which may provide her with little sperm due to the presence of the plug.

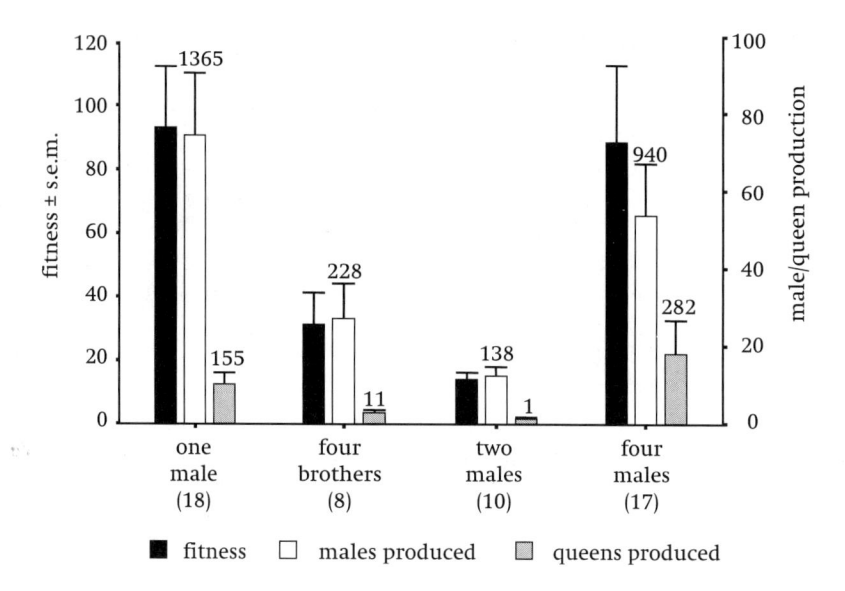

Figure 4.2. Reproductive success of colonies of *B. terrestris* according to treatment. The foundress queens had been artificially inseminated with sperm from one, two, or four unrelated males, or four brothers. Colonies were then placed out in the field in Switzerland. Fitness is defined as the number of queens multiplied by two, plus the number of males. Numbers above bars indicate the total number of sexual organisms sampled. Number of colonies is given in parentheses at the bottom. From Baer and Schmid-Hempel (2001).

There are other differences between honeybees and bumblebees that may influence the strength of selection pressures operating on queens with regard to how many times they mate. Bumblebee nests are much smaller in size and of shorter duration compared to honeybee nests, both of which will tend to make bumblebees less prone to epizootics than honeybees. Honeybee colonies can persist for many years, and during this period it is inevitable that some of the thousands of workers will bring pathogens back to the nest. Conversely, bumblebee nests generally last for only a few months, and the period of intense worker activity may last for only a few weeks. With luck, a bumblebee nest may entirely escape attack by a serious pathogen before the new reproductives have been reared. Thus the need for genetic variability within colonies may be less. There are a small number of tropical bumblebees that have large, perennial nests similar to those of honeybees (Michener and Laberge 1954; Dias 1958; Michener and Amir 1977; Brian 1983). It would be interesting to examine whether these species are also monogamous.

The recent development of artificial insemination techniques in bumblebees makes it possible to inseminate bumblebee queens with sperm from more than one male, and study the consequences (Baer and Schmid-Hempel 2000). Baer and Schmid-Hempel (2001) inseminated *B. terrestris* queens with sperm from one, two or four unrelated males, or four brothers, and placed the resulting colonies out in the field in Switzerland. Genetic diversity of the colony was negatively related to the prevalence and intensity of infection with the trypanosome gut parasite *Crithidia bombi*, but nevertheless colonies produced by singly-mated queens had the highest reproductive success (Figure 4.2). Colony fitness appeared to follow a U-shaped function, being lowest when queens were mated twice. The mechanism underlying this result has not yet been established, but Baer and Schmid-Hempel (2001) speculate that low levels of multiple mating may lead to high levels of conflict between workers of different patrilines within the nest, reducing colony fitness. Whatever the cause, it seems that bumblebee queens may be constrained by an adaptive valley, beyond which high fitness could be achieved. The additional constraints on multiple mating imposed by males through prolonged mating duration and use of sperm plugs may make it very difficult for bumblebee queens to escape monogamy.

B. hypnorum is one of very few bumblebee species in which multiple mating by queens regularly occurs. Although many queens are monogamous, some mate two or more times (Schmid-Hempel and Schmid-Hempel 2000; Paxton *et al.* 2001). Molecular studies by Paxton *et al.* (2001) indicate that one male predominates fathering of the offspring (mean 69%), perhaps due to a partially effective sperm plug. It would be interesting to examine the consequences of multiple mating for colony fitness and social organization in this species.

5
Natural enemies

Bumblebees are attacked in various stages of their life cycle by a diverse range of predators, parasites and parasitoids. The importance of these organisms is perhaps best illustrated by the vigor of bumblebees when they are freed from their natural enemies. In New Zealand four bumblebee species were introduced in 1885 from the UK, and only three of their many parasites were accidentally introduced with them (Donovan and Wier 1978). Two of these bumblebee species (*B. subterraneus* and *B. ruderatus*) are now extinct or nearly so in the UK. In contrast all four species are flourishing in New Zealand, and they often occur at extraordinary densities, far greater than those observed in their natural range, suggesting that elsewhere their numbers are held in check by natural enemies.

In general, spectacularly little is known about the biology of most bumblebee parasites and parasitoids. In particular, the microorganisms and mites associated with bumblebees have received very little attention, and no doubt many have yet to be discovered. For those that have been identified and named, in the vast majority of cases almost nothing is known of their distribution, host range, and the impact that they have on the population dynamics of their hosts. Alford (1975) provides detailed descriptions of the life cycles of some species, focussing particularly on the UK. More recently, Schmid-Hempel (1998) provides an excellent review of the parasites of bumblebee and other social insects. These reviews serve primarily to illustrate the enormous gaps in our knowledge; there is great scope for further work. What follows is a description of the biology of the better known and more abundant natural enemies of bumblebees, but the list is far from comprehensive.

5.1 True Predators

Foraging bumblebees are generally said to have few 'true predators', organisms that kill and consume many prey during their lives (*sensu* Thompson 1982). For this reason it has often been argued that foraging bumblebees are not constrained in their behaviour by predation (Pyke 1978*a*) (e.g. they do not spend time looking around for predators in the way that, say, an antelope might). Bumblebees are among the most obvious warningly coloured organisms in northern temperate zones. There is clear evidence for colour pattern convergence among groups of species in North America, Europe, and Kashmir, presumably a result of Müllerian mimicry (Plowright and

Owen 1980; Williams 1991) (Plate 5). Very often, species with near-identical patterns are rather distantly related (notably, *Psithyrus* species are often similarly coloured to their hosts). Male bumblebees do not have a sting, so are presumably automimics of females of their own or other species (Mallet 1999), although interestingly males often differ quite obviously from females in their colour pattern. Perhaps there is a conflict between sexual selection and mimicry.

In temperate regions the main true predators of foragers are probably birds and spiders. Shrikes are said to be particularly partial to bumblebees (Owen 1948). Grönlund *et al.* (1970) found that bumblebees may make up to 40% of the total food intake of the great grey shrike, *Lanius excubitor* in the autumn in Finland. However, shrikes are generally uncommon birds. Spotted flycatchers (*Muscicapa striata*) occasionally take bumblebees, removing the sting by wiping the bee against a branch (Davies 1977). In southern Europe bee-eaters (*Merops apiaster*) do likewise. However, most other insectivorous birds avoid bumblebees.

Most spider webs are too flimsy to catch bumblebees, but Plath (1934) observed that the North American species *Argiope aurantia* frequently caught bumblebees. The larger crab spiders such as *Misumenia vatia*, which do not spin a web but rather wait on flowers for their prey, are also capable of catching bumblebees (Plath 1934), but the rates of predation are low (Morse 1986b). Perhaps the main arthropod predators of foraging bumblebees are the robber-flies (Diptera: Asilidae). Robber-flies are active fliers that catch flying prey in the air with their powerful legs. The larger robber fly species are capable of taking smaller bumblebees, and some species such as *Proctacanthus hinei* and *Mallophora bomboides* prey extensively on bumblebees (Bromley 1934). Interestingly, *M. bomboides* is also a Batesian mimic of its main prey, *B. americanorum* (Brower *et al.* 1960). Published records of predation on bumblebees by robber-flies appear to be confined to North America (Brown 1929; Fattig 1933; Bromley 1936, 1949). Overall it seems probable that predation on foraging bumblebees is rare.

In contrast, bumblebee nests are attacked by a number of predators which may have a significant impact on their populations. In the UK, nests are frequently destroyed by badgers (*Meles meles*), which entirely consume the brood, comb, and most of the adult bees (Alford 1975). In North America skunks (*Mephitis mephitis*) are similarly destructive (Plath 1923b, 1934). Other nest predators are said to include foxes (*Vulpes vulpes*) (Southern and Watson 1941), moles (*Talpa europea*), weasels (*Mustela nivalis*) (Sladen 1912), shrews (*Sorex* spp.) and voles (*Clethrionomys* and *Microtus*) (Alford 1975). In Iceland, bumblebee nests may be a major food source for mink (*Mustela vison*) (Prys-Jones *et al.* 1981). Newman (1851) estimated that two-thirds of bumblebee nests in England were destroyed by the field mouse (*Apodemus sylvaticus*). However, in general we have very little quantitative data on predation rates by any of these organisms.

One predator which is undoubtedly of great importance is the wax moth, *Aphomia sociella* (Lepidoptera: Pyralidae). This species only occurs in the nests of bumblebees, and was said by Hoffer (1882–3) to be one of their most serious enemies. The moth lays batches of eggs in the nest, and the gregarious larvae indiscriminately consume the comb, larvae and pupae. They spin silken tunnels which presumably protect them from the adult bees. The bees appear to have no effective defense against them, and in heavy infestations the nests are entirely destroyed (D.G. pers. obs.). The larvae overwinter in a ball of tubular cocoons near the destroyed nest. They pupate in the spring, and give rise to adults from June onwards (Alford 1975). This moth appears to be particularly abundant in gardens in southern England, where I have found infestation levels in nests of *B. terrestris* of up to 80% (D.G. unpublished data). Since much lower infestation levels (~20%) were found in nests situated in farmland, this suggests that bumblebee populations may be higher in gardens than elsewhere. This moth is known to be an important predator in mainland Europe (Hoffer 1882–3; Hasselrot 1960; Pouvreau 1967), and has been introduced to North America (Forbes 1923).

Interestingly, bumblebee species vary greatly in their enthusiasm for nest defence. Some species, including *B. terrestris*, *B. muscorum*, and *B. hypnorum* are notoriously aggressive (Alford 1975; Schmid-Hempel 1998). They readily attack intruders near to their nest, which they will bite and sting simultaneously. Often workers will pursue intruders for some distance (D.G. painful pers. obs.). In contrast species such as *B. pratorum* and *B. pascuorum* are remarkably docile. Their nests can be destroyed yet they will make little effort to defend them. This variation in behaviour does not follow taxonomic boundaries (e.g. *B. pascuorum* and *B. muscorum* are close relatives), and remains unexplained.

5.2 Parasitoids

Parasitoids are specialized organisms that develop on or in the body of their host, and successful development of the parasitoid invariably causes the death of the host (thus distinguishing them from parasites). They belong almost exclusively to the Hymenoptera and Diptera. A great deal is known about some parasitoids, notably the Hymenopteran parasitoids that attack agricultural pests. Rather less is known about parasitoids of social insects.

5.2.1 *Conopidae (Diptera)*

At least four genera of conopids attack bumblebees: *Conops*, *Myopa*, *Physocephala*, and *Sicus* (Smith 1959, 1966, 1969). Conopids are parasitoids of adult bees, both workers and males (Postner 1952; Alford 1975). In Europe, queens fly too early to be attacked, but in Canada, queens of late-emerging species are attacked by *Physocephala texana* (Hobbs 1965b, 1966a,b). The life cycle of conopids is described in detail by Alford (1975). The adult fly waits at flowers for foraging bees, and inserts a single egg through the intersegmental membrane into the abdomen of the host. Remarkably, the female fly has no hardened ovipositor for penetration of the host cuticle. Perhaps for this reason *Psithyrus* are never attacked, for they have a tougher exoskeleton than other *Bombus*. Bees are attacked from June to August.

Once inside the host the parasitoid egg rapidly hatches. The larvae consume haemolymph during their first two instars, but in the third and final instar switch to feeding on host tissues within the abdomen and, in some conopid species, they also feed upon the contents of the thorax. This leads to the death of the host bee about 10–12 days after infection (Müller and Schmid-Hempel 1992b). Shortly afterwards the parasitoid pupates, remaining within the abdomen of the host. The adult fly emerges the following summer.

The behaviour of the host changes once infected. Workers of B. terrestris spend less time in their nest, and tend to stay outside the nest at night. They also actively seek out cold microclimates (Müller and Schmid-Hempel 1993a). This behaviour may have an adaptive explanation; by doing so they maintain a lower mean body temperature, which slows the development of the parasitoid and thus increases host longevity. Host workers continue to forage while parasitized, although they have a reduced capacity to carry nectar since the presence of the parasite constricts the volume of nectar that the honey stomach can contain (Schmid-Hempel and Schmid-Hempel 1991). Thus prolonging their life will benefit the colony. Conversely, in late stages of infection it seems that the parasitoid manipulates the behaviour of the host. Before death, parasitized bees tend to bury themselves, and the parasitoid is more likely to survive the winter when underground (Müller 1994).

The incidence of parasitization by conopids can be high but is very variable, ranging from zero to 70% (de Meijere 1904; Cumber 1949c; Postner 1952; Schmid-Hempel et al. 1990; Schmid-Hempel and Müller 1991; Schmid-Hempel and Schmid-Hempel 1996a). Interestingly, the incidence of parasitization has been found to vary according to colony size; Müller and Schmid-Hempel (1993b) found that workers of B. lucorum from larger colonies were more likely to be parasitized in Switzerland, but Macfarlane and Pengelly (1974) found the opposite effect for bumblebees in Canada. Since parasitization occurs at flowers, it is unclear why colony size should influence the prevalence of infection. Heavy infestation levels do impact on fitness of bumblebee nests, for they result in the rearing of smaller queens (Müller and Schmid-Hempel 1992a) which are more likely to die in the winter (Holm 1972; Owen 1988), and also in a switch from production of queens to production of males (Müller and Schmid-Hempel 1992b).

Host selection by conopids is a subject that has recently attracted attention. Only one parasitoid ever emerges from a host, so one would expect strong selective pressure on female conopids to avoid laying eggs in hosts that are already parasitized. However, hosts are frequently multiply parasitized (Clausen 1940; Schmid-Hempel and Schmid-Hempel 1989, 1996b). When this occurs there must be fierce competition resulting in the death of all but one parasitoid. Those few conopids that have been studied do not appear to be host-specific, although they do exhibit preferences for particular host species. For example in Switzerland Schmid-Hempel and Schmid-Hempel (1996a) found that *Physocephala rufipes* attacked B. pascuorum, B. lapidarius and, occasionally, B. terrestris/ lucorum, while the sympatric conopid *Sicus ferrugineus* preferentially attacked B. terrestris/ lucorum and B. pascuorum but was never found in B. lapidarius. S. ferrugineus also preferentially parasitized larger workers. Further host records are to be found in de Meijere (1912), Freeman (1966) and Schmid-Hempel (1994). Thus there is considerable potential for both intra and interspecific competition within hosts, and the choice made by the female fly is vital to the fitness of her offspring.

5.2.2 *Sarcophagidae (Diptera)*

Various *Brachicoma* spp. are parasitoids of the brood of bees and wasps. Several species commonly attack bumblebees, notably *Brachicoma devia* in the UK and *Brachicoma sarcophagina* in North America (Alford 1975). The host range of *Brachicoma* spp. appear to be broad, spanning different families, but they do appear to exhibit preferences; for example *B. sarcophagina* was most frequently recorded from *B. bimaculatus* and *B. fervidus* (Townsend 1936). The adult fly must enter bumblebee nests to deposit her offspring, and presumably has means of overcoming or avoiding the nest defences. She is viviparous, depositing young larvae directly on to bumblebee larvae. The fly larvae do not feed until their host spins a cocoon for pupation. They are ectoparasitoids, slowly consuming their host from the outside. A bumblebee larvae can support up to four parasitoids (Alford 1975). Once fully developed, the larvae drop to the floor of the nest and pupate among the nest debris. There may be several generations during the summer.

Other known bumblebee parasitoids within the Sarcophagidae include *Boettcharia litorosa*, *Helicobia morionella*, *Sarcophaga* spp. and *Senotainia tricuspis* (Macfarlane and Pengelly 1974, summarized in Schmid-Hempel 1998).

5.2.3 *Braconidae (Hymenoptera)*

Syntretus splendidus is a gregarious endoparasitoid of adult bumblebees, including queens, workers, males and the *Psithyrus* species *B. (P.) vestalis* (Pouvreau 1974; Alford 1975; Schmid-Hempel et al. 1990). Egg-laying has never been observed, but is thought to take place during May and June when bees are foraging on flowers as in conopids. The female may lay up to 70 eggs in queens, but usually less than 20 in workers (Alford 1968, 1973). A variety of host species are attacked, including *B. terrestris*, *B. lucorum* and *B. pascuorum* (Alford 1968). After hatching, the eggs develop through five larval instars within the thorax or abdomen of the host. In the fifth instar they emerge from the abdomen and burrow into the soil to pupate.

Parasitized bees continue to forage and behave normally, although the ovaries of queens degenerate so that egg-laying ceases. The offspring reared by parasitized queens are significantly smaller than is usual (Alford 1968). After emergence of the parasitoids, the bee dies. The distribution and importance of braconid parasitoids of bumblebees is unknown. Most studies of *S. splendidus* have been carried out in the UK, but similar parasitoid larvae have been found in bumblebees in North America (Plath 1934) and Sweden (Hasselrot 1960), suggesting that they are widespread.

5.2.4 *Mutilidae (Hymenoptera)*

Mutilla europaea is a parasitoid of bee larvae, attacking bumblebees, honeybees, and probably other bee species. This is a rare insect in the UK, and little is known of the details of its biology. The female is wingless and resembles a large hairy ant in appearance. She invades bee nests and lays her eggs in the pupal cocoons. The larvae consume part or all of the host, and pupate inside a cocoon spun within the pupal cocoon of the host (Alford 1975).

5.3 Parasites and Commensals

The parasites associated with social insects have been authoritatively reviewed by Schmid-Hempel (1998). Numerous organisms from diverse taxa are found associated with bumblebees and their nests. Some probably have no impact on their hosts (commensals) while others are major sources of mortality. With a few notable exceptions such as the nematode *Sphaerularia bombi*, very little is known about the biology of these organisms. This is particularly true of micro-organisms; few studies have examined the microorganisms associated with bumblebees (or indeed those associated with most other insects) and we have very little idea of their importance to bumblebee population dynamics.

5.3.1 *Viruses*

Entomopox-like viruses have been found in workers of the bumblebees *B. impatiens*, *B. pennsylvanicus*, and *B. fervidus* in North America, although no adverse effects were found in infected individuals (Clark 1982). These viruses were most frequent in the salivary glands, and it seems likely that they are transmitted by ingestion of contaminated food.

5.3.2 *Prokaryotes (bacteria and others)*

It seems almost certain that many bacteria are associated with bumblebees and their nests, either as parasites or commensals, but there have been exceedingly few studies. In contrast, the bacteria *Melissococcus pluton* and *Paenibacillus larvae* which infect honeybees have been studied in considerable detail (Schmid-Hempel 1998). These organisms cause substantial mortality, demonstrating that bacterial diseases can be important in bees.

One particularly interesting group of prokaryotes associated with bumblebees are the Spiroplasmataceae. These bacteria-like organisms cause systemic infections in plants, and frequently occur on the surface of flowers. They have also been identified in various insects, including honeybees, the bumblebees *B. impatiens* and *B. pennsylvanicus* (Clark et al. 1985), and the solitary bees *Osmia cornifrons* and *Anthophora* sp. (Raju et al. 1981). In insects they occur in the gut and haemolymph.

These organisms may primarily be sexually transmitted diseases of plants that employ bees as vectors to move between hosts (Durrer and Schmid-Hempel 1994). In honeybees infected with *Spiroplasma melliferum* death occurs after about 1 week (Clark 1977). Little is known of the pathological effects of these organisms on other bees.

5.3.3 Fungi

A range of generalist fungal pathogens including *Cordyceps*, *Paecilomyces*, and *Beauveria* were recorded from UK bumblebees by Leatherdale (1970). Various other fungi that have been occasionally recorded from bumblebees include *Aspergillus candidus*, *Cephalosporium* sp., *Hirsutella* sp., *Paecilomyces farinosus*, and *Verticilium lecanii* (summarized in Schmid-Hempel 1998). As far as is known, none of these fungi regularly cause significant mortality in bumblebees. The yeasts *Candida* and *Acrostalagmus* do appear to be widespread in bumblebees; they were found in about 30% of queens examined by Skou et al. (1963), and they appeared to trigger abnormally early emergence from hibernation. It is possible that these yeasts are important causes of overwintering mortality (Schmid-Hempel 1998).

5.3.4 Protozoa

Several protozoans infect bumblebees. The primary route of transmission is through ingestion of spores; infected hosts release further spores from their gut. Protozoan parasites are probably widespread, although most studies have been localized. For example the flagellate trypanosome *Crithidia bombi* is frequent in the guts of bumblebees in Scandinavia (Schmid-Hempel 1998). In Switzerland, colonies of *B. terrestris* usually contract an infection within two weeks of the start of foraging activity (Imhoof and Schmid-Hempel 1999), and by the end of the season almost all nests become infected (Schmid-Hempel 1998). The microsporidian *Nosema bombi* is also seemingly common; Skou et al. (1963) found that 18 of 99 *Bombus* queens sampled in Denmark were infected, while Fisher (1989) found it in 10% of spring queens and 61% of mature colonies in New Zealand. It seems that colony size can influence the likelihood of infection, for Schmid-Hempel (1998) found that workers of *B. terrestris* from larger colonies were more likely to be infected; presumably colonies with more foragers are more likely to contract infections.

The effects that protozoa have upon their hosts are variable. *C. bombi* has high infectivity but low virulence. Infection spreads rapidly within nests through contact with contaminated nest material. Infection can also be contracted on flowers by drinking nectar contaminated by other bees, and so the protozoan can be transmitted horizontally between colonies (Durrer and Schmid-Hempel 1994). In laboratory conditions with unlimited food, effects on the host are generally small, including slight reductions in ovary size and overall colony growth (Shykoff and Schmid-Hempel 1991c). Reduction of ovary size of workers through infection with *C. bombi* may actually benefit the foundress, for infected workers are less likely to lay their own eggs (Shykoff and Schmid-Hempel 1991c). In contrast, when hosts are stressed through starvation, infection with *C. bombi* increases mortality by 50% compared to control groups (Brown *et al.* 2000). Since starvation is more likely during the early stages of colony development, when fewer workers are available to forage, this may be when the parasite has most impact (Schmid-Hempel 2001).

Other protozoans exhibit high virulence even in unstressed hosts. The neogregarinid *Apicystis bombi* is thought to have a major impact upon infected hosts, destroying the fat body and severely reducing colony reproduction (Durrer and Schmid-Hempel 1995). Infection with *N. bombi* may also reduce host fitness; Skou *et al.* (1963) found that infected queens entered hibernation late and left hibernation early, and this is likely to influence the probability of successful nest-founding. This parasite can spread rapidly within colonies, infecting both adults and larvae, but mortality seems to occur late in the season. Fisher (1989) did not find any significant effect of infection on colony growth. Contrary to expectation, one study has found that infected colonies actually produced more new queens and males than those that were uninfected (Imhoof and Schmid-Hempel 1998b).

Clearly the interactions between bumblebees and their parasites are complex. It seems likely that much of the variation between studies results from differences in the particular strain of parasite used and host population tested. When exposed to hosts from different regions with which it has not coevolved, *C. bombi* causes greater mortality (Imhoof and Schmid-Hempel 1998a). Similarly, Schmid-Hempel and Loosli (1998) demonstrate that susceptibility to *N. bombi* varies greatly between colonies of *B. terrestris*, most probably the result of host–parasite genotype–genotype interactions. It seems probable that host genotypes that are resistant to locally abundant strains of parasite will enjoy high reproductive success and increase in frequency, but that parasites will rapidly adapt to more abundant host genotypes so that susceptibility of host strains and virulence of parasites are in constant flux (Schmid-Hempel 2001).

Recent evidence suggests that the breeding system of most bumblebee species makes them particularly susceptible to intra-colony epizootics of parasites such as protozoans. Queens of most bumblebee species mate only once, so that workers are all full siblings (Estoup *et al.* 1995). Due to the haplodiploid genetics of Hymenoptera, this means that all workers within a colony are 75% related to each other. Parasites and pathogens probably act with positive frequency dependence, so that rare host genotypes are less likely to be infected (Hamilton 1980). Thus genetic variability within a colony of a social organism is likely to reduce the impact of parasites (Tooby 1982; Hamilton 1987; Sherman *et al.* 1988). Experimental tests with bumblebees support this hypothesis; infections of *C. bombi* spread more slowly among groups of unrelated workers than among related workers (Shykoff and Schmid-Hempel 1991a,b). In an elegantly simple experiment, Liersch and Schmid-Hempel (1998) manipulated the genetic variability in colonies of *B. terrestris* by moving brood between nests; when placed under field conditions, colonies with artificially enhanced variability suffered from fewer parasites.

Genetic variability can be greatly increased if queens mate with several males, but most bumblebee queens do not do so. Why not? Mating takes up to an hour (Alford 1975) and during this period the mating pair cannot fly. They are presumably vulnerable to predation at this time, so it may pay a queen to mate only once. Also, males place a mating plug in the reproductive tract of the female which is at least partially effective in preventing further matings (see Chapter 4). However, Schmid-Hempel (1998) argues that monogamy may be the best strategy for the female. He demonstrates that, because reproductive success of bumblebee colonies increases disproportionately with colony size, selection by parasites may actually favour monandry. A genetically diverse colony can expect a low prevalence of parasites. In contrast, a genetically uniform colony can expect either to be wiped out by an epizootic, or to have very low rates of parasitism (if by chance it avoids contact with parasites suited to its genotype). If the latter occurs, it will enjoy a very high reproductive success. Essentially, he suggests that queens are opting for a high risk strategy because the reproductive rewards from a low-risk strategy are poor.

Newly developed techniques for artificial insemination of bumblebees (Baer and Schmid-Hempel 2000) make it possible to test this hypothesis. Baer and Schmid-Hempel (2001) inseminated queens of *B. terrestris* with sperm from one, two or four unrelated males or four brothers. Once colonies were established they were placed out in the field, and their fitness and the prevalence of infection with *C. bombi* subsequently compared. As predicted, colonies with greater genetic variation had lower prevalence and intensity of infection (Figure 5.1). However, intriguingly, colony fitness, as measured by the number of males and queens produced, was highest in colonies with singly mated queens. This has not been adequately explained, but may be due to conflict between workers of different patrilines within the nest (see Chapter 3). This suggests that selection for resistance to parasites does favour multiple matings, but that other factors constrain the evolution of polyandry.

5.3.5 *Nematodes*

One of the best-known parasites of bumblebees is the nematode worm *Sphaerularia bombi*. This parasite is unusual in that it only attacks queens (and female *Psithyrus*), a strategy that inevitably restricts it to a very small proportion of the host population. The life cycle is described and illustrated in detail by Alford (1975). A mated nematode female enters through the cuticle of a queen while she is hibernating in the soil, and takes up residence in the haemocoel. By the time the queen emerges from hibernation the uterus of the nematode has everted and greatly enlarged, so that it dwarfs the rest of her body. Hosts are often multiply infected with several nematodes (Alford 1969; Poinar and van der Laan 1972). Infection rates can be high; Schmid-Hempel *et al.* (1990) reports an average of 12%, but on occasion far higher rates have been recorded (Alford 1975). The female nematode releases eggs into the haemocoel, which rapidly hatch. She may produce up to 100 000 offspring. The juveniles migrate to the gut and are egested with the faeces. They reach adulthood and mate in the soil, and so complete the cycle.

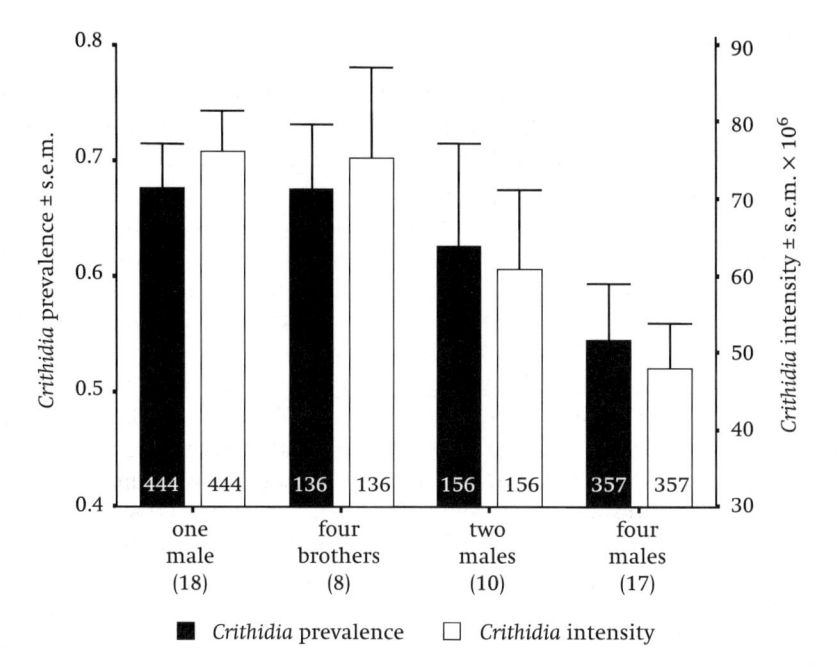

Figure 5.1. Intensity (the average number of cells found in the gut of workers per colony) and prevalence (the proportion of parasitized workers in a colony) of *Crithidia bombi* in colonies of *Bombus terrestris*. The foundress queens had been artificially inseminated with sperm from one, two, or four unrelated males, or four brothers. Both intensity and prevalence of infection differ significantly between treatments ($p < 0.0001$ and $p = 0.025$, respectively). Numbers within the bars indicate the number of workers checked per group. Number of colonies is given in parentheses at the bottom. From Baer and Schmid-Hempel (2001).

One of the more intriguing aspects of this parasite is its ability to influence the behaviour of its host. In a healthy queen the corpora allata swells in the spring, releasing hormones that stimulate the development of the ovaries. In a parasitized queen this does not occur. Rather than attempting to found a nest, the queen investigates overwintering sites and so contaminates them with parasites (Lundberg and Svensson 1975). Any queens remaining on the wing by May or June are generally infected, since healthy queens will by this stage have founded a nest. Infected queens become increasingly sluggish and eventually die in spring or early summer; they never manage to reproduce. Occasionally the parasite must pass through two generations in a single season for young queens have been found in late summer containing well developed adult nematodes (Alford 1975).

It seems that infected queens do not travel far. In New Zealand, *S. bombi* was presumably introduced accidentally with the first bumblebee releases in 1885. By 1990, the parasite had spread by only about 40 km, while the bumblebees had colonized the whole of New Zealand within a few years of their release (Macfarlane and Griffin 1990).

5.3.6 Mites (Acarina)

At least 15 genera of mites are associated with bumblebees (Alford 1975; Eickwort 1994; Schmid-Hempel 1998). The most familiar of these are mites of the genus *Parasitellus* (Mesostigmata; Parasitidae), which are very often to be seen attached to the bodies of adult bumblebees, particularly queens. These mites are only ever found in close association with bumblebees (Richards and Richards 1976; Schousboe 1987; Schwarz et al. 1996). However, they do not feed directly upon bumblebees, but are phoretic, using the adult bees for transport between nests. This is a common phenomenon; mites have poor locomotory abilities, but with their small size they can easily attach themselves to larger organisms and so gain a free ride (Evans 1992). *Parasitellus* species are thought to feed upon wax, pollen, and other small arthropods that are found in bumblebee nests (Richards and Richards 1976). Only the deutonymph stage is phoretic, colonizing new nests by transferring from workers to flowers, and then awaiting the arrival of another worker (Schwarz and Huck 1997). The prevalence of *Parasitus* spp. is generally high. Schousboe (1987) found that 15–28% of spring queens of *B. terrestris/lucorum* were infested in Denmark. Schwarz et al. (1996) recorded infestation levels of 22% on spring queens of *B. pascuorum* and 46–49% on *B. terrestris*, *B. lucorum*, and *B. lapidarius* in Switzerland. Comparable estimates from Corbet and Morris (1999) were 57% for *B. terrestris*, 83% for *B. pacuorum*, and 100% for *B. hortorum* in England. With this level of prevalence at the beginning of the season, it is not surprising that the vast majority of bumblebee nests become infested by the end of their growth (Huck et al. 1998).

Because these mites do not feed upon the bees themselves, it is debatable whether they have a negative impact. However, infestation levels can be high; Huck et al. (1998) report up to 165 deutonymphs on a single *B. lapidarius* queen. It seems inevitable that loads of this magnitude must hamper a queen's ability to fly, and so her ability to find food, a mate and a hibernation site.

A diversity of other mites are found on or in bumblebees, including the tracheal mite *Locustacarus buchneri* (Skou et al. 1963), *Scutacarus* spp. (Schousboe 1986), *Pneumolaelaps* spp., *Hypoaspis* spp. (Hunter and Husband 1973), and *Kuzinia* spp. (Goldblatt 1984). It seems probable that many more await identification. Some, such as *L. buchneri*, are truly parasitic, feeding directly on their host, and they undoubtedly reduce host fitness (Husband and Sinha 1970). However, most are not parasitic. For example *Scutacarus acarorum* feeds on fungi (Schousboe 1986), and *Kuzinia laevis* on pollen and fungi (Chmielewski 1969). They are all probably phoretic to varying degrees, and a range of species can be found on flowers visited by bumblebees (Schwarz and Huck 1997). One very small species, *Scutacarus acarorum*, is actually phoretic on larger mites such as *Parasitellus*, even while their hosts are phoretic on bumblebees (Schwarz and Huck 1997), inevitably bringing to mind 'A flea hath smaller fleas that on him prey; and these have smaller fleas to bite 'em, and so proceed ad infinitum' [Jonathan Swift 1733].

Because bumblebee nests are short lived, mites must attach themselves to new queens at the end of the season (Stebbing 1965; Richards and Richards 1976). They then overwinter with the queen in her underground hibernaculum, and infest the nest that she founds in the spring. In choice tests, deutonymphs of *Parasitellus fucorum* exhibited a strong preference for queens over males, and readily transferred from males to queens, which clearly makes sense since males do not go in to nests (Huck *et al.* 1998). In mature nests of *B. terrestris* the vast majority of deutonymphs are attached to new queens (D.G. unpublished data). Similarly, the phoretic instars of *Parasitellus, Pneumolaelaps,* and *Scutacarus* are all found more frequently on queens than on workers (Hunter and Husband 1973; Richards and Richards 1976; Schousboe 1986, 1987).

5.3.7 *Other commensals*

If a bumblebee comb is lifted a wriggling mass of insect larvae is usually revealed. These organisms scavenge upon detritus, dead brood, wax, pollen, and adult bee faeces. Many of these organisms also occur elsewhere (e.g. in bird's nests, or in the faeces of mammals). The majority have little or no adverse effects on their hosts, and may even be helpful in disposing of waste. In general the most abundant of these scavengers are dipteran larvae (maggots). One of the most frequent denizens of bumblebee nests is *Fannia canicularis* (Muscidae). *Volucella bombylans* (Syrphidae) also deserves particular mention for this fly is an obligate bumblebee nest commensal and the adult fly is a splendid and convincing bumblebee mimic (Evans and Waldbauer 1982). It is also polymorphic, with one morph mimicking bees with yellow stripes and a white tail (such as *B. hortorum*) and another mimicking black bumblebees with a red tail (such as *B. lapidarius*). If killed by the bees when attempting to enter a nest, the female fly immediately lays her eggs, which generally go unnoticed by her assailants (Sladen 1912).

Other nest scavengers include several species of lepidopteran larvae including *Endrosis sarcitrella* (Oecophoridae), *Vitula edmandsii* and *Ephestia kühniella* (Pyralidae) and an as yet unidentified member of the Tineidae (Smith 1851; Davidson 1894; Frison 1926; Alford 1975; Whitfield and Cameron 1993; Whitfield *et al.* 2001). In contrast with *A. sociella*, it seems that most of these species inflict little or no damage on the nest. Lastly, all stages of the life cycle of the coleopteran *Antherophagus nigricornis* can be found in bumblebee nests (Alford 1975). This beetle is phoretic; the young adult beetle climbs on to a flower, and then hitches a ride to a nest by clinging to the tongue, antennae or leg of a visiting bumblebee (von Frisch 1952). The adult beetle and its offspring feed on nest debris.

These nest commensals themselves support a range of parasitoids including ichneumonid and braconid wasps which may often be found in bumblebee nests (reviewed in Alford 1975; see also Whitfield and Cameron 1993; Whitfield *et al.* 2001).

5.4 The Immune System of Bumblebees

Bumblebees do have a defence against internal parasites and pathogens; as in most insects, they have an immune response (Gupta 1986). Small foreign particles are phagocytosed by haemocytes. Larger foreign bodies are also attacked by haemocytes in large numbers, and may eventually be entirely enclosed. The haemocytes subsequently melanise, forming a tough capsule around the intruder and isolating it from the host tissue. In bumblebees, foreign bodies (including parasitoid eggs) are rapidly encapsulated in this way (Schmid-Hempel 1998; Allander and Schmid-Hempel 2000; Moret and Schmid-Hempel 2000). Encapsulation may kill the parasitoid, but we do not know how often bumblebees are successful in doing this. Activation of the immune response is known to be costly. Moret and Schmid-Hempel (2000) stimulated the immune response of workers of *B. terrestris* by implanting non-pathogenic and non-toxic latex beads or lipopolysaccharides (molecules that are normally found on the surface of bacteria). When maintained on a near-starvation diet, worker mortality was substantially increased by this procedure, and it was concluded that it was the cost of melanization that directly caused mortality. A similar approach has also revealed that when exposing workers to two consecutive challenges, the second receives a weaker encapsulation response than the first, suggesting that the immune response is readily depleted (Allander and Schmid-Hempel 2000).

The encapsulation response has been found to be very variable between colonies of *B. terrestris* (Schmid-Hempel 1994; König and Schmid-Hempel 1995; Schmid-Hempel and Schmid-Hempel 1998). This may in part reflect environmental effects; colonies with reduced resources are likely to exhibit a weaker immune response. Also, there is likely to be genetic variability in immune response between bumblebee populations as a result of variation in their past exposure to pathogens.

5.5 Social Parasitism

The success of social insects is largely due to their ability to accrue resources efficiently through division of labour, and to store these resources within the nest. A number of organisms have evolved methods for diverting the efforts of social insects to their own ends, so that the parasites benefit from the resources gathered, and are often directly cared for by their hosts. This is known as social parasitism. Very often social parasites are closely related to their hosts, for they are thus better equipped with the chemical armoury necessary to subvert the efforts of their host to their own ends. Bumblebees suffer from two sorts of social parasites; other non-*Psithyrus* bumblebees (either of their own or a different species) and cuckoo bees (subgenus *Psithyrus*). These are considered in turn, although in many respects the details are similar.

5.5.1 Nest usurpation

Bumblebee queens vary greatly in their time of emergence from hibernation, even within species, perhaps as a result of their choice of overwintering site or their body condition. Late-emerging queens, when searching for nest sites, will often find that suitable sites are already occupied. In this situation they may attempt to take over an established nest for themselves (Alford 1975). Species that tend to emerge late, such as *B. rufocinctus*, are particularly prone to this behaviour (Hobbs 1965*b*). In years when queens are abundant, nest usurpation may become very frequent (Bohart 1970). Sladen (1912) describes one nest of *B. terrestris* which contained 20 dead queens, presumably the foundress queen and successive usurpers. The process of usurpation is described in detail by Alford (1975). The foundress and intruder may avoid each other for some time, but a fight to the death eventually ensues. The fights are usually brief, being concluded when one queen successfully stings the other. If the intruder is successful she will continue to care for the brood in the normal manner. Usurpation becomes rarer as colonies grow in size, and is never observed once the second batch of brood have emerged.

There are very few data as to how frequently nest usurpation is successful, although the fact that the strategy persists suggests that it must sometimes succeed. However, Paxton *et al.* (2001) used microsatellite analysis of workers genotypes to demonstrate the presence of workers that were unrelated to the queen in 6 of 11 nests of *B. hypnorum*. They concluded that these workers were probably the offspring of a previous queen that was usurped (the alternative explanation being that these workers had drifted between nests).

Usurpation only occurs within species of the same subgenus (Hobbs 1965*b*). Thus for example *B. terrestris* will often attempt to usurp its sister species, *B. lucorum*, which tends to emerge slightly earlier. It is easy to see how this behaviour could eventually evolve into obligate usurpation. In arctic North America, *B. hyperboreus* frequently usurps *B. polaris* (Milliron and Oliver 1966; Richards 1973). Because the season is short, the usurping queen does not rear any workers of her own, but rears only reproductives. Since no worker *B. hyperboreus* are reared, the life cycle then becomes identical to that of *Psithyrus* (see below). There is one *Bombus* species outside of the subgenus *Psithyrus* which is suspected to have adopted a obligate parasitic lifestyle. No workers of *B. inexspectatus* have been recorded, and it is thought that this species may be an obligate parasite of its close relative *B. ruderarius* (Yarrow 1970).

5.5.2 *Cuckoo bees* (Psithyrus)

Cuckoo bees (subgenus *Psithyrus* spp.) were for many years placed in a separate genus to the 'true' bumblebees (Williams 1994). In all probability they evolved from social *Bombus* via nest usurpation as described above. Most recent authors agree that they have a monophyletic ancestry (Plowright and Stephen 1973; Pekkarinen *et al.* 1979; Ito 1985; Williams 1985*a*, 1994; Pamilo *et al.* 1987). *Psithyrus* do not have pollen baskets and are unable to produce wax, and so they now have an obligate dependency on social bumblebees.

Female *Psithyrus* emerge from hibernation later than their hosts, and spend some time foraging on flowers while their ovaries develop. They then search for nests of their host species, probably at least in part using scent (Frison 1930). The female *Psithyrus* will enter a bumblebee nest and attempt to dominate or kill the foundress. *Psithyrus* females have a more powerful sting and mandibles than their hosts, and they have a generally thicker exoskeleton. They are thus at a distinct advantage in conflicts with the foundress queen, and if they attack a colony before the second batch of workers have been produced they usually prevail. For example, Frehn and Schwammberger (2001) placed *B. (P.) vestalis* queens into each of three young *B. terrestris* nests: the host queens died after 6 h, 2 and 6 days, respectively. Sometimes the host queen will retreat and become subservient to her usurper (Hobbs et al. 1962; Hobbs 1965a). She behaves much like a worker; if the nest is disturbed she engages in active nest defence, rather than hiding within the comb as she would normally do.

Although bumblebee queens are unable to repel *Psithyrus*, a large group of workers may do so. When *Psithyrus* attempt to invade large nests they are usually fiercely attacked by a number of workers and may be killed. Interestingly, most *Psithyrus* species are able to parasitize host species from more than one subgenera (Sakagami 1976), and they are notably less specific than *Bombus* species which are only able to usurp nests of their own species or very close relatives (Alford 1975). Perhaps their physical strength means that they do not need to closely match the chemistry of their hosts to successfully invade. In the UK at least, *Psithyrus* often resemble their hosts in coloration. Most authors agree that this is probably not to aid entry in to the nest, but that the *Psithyrus* and their hosts are members of Müllerian mimicry groups (Alford 1975; Prys-Jones and Corbet 1991).

If her take-over attempt is successful, the *Psithyrus* female will lay eggs which will be reared by the bumblebee workers as their own (Weislo 1981; Fisher 1987). This form of social parasitism is known as inquilinism (Hölldobler and Wilson 1990). Since *Psithyrus* do not have a worker caste, all of the offspring are males or future breeding females. The invading *Psithyrus* may eat host eggs and young larvae, but older ones are allowed to develop to add to the work force. Nests that have been invaded produce few or no host queens or males, although workers do lay eggs and a few of the resulting male offspring may survive (Frehn and Schwammberger 2001). The *Psithyrus* queen presumably attempts to prevent this; she chases and mauls the workers, particularly those with active ovaries (Fisher 1988; Frehn and Schwammberger 2001).

Nests may be easier to appropriate when they are small, but then a smaller workforce will be available for rearing the offspring of the *Psithyrus* female (Fisher 1984). In nests with few workers, few *Psithyrus* are reared and they tend to be smaller in size than usual (Alford 1975). Thus there is likely to be a trade-off between the ease with which nests can be taken over and the benefits to be accrued from doing so. Müller and Schmid-Hempel (1992b) found that female *B. (P.) bohemicus* preferentially attacked the largest nests of *B. lucorum*. Little information is available on the frequency with which bumblebee nests repel invasion. *Psithyrus* may target large nests simply because they are easier to find (presumably they produce a stronger odour).

Bumblebee queens may also face a trade-off with regard to the optimum time at which to leave hibernation and found a nest. Müller and Schmid-Hempel (1992b) found that early-founded nests of *B. lucorum* were most frequently attacked by *B. (P.) bohemicus*, but early-founded nests have a higher expected reproductive output if they are not taken over by *Psithyrus*. Thus we might expect that queens should nest early in areas where *Psithyrus* are scarce, but nest later when they are common.

The frequency of invasion of bumblebee nests by *Psithyrus* is highly variable both between localities and years. Alford (1975) considered attacks by *Psithyrus* to be generally rare, but high infestation levels have been recorded. Awram (1970) found that more than 50% of nests of *B. pratorum* were taken over by *B. (P.) sylvestris*, while Sladen (1912) reports rates of 20–40% for invasion of *B. lapidarius* nests by *B. (P.) rupestris* (both studies were carried out in England). More recently Müller and Schmid-Hempel (1992b) recorded rates of attack of *B. lucorum* by *B. (P.) bohemicus* approaching 30% in Switzerland, although these attacks were not necessarily successful.

In most other respects the life cycle of *Psithyrus* is rather similar to that of their hosts. Mating occurs in mid to late summer, and only females hibernate. Males are far more frequently seen that females, and they are very commonly observed feedingly sluggishly on flower heads of thistles (*Cirsium* and *Carduus* spp.) and knapweeds (*Centaurea* spp.) in July and August.

6
Foraging economics

'Time is honey'
(Bernd Heinrich 1996)

Insect foraging behaviour in particular is an area in which knowledge has advanced rapidly in recent years, and much of this research has focussed on bumblebees. There are a number of reasons why bumblebees are excellent organisms for studies of foraging behaviour: they are abundant in the northern hemisphere where most researchers are based; they are conspicuous, docile, and easily observed without causing interference; and they forage ceaselessly, even under cool, cloudy conditions when other insects are inactive. Furthermore, studies of pollinator behaviour are of particular interest because the majority of flowering plants rely upon insects to mediate pollen transfer. Thus it is the behaviour of insects which determines which flowers will set seed and which will not, and which governs the pattern of transfer of gametes between plants. Aside from their economic and ecological importance as pollinators, bumblebees have become popular vehicles for examining the assumptions and predictions of foraging models and the interplay between learning, memory constraints and foraging efficiency in a complex and unpredictable environment (e.g. Heinrich 1979*a*; Cresswell 1990; Dukas and Real 1993*a,b,c*; Dreisig 1995).

Bees and a number of other insect groups, including butterflies and moths (Lepidoptera), some flies (Diptera) and beetles (Coleoptera), depend for their sustenance upon pollen or nectar rewards provided by flowers. Both nectar and pollen have much to recommend them as food for insects. Nectar provides sugars and water necessary to sustain an active adult insect, while pollen is a rich source of protein. Neither are protected by toxins or physical defences, as are most plant tissues. Proteins are particularly important during growth of immature insects, but it is generally only adult winged insects that specialise in visiting flowers. Immature stages do not have the mobility necessary to gather such carefully rationed and sparsely scattered resources. Of course, bees have overcome this problem; the larvae feed upon pollen and nectar collected by the adults.

Even for insects capable of flight, efficient collection of floral rewards is problematic. The distribution of rewards is unpredictable in time and space; individual plants and plant species open their flowers at different times of the day and flower at different times of the year (Waser 1982*b*; Zimmerman and Pyke 1986; Real and Rathcke 1988). Rewards per flower vary greatly between plants of a single species and between flowers on a single plant due to genetic and environmental influences on reward production rates and also in response to the pattern of depletion of rewards by foragers (e.g. Pleasants and Zimmerman 1979, 1983; Zimmerman 1981a,b; Brink 1982; Thomson et al. 1982; Pleasants and Chaplin 1983; Cruden et al. 1984; Zimmerman and Pyke 1986; Real and Rathcke 1988; Mangel 1990; Waser and Mitchell 1990; Gilbert et al. 1991). At any one time, many flowers may be empty (Wetherwax 1986; Real and Rathcke 1988; Cresswell 1990; Waser and Mitchell 1990). To add to the difficulties many plant species hide their floral rewards within complex flowers so that only insects with an appropriate morphology can enter them; for example, flowers of broom, *Cytisus scoparius*, produce an abundance of pollen, but it can only be accessed by heavy insects such as bumblebees that have sufficient weight to depress the keel of the flower, so revealing the stamens. Very often, the nectaries are located at the bottom of a narrow tube so that efficient nectar extraction necessitates a proboscis which in length roughly matches or exceeds the depth of the tube (e.g. Inouye 1978, 1980a; Pyke 1982). Thus many of the flowers which a forager encounters may have rewards which are at least partially inaccessible. Even if the forager possesses a suitable morphology, learning to handle flowers with complex structure takes time (Kugler 1943; Schremmer 1955; Weaver 1957, 1965; Macior 1966; Heinrich 1976b; Laverty 1980, 1994a; Waser 1983; Schmid-Hempel 1984; Lewis 1986). Also it seems that insects are unable to retain effective handling skills while foraging among several plant species with different flower structures (Heinrich et al. 1977; Lewis 1986; Woodward and Laverty 1992). Thus insects must make economic decisions while faced with a bewildering array of flowers of varying abundance, structure, colour and reward, incomplete knowledge as to which flowers contain rewards and as to how to extract these rewards, and limitations on their ability to simultaneously remember handling skills for a range of different flowers (Wells and Wells 1986).

Making the wrong decisions is particularly costly for bumblebees because, for them, flight is energetically very costly. Many of the details of bumblebee flight have been revealed by a series of ingenious experiments performed by Charles Ellington and coworkers. They persuaded bumblebees to fly in place against an air stream of variable velocity in a wind tunnel, using moving visual cues to convince the bees that they were making forward progress. They were thus able to film bees flying at up to 4.5 m s^{-1} with a stationary camera. To summarize their results very briefly, bumblebees beat their wings at about 160–200 Hz in flight, with larger bees tending to have slightly lower frequencies than smaller bees (Dudley and Ellington 1990; Hedenström et al. 2001). The frequency and amplitude of wingbeats does not appear to change whether hovering or moving forwards up to a speed of about 4.5 m s^{-1} (Dudley and Ellington 1990).

To maintain such high wingbeat frequencies requires considerable energy expenditure. Using thermal balance analysis, Heinrich (1975a) calculated that flying bumblebees consume about 80–85 ml O_2/g/h, or roughly 0.04 Watts per worker. This agrees reasonably well with analysis of oxygen use of bees flying in a sealed wind tunnel, which suggest consumption levels between 40 and 70 ml O_2/g/h

(Ellington et al. 1990). Interestingly, as with wing beat frequency, the metabolic rate of bumblebees does not seem to vary with flight speed, at least within the range 0–4 m s^{-1}. More recently, Ellington's group pioneered the use of doubly labelled water to quantify the metabolic rate of flying bumblebees (Wolf et al. 1996). This has the advantage that it can be used for free flying bees under field conditions (Wolf et al. 1999). This approach revealed great variation in the metabolic costs of individual B. terrestris flying in windy conditions, which are as yet unexplained. Predictably, the average metabolic costs of flying in windy conditions were higher than when flying within the shelter of a greenhouse. To obtain sufficient oxygen for respiration during flight, bumblebees cannot rely on diffusion of oxygen through the trachea. Instead they actively pump air in and out of internal body sacs by contracting and extending the abdomen (Heinrich 1979b; Komai 2001).

The estimates for the metabolic costs of flight suggest that flying bumblebees have one of the highest metabolic rates recorded in any organism, being 75% higher than that of hummingbirds. To illustrate the magnitude of their metabolic rate, Heinrich compares a flying bumblebee to a jogging human male (Heinrich 1996). The human burns the energy in a Mars bar in roughly 1 h. A bumblebee of equivalent mass would burn the same energy in just 30 s.

Thus for bumblebees, profitable foraging is a challenge. The rewards they must gather are sparsely and to some extent unpredictably distributed, yet to gather them they must expend considerable energy in flight. If the time taken to locate and handle each flower is too long, or the reward too small, then foraging will result in a net loss of resources. A queen that makes poor decisions will quickly starve, while an inefficient worker will drain the resources gathered by her nestmates. It is perhaps not surprising that bumblebees have evolved an array of behaviours to improve (if not maximize) foraging effiency.

The foraging behaviours of bumblebees can be conveniently considered as operating on a variety of spatial scales. At the landscape level, we can examine how far they travel from their nests to obtain forage, and how they exploit the patchy nature of floral resources. We might also examine which flower species they choose to feed upon, and within flower species which individual flowers they visit (for they are selective even at this fine scale). Bumblebees do not forage in isolation, but are influenced by the behaviour of siblings, other conspecifics, other bumblebee species, and other flower visiting insects. All of these factors influence the foraging success of bumblebees, and have implications for the efficacy of crop pollination and the reproductive success and population structure of wild plant populations. I shall examine them in turn.

7
Foraging range

Both solitary and social bee species provision their broods by central place foraging, in which they accumulate reserves of pollen and sometimes nectar in a nest. The foraging range of bees is a fundamental aspect of their ecology, for it determines the area of habitat which a nest can utilize (Bronstein 1995; Westrich 1996). For example, if we are to study the effects of habitat management on bee populations, the appropriate scale of experimental plots depends on the area over which the occupants of each nest forage. To encourage bees, nest sites and forage sites must be sufficiently close together that foragers can economically visit the forage. Similarly, when bees are used for crop pollination their foraging range will determine the area of crop that workers from a nest may visit, and the distances over which pollen might travel (Free 1970; Procter and Yeo 1973; Corbet et al. 1991). This is particularly relevant when transgenic crops are being grown.

Flight ranges are known to vary among bee species. Quantifying the foraging range of honeybees is relatively easy because the 'waggle' dance of a returning forager describes the distance and direction of the food source (von Frisch 1967; Seeley 1985a). Studies of this sort indicate a foraging range of 1–6 km, more rarely up to 20 km (von Frisch 1967; Visscher and Seeley 1982; Seeley 1985b; Schneider and McNally 1992, 1993; Waddington et al. 1994; Shwarz and Hurst 1997). Honeybees are clearly not 'doorstep foragers'. Little is known of the foraging range of most other bee species, but those estimates that are available suggest that honeybees are unusual, and that most bees forage over much shorter distances. For example, *Melipona fasciata* travels up to 2.4 km (Roubik and Aluja 1983) and members of the Trigonini over 1 km (Roubik et al. 1986). Solitary bee species are generally thought to travel only a few hundred metres at most (Schwarz and Hurst 1997). The maximum recorded foraging distance is 24 km for euglossine bees (Janzen 1971).

It has long been assumed that bumblebees forage as close to their nest as possible, for intuitively this would seem to be the sensible strategy (Braun et al. 1956; Free and Butler 1959; Crosswhite and Crosswhite 1970; Teräs 1976; Heinrich 1976b; Bowers 1985a; Free 1993). This view is strengthened by optimality models which predict that choosing the closest foraging sites is the best strategy, all else being equal (Heinrich 1979b) (but of course the same arguments ought to apply to honeybees, yet, they travel great distances). Remarkably, despite the wealth of literature on bumblebee foraging behaviour there are rather little hard data available on bumblebee foraging ranges (Bronstein 1995; Osborne et al. 1999; Cresswell et al. 2000).

They are unexpectedly hard to quantify. Four alternative means have been attempted, each with limited success.

7.1 Studies with Marked Bees

This is the obvious approach; one simply marks the bees, and then searches the flowers in the surrounding area to see where the bees go to forage. If bees are marked while foraging this is not particularly informative. Bowers (1985a) marked *B. flavifrons* while foraging in discrete meadows surrounded by forest in Utah. Subsequent recaptures were all in the same meadows in which the bees had been marked, leading to the conclusion that bees did not move between meadows. However, the location of the nests were not identified. One would expect individual bees to return to the same sites repeatedly (see Thomson *et al.* 1982, 1987, 1997), but this does not mean that their nest was situated nearby.

It is far better to mark the bees at their nest, and then search for them. Unfortunately, this is also fraught with difficulties. Where forage is abundant and available close to the nest (e.g. if the nest is in a flowering crop), a small proportion of marked bees have been observed foraging, and these tended to exhibit a leptokurtic distribution with most foraging close to the nest (Saville 1993; Schaffer and Wratten 1994). Where nests have been situated in a more typical fragmented habitat with small, scattered sources of forage, very few marked bees have been recorded (Kwak *et al.* 1991a; Dramstad 1996; Saville *et al.* 1997). For example, Dramstad (1996) found that few *B. terrestris/lucorum* foraged on sources of suitable forage located within 50 m of their nest, and searches up to 300 m located very few marked bees. Dramstad concluded that bumblebees do not forage close to their nests, but exactly where they do forage was not clear. Schaffer (1996) suggested that the leptokurtic distribution observed in earlier studies was an artefact of the experimental design. Observer effort was always biased towards searching areas close to the nest for the simple reason that the area to be searched increases as the square of the distance from the nest. To properly evaluate the distribution of marked bees within even a moderate distance of the nest such as 1 km would require 3.1 km^2 to be thoroughly searched; a formidable task. Since more recent studies (see below) suggest that the actual foraging limit may be much greater than this, the mark-reobservation method is of little use without a huge team of researchers.

7.2 Homing Experiments

It has long been known that the Aculeate Hymenoptera possess homing abilities. The pioneering entomologist Jean-Henri Fabre demonstrated that the solitary sphecid wasp *Cerceris tuberculata* and the gregarious bee *Chalicodoma muraria* could return to their nests when transported several kilometres in darkened boxes (Fabre 1879, 1882). Similar experiments have since been performed on a range of solitary and social species (reviewed by Wehner 1981; more recently Ugolini 1986; Ugolini et al. 1987; Schöne et al. 1993a,b; Southwick and Buchmann 1995; Chmurzynski et al. 1998; Capaldi and Dyer 1999). Most of these studies have examined homing from distances ranging from 100 m to 3–4 km, and all have found that at least a proportion of the released insects return. The greatest distance over which Hymenoptera have been found to successfully return home is 23 km in the Euglossine bee *Euplusia surinamensis* (Janzen 1971). Recently, homing experiments have also been carried out on the bumblebee *B. terrestris* (Goulson and Stout 2001). The maximum distance from which a bee successfully returned to its nest was 9.8 km, with a clear decline in the probability of a bee returning as the distance over which it was displaced increased (Figure 7.1). Interestingly, no relationship was apparent between the displacement distance and the time to return to the nest. Times from release to recapture varied between 6 h (from 2 km) to 9 days (from 3.5 km). Notably, one bee that returned to its nest after being displaced by 4.3 km was observed on a subsequent occasion gathering nectar at the release site, which contained a large patch of nectar-rich flowers.

This is all very interesting, but what does it tell us about the natural foraging range of bumblebees? This is a matter for debate. Prolonged homing times have been found in studies of other species. For example, of 374 *Anthophora abrupta* displaced up to 3.2 km from their nest, homing times varied from 20 min to 50 h (Rau 1929). This may provide a clue as to how Hymenoptera locate their nests. *B. terrestris* is capable of flying at speeds of up to 15.7 km h^{-1} even when laden with a harmonic radar aerial (Osborne et al. 1999), and so could theoretically return from up

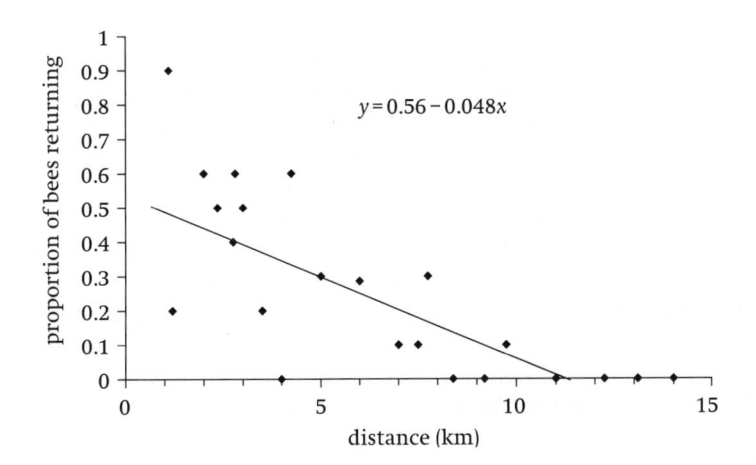

Figure 7.1. The proportion of marked *B. terrestris* returning to their nest after artificial displacement over distances up to 15 km. In total, 220 bees were displaced in batches of 10, so that each point marked is based on 10 bees. There was a clear negative relationship between the proportion of bees returning and the distance of the release site (linear regression, $r^2 = 0.55$, $p < 0.001$). From Goulson and Stout (2001).

to 15 km within 1 h. Pigeons have prodigious homing abilities, and do so rapidly, flying more-or-less directly from the release site to their loft. Their homing abilities have been the subject of intensive studies, and it is thought that they possess some sort of coordinate system which enables them to determine their location relative to home, perhaps based on olfactory or magnetic stimuli (Witschko and Witschko 1988; Wallraff 1990; Able 1994). They may also be able to detect information as to the direction of transport during the outward journey, and simply reverse their path (Wallraff 1980; Wallraff et al. 1980). It seems highly unlikely that Hymenoptera possess either of these abilities. If they did, we would expect them to return swiftly to their nests. Most recent authors have concluded that a third mechanism is most likely. Displaced insects are thought to use a systematic search for familiar landmarks, and then use these to locate their nest (reviewed in Wehner 1989). Desert ants (*Cataglyphis* spp.) engage in systematic searches when displaced to unfamiliar terrain (Wehner 1996). Honeybees are known to use visual landmarks to aid navigation between the nest and forage (Wehner 1981; Kastberger 1992; Dyer 1996), and use a sun compass to relate the positions of landmarks and the nest (Wehner 1994). Honeybee homing is better when prominent horizon landmarks are present (Southwick and Buchmann 1995). Searching for familiar landmarks could lead to protracted homing times, and explain why from more distant sites many bees fail to return.

This hypothesis is consistent with the behaviour of released bumblebees; upon release, they immediately begin to circle the release site (Goulson and Stout 2001). The radius of the circle gradually increases until the bees are lost from sight. Thus, it seems probable that the homing mechanism used by Hymenoptera is a systematic search of an expanding circle surrounding the release site until a familiar landmark is recognized. If this is the homing mechanism in use, then we might expect all bees released within their home foraging range to successfully return to the nest, and to do so rapidly, while bees released beyond their range should return slowly if at all. If all bees had a similar knowledge of the environment, this would lead to a stepped (non-linear) response between distance and both the proportion of bees returning and the speed with which they do so. With a large data set it may be possible to test for such relationships. However, individual bees are likely to vary in their ability to home according to their foraging experience (they typically live for only a few weeks) and also to the particular directions that they have previously explored, which will determine the number and distribution of familiar landmarks. Rau (1929) found that homing success in *A. abrupta* was strongly related to age, with older (and presumably more experienced) bees being much more likely to return to the nest. Such variability may obscure relationships between displacement distance and success in homing.

It seems improbable that a bee released 9.8 km from its nest could find familiar landmarks unless its home range were several kilometers in radius (Goulson and Stout 2001). Since one marked bee which returned to its nest from 4.3 km was subsequently seen foraging at the site where it was released, bumblebees are clearly capable of remembering the location of forage at such distances and successfully navigating too and from these patches. Whether bumblebees locate forage at such distances from their nests naturally remains to be determined. Since many bumblebees successfully returned from considerable distances (>5 km), it seems likely that *B. terrestris* naturally forage over several kilometers. However, we must be cautious is interpreting homing experiments because we cannot directly observe the bees during most of their journey home. We do not know the size of the area that a bumblebee can search for familiar landmarks (if indeed that is what they do), and so we can only speculate as to the likely foraging area.

7.3 Harmonic Radar

Harmonic radar is an ingenious system for tracking insect movements that was developed by the Natural Resources Institute in conjunction with the Institute of Arable Crops Research, Rothamsted (Riley *et al.* 1996, 1998; Osborne *et al.* 1997). The technique involves attaching a 16 mm vertical aerial-like transponder to the thorax of the bumblebee (or other insect) (Plate 6). This captures the energy in radar emissions sent out from a base unit and re-emits the energy at a higher frequency, providing a signal that can be tracked as the bee flies. The transponder is very light in comparison with the weight of a typical bumblebee forager (6–7% of the insects' body mass), and is much lighter than the normal foraging load which can reach 90% of the body mass (Heinrich 1979*b*). As far as can be ascertained the transponder does not seem to interfere greatly with the behaviour of the bumblebee, although bees with transponders do take longer to complete foraging trips than usual, perhaps because the transponder impedes their handling of flowers (Osborne *et al.* 1999). These studies have provided some fascinating insights into the flight paths of the bumblebee *B. terrestris*. For example, they have revealed that bumblebees can compensate for cross winds and manage to fly directly between the nest and patches of forage by flying at an angle to their intended course (Riley *et al.* 1999). Unfortunately this technique has a major limitation with respect to determining foraging range, for bees can only be detected up to 700 m, and then only if they remain within a direct line of sight of the radar equipment. Osborne *et al.* (1999) found that very few foragers flew less than 200 m from their nest, even though there were patches of suitable forage just 50–100 m from the nest. In separate studies in June and August, Osborne *et al.* (1999) recorded mean maximum distances from the nest of 339 m (range 96–631 m) and 201 m (range 70–556 m), respectively (Figure 7.2). However, many of these bees flew beyond the range of the radar, being lost behind hedges or other landscape features, so these do not represent the actual foraging ranges; 13 of 35 tagged bees flew beyond radar range in June, and 14 of 30 in August. This study confirms that bumblebees do not often forage on the closest available patches of forage, but without improvements to the technology the harmonic radar approach cannot tell us what the true distribution of foraging ranges is, or what the limit to foraging range might be.

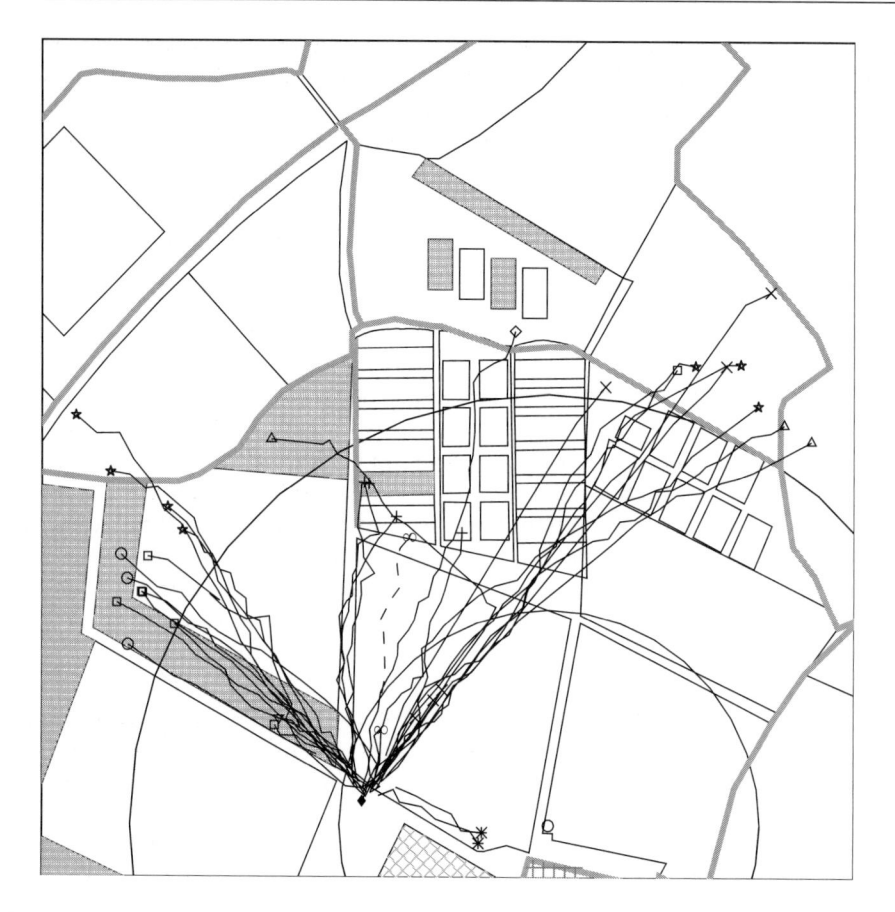

Figure 7.2. Harmonic radar tracks of the bumblebee, *B. terrestris*, flying away from the nest. Rings are at 200 and 400 m from the radar. Shaded areas are patches of forage. Thick lines are hedges. Each symbol denotes an individual bee. Thirty five outward tracks are shown, made by 9 bees. One bee (normally marked as +) stopped for 5 min 54 s and then continued. The continuation is plotted as _ _ _ _∞. The ends of tracks are where bees were lost from the radar, and do not necessarily indicate where the bee stopped to forage. From Osborne *et al.* (1999).

7.4 Modelling Maximum Foraging Range

The economics of bumblebee foraging have been the subject of considerable research, primarily by Bernd Heinrich (reviewed in Heinrich 1979*b*). Detailed information is available on the energetic cost of bumblebee flight, on the amount of forage that they can carry, the time it takes them to handle flowers of different species, and the energetic rewards that they obtain per flower. Cresswell *et al.* (2000) combined these data to estimate the maximum distance that a bumblebee could travel to reach a patch of nectar-rich flowers and return with a net profit.

The most accurate estimates of flight speeds are provided by recent studies using harmonic radar, which arrived at an average speed of 7.1 m s^{-1} for *B. terrestris* (Riley *et al.* 1999). The energetic costs of flight have been estimated from oxygen consumption rates to be 1.2 kJ h^{-1} (Ellington *et al.* 1990). Estimates of the volume of the honey stomach (which determines the maximum volume of nectar that a bee can carry) vary from 60 to 100 μl (Allen *et al.* 1978; Heinrich 1979b), and vary greatly according to the size of the worker (Goulson *et al.* 2002b). Nectar concentrations vary greatly between flower species, and also with time of day, but commonly fall within the range 40–60% (about 1–1.5 M sucrose or equivalent) (Cresswell *et al.* 2000). The metabolism of nectar sugars yields 16.7 kJ g^{-1} (Heinrich 1979b). Assume that a worker leaves the nest with an empty honey stomach, and occasionally stops to feed at flowers to fuel the outward flight. If it returns with any nectar, it has thus made a profit. The limit to the distance from which the bee can return with a net profit is simply given by the time taken to burn up all of the sugars in the honey stomach on the return flight. For a 40% nectar solution and a honey stomach capacity of 80 μl, the maximum range is about 10 km (Cresswell *et al.* 2000). If the nectar is more concentrated, or if the bees concentrated the nectar within their honey stomach as they foraged, then the range could be larger.

The model is concerned only with nectar collection, and so tells us nothing about the economics of trips to gather pollen, or trips where both nectar and pollen are collected. Since pollen is the only source of protein for larval growth, if pollen is in short supply near to the nest it is conceivable that workers could engage in flights of more than 10 km to obtain it. Such flights would result in a net loss of energy, but could be fuelled by visits to nearer nectar sources on the outward and return journeys. Unfortunately, the vast majority of studies of the economics of bumblebee foraging have focused exclusively on nectar collection, and so we have a very poor understanding of the energetics of pollen collection.

7.5 So Why do not Bumblebees Forage Close to their Nests?

Cresswell's model makes two other interesting points. That bees should choose the forage patch closest to the nest, given a choice of equally rewarding patches; but that because flight is swift relative to the time spent within patches of flowers, the more distant site does not have to be much more rewarding to be the better option. Thus for example, if given the choice between a patch of flowers immediately adjacent to the nest, and a more distant patch in which average nectar rewards are twice as high, it may be worth flying up to 4 km further to reach the more distant patch.

If this model is approximately correct in its assumptions, then it may explain why bumblebees often do not seem to visit patches of apparently suitable forage close to their nests. Even in the most fragmented and impoverished habitat there is likely to be a considerably better patch located within a radius of, say, 5 km (an area of 78 km^2!).

A second and compatible hypothesis to explain why most individual bumblebees do not forage close to their nest is that it minimizes intra-colony competition (Dramstad 1996). Colonies of some species such as *B. terrestris* grow to contain up to 400 workers (Alford 1975), and a work force of this size would quickly deplete resources close to the nest; Heinrich (1976a) found that bumblebees could remove 94% of the standing crop of nectar within an area. Since it is the cumulative foraging success of all foragers that determines colony success, one would predict that there is a trade-off between travel time and competition. Patches that are near the nest should always receive more visits than those that are further away, but the difference need only be slight if travel is rapid (Dukas and Edelstein-Keshet 1998; Cresswell *et al.* 2000). Given that each individual bee may visit hundreds, sometimes even thousands, of flowers in a foraging bout, one might expect only a very few bees to visit the nearest patches if the patches are small. This might explain the very low numbers of observed visits to patches close to the nest described by Dramstad (1996) and Osborne *et al.* (1999).

An alternative potential explanation for the apparent tendency of bumblebees to forage far from their nest relates to predation. It has been suggested that colonial food provisioners such as bumblebees may avoid foraging close to their nests so as to avoid attracting predators or parasites to the nest (Dramstad 1996). Bumblebee nests are attacked by a range of organisms including queens of their own species (reviewed in Alford 1975). Attacks by cuckoo bees (*Psithyrus* spp.) can be particularly frequent; Sladen reported that 20–40% of *B. lapidarius* nests were taken over, while Awram (1970) found that more than 50% of nests of *B. pratorum* were infested by *B. (P.) sylvestris*. Although these frequencies may be atypical, they nonetheless suggest that selection pressure to avoid attack should be powerful. Dramstad (1996) suggests that a high concentration of workers close to the nest might attract *Psithyris* queens and other enemies. It is not known how *Psithyris* females locate the nests of their hosts. If foragers remained close to their nests then the chemical cues deposited on flowers to aid foraging (Goulson *et al.* 1998b) could potentially provide a 'scent magnet' to cuckoo bees (Dramstad 1996). Similarly, if foragers were all concentrated in the area close to the nest then they might attract aggregations of predators; Gentry (1978) found that aggregations of pollinators on the flowers of tropical trees attracted large numbers of bee-eating birds. Little is known of the risk of predation faced by foraging bumblebee. Many researchers have considered predation on foraging bumblebees to be rare and of minimal ecological importance (Brian 1965a; Pyke 1978d; Zimmerman 1982; Hodges 1985b), but others suggest predation rates are significant (Rodd *et al.* 1980; Goldblatt and Fell 1987). There are also likely to be big differences between regions; for example, the UK has very few bee-eating birds, but they are common in southern Europe.

The difficulty with these hypotheses concerning the risk of attack by predators, parasites, or inquilines is that they are difficult to test since the available evidence suggests that bumblebees do not forage close to their nests. It is not easy to demonstrate that this is a result of 'the ghost of predation past'. A comparison of the foraging ranges of species with and without these enemies could in theory provide an insight, if such species exist, but our sketchy knowledge of both the foraging ranges and natural enemies of different bumblebee species prevents a meaningful comparison at present.

Interestingly, there is now some evidence (albeit rather flimsy) that bumblebee species may differ markedly in foraging range and that this correlates with colony size. Experiments with bees marked at the nest and anecdotal observations suggest that species such as *B. pascuorum, B. sylvarum, B. ruderarius,* and *B. muscorum* are 'doorstep foragers', mostly remaining within 500 m of their nests whilst *B. lapidarius* forages further afield (mostly <1500 m), and *B. terrestris* regularly forage over more than 2 km (Walther-Hellwig and Frankl 2000). These experiments suffer from a reduced intensity of sampling at greater distances from the nest, and also rather small sample sizes. Nonetheless, this cannot explain the differences that were found between species. These differences appear to correspond to the known nest sizes of these bee species; *B. terrestris* and *B. lapidarius* nests grow to a large size (100–400 workers), while those of *B. pascuorum, B. sylvarum, B. ruderarius,* and *B. muscorum* (all commonly known as carder bees, and belonging to the subgenus Thoracobombus) are generally small (20–100 workers) (Alford 1975). In comparison, honeybees which have colonies consisting of many thousands of workers, forage over great distances (up to 20 km; Schwarz and Hurst 1997). More quantitative data is required to test whether bumblebee species do differ in foraging range. Up until the recent work of Walther-Hellwig and Frankl (2000), studies have implicitly assumed that foraging range is similar in all species. If it is not, then this might represent a fundamental difference between the ecology of bumblebee species which has clear implications for management of bumblebees for pollination or conservation purposes.

8
Exploitation of patchy resources

Within the foraging range of a bee, there will probably be many different patches of flowers, varying in size and in the plant species of which they are composed. Each individual bee has to choose which patch(es) to exploit. One might naively predict that, all else being equal, large patches should be favoured over small ones, since this would minimize traveling time between flowers. However, if all bees adopted this strategy, then large patches would become overrun with bees, and flowers in small patches would contain much more nectar because they would never be visited. Whatever size of patch a bee chooses, it must then decide how long to stay. The longer it stays in a patch, the less resources the patch will contain as the bee uses them up (unless the patch is so large that it produces rewards faster than the bee can gather them). At some point, the bee would probably be better served by going to find another patch.

One approach to understanding forager behaviour when exploiting patchy resources which has proved to be fruitful is the use of optimality models. Although optimality models have in the past received much criticism (e.g. Pierce and Ollason 1987), they remain a valuable starting point for generating hypotheses to explain behaviour, and also provide a means of testing our understanding of behaviour. Optimal foraging models generally assume that foragers maximize their rate of resource acquisition (Charnov 1976). This is a reasonable assumption for workers of social insects such as bumblebees since they are freed from many of the constraints which are likely to affect the behaviour of other foragers, (Pyke 1978a), and so optimality models have frequently been applied to bumblebees (e.g. Pyke 1978a,b, 1983; Hodges 1981, 1985a; Best and Bierzychudek 1982; Zimmerman 1982, 1983; Cibula and Zimmerman 1987; Dreisig 1995; Goulson 1999, 2000b). The assumption of maximized rate of resource acquisition is less reasonable for insects such as butterflies which intersperse nectaring with activities such as searching for mates or oviposition sites, and so regularly indulge in longer flights than do most foraging bumblebees (Schmitt 1980; Waser 1982a; Goulson et al. 1997a).

Two optimal foraging models are particularly relevant to bumblebee foraging among patches of flowers, the ideal free distribution (Fretwell and Lucas 1970) and the marginal value theorem (Charnov 1976). I shall examine the predictions of these two models in turn.

8.1 The Ideal Free Distribution

Flowers typically exhibit a patchy distribution at a number of levels; flowers are often clustered into inflorescences, several flowers, or inflorescences may be clustered on each plant, and the plants themselves are likely to be patchily distributed. According to the ideal free distribution model, the evolutionary stable strategy for foragers exploiting a patchy resource is to equalize the rate of gain of reward in all patches by matching the proportion of foragers in each patch to the rate of reward production in the patch (Fretwell and Lucas 1970). Applying the ideal free distribution model, we predict that the ratio of foragers to flowers should be independent of patch size (assuming that reward production per flower does not vary with patch size). If we incorporate travel time between patches, we would expect the proportion of foragers to flowers to increase with patch size, so that foragers in small patches receive a higher reward per time within the patch but spend more time moving between patches; overall, the rate of reward received by all foragers is equal. For social organisms, the distribution of foragers among patches may also be influenced by the location of nests; if we take into account travel time from the nest then we would predict a higher proportion of foragers close to nests (Dukas and EdelsteinKeshet 1998). However, since travel time is often likely to be negligible compared to time spent within patches (Dreisig 1995), the ratio of foragers to flowers should generally remain more or less independent of patch size and nest locations.

So do bumblebees achieve an ideal free distribution? An ideal free distribution can be achieved by non-random searching or by non-random choice of patches (i.e. a preference for large patches) (Dreisig 1995). Both are exhibited by bumblebees when visiting flowers.

8.1.1 Search patterns within patches

Foragers which adopt a non-random search pattern can achieve a higher reward per time than individuals which are searching randomly, so that non-random searching should predominate where it is possible (although if all foragers have non-random search strategies then rewards per time are the same as when all foraging is random (Possingham 1989)). At least two forms of non-random spatial searching have been identified in bumblebees visiting flowers: trap-lining along established routes, and systematic searching (which does not require a prior knowledge of the area).

Traplines of various lengths have been identified in butterflies (Gilbert 1975) and in a variety of bees, including euglossines (Janzen 1971; Ackerman et al. 1982), flower bees (*Anthophora* spp.; Kadmon 1992), honeybees (Ribbands 1949), but most frequently in bumblebees (Manning 1956; Heinrich 1976b; Thomson et al. 1982, 1987, 1997; Corbet et al. 1984; Williams and Thomson 1998). Bees possess impressive navigational abilities and are able to remember the relative positions of landmarks and rewarding flower patches (Southwick and Buchmann 1995; Menzel et al. 1996, 1997) which is no doubt valuable in following traplines. For example honeybees are able to integrate movement vectors; after a series of movements between patches they are able to plot a direct route home, thus avoiding the need to backtrack (Menzel et al. 1998). It seems likely that bumblebees also have this ability. Traplining along a regular route enables the forager to learn which flowers or patches are most rewarding, and also to avoid visiting flowers which it has recently depleted. Bumblebees seem to show very strong fidelity to sites at which they have previously found a reward, visiting the same patches over and over again on successive days or weeks (Bowers 1985a; Dramstad 1996; Saville et al. 1997; Osborne et al. 1999; Osborne and Williams 2001). Visiting the same group of flowers at regular intervals may discourage competitors since any new forager attempting to exploit the same flowers may not know which flowers have been most recently depleted and so will initially receive a lower rate of reward than the resident forager (Corbet et al. 1984; Possingham 1989).

Even without prior knowledge of the distribution of rewarding flowers, foragers can improve their efficiency compared to a strategy of random searching by using a systematic spatial search pattern so that they avoid encountering areas where they have recently depleted rewards (Bell 1991). For example, various bee species and also Lepidoptera are able to remember their direction of arrival at a flower, and tend to continue in the same direction when they leave (reviewed in Waddington and Heinrich 1981; Pyke 1983, 1984; Schmid-Hempel 1984, 1985, 1986; Cheverton et al. 1985; Dreisig 1985; Ginsberg 1985, 1986; Ott et al. 1985; Plowright and Galen 1985; Soltz 1986; Schmid-Hempel and Schmid-Hempel 1986; Kipp 1987; Kipp et al. 1989; for exceptions see Zimmerman 1979, 1982). Bumblebees are even able to accomplish this correctly when the flower is rotated while they are feeding on it, provided that there are landmarks available by which they can keep track of their orientation relative to their direction of arrival (Pyke and Cartar 1992). Recent evidence suggests that they also may be able to use the earth's magnetic field to orientate themselves (Chittka et al. 1999b).

Superimposed on the general tendency for foragers to exhibit directionality, they may also adjust their turning rates and movement distances according to the size of rewards so that they quickly leave areas with few flowers or unrewarding flowers, and remain for longer in patches which provide a high reward or where flowers are dense (Pyke 1978a; Pleasants and Zimmerman 1979; Heinrich 1979a; Thomson et al. 1982; Rathcke 1983; Real 1983; Cibula and Zimmerman 1987; Kato 1988; Cresswell 1997) Short flights and frequent turns entail the risk of revisiting flowers, but this is presumably more than offset by the benefits of remaining within a patch containing many or highly rewarding flowers (Zimmerman 1982). Bumblebees also tend to remain longer in patches with 'landmarks' (features protruding above the herb layer), perhaps because the landmarks facilitate a systematic search (Plowright and Galen 1985). If bumblebees do encounter flowers that they have already emptied, they have a further trick up their sleeve. They are able to distinguish and avoid entering flowers that have recently been visited by detecting the scent of previous visitors (although they still incur a small time penalty due to the time it takes to detect the scent) (Núñez 1967; Wetherwax 1986;

Giurfa and Núñez 1992b; Giurfa 1993; Giurfa et al. 1994; Stout et al. 1998; Goulson et al. 1998b) (see Chapter 11).

Systematic search patterns are also evident in the movements of bees between flowers on the same inflorescence. Many plants present flowers in a vertical raceme, which bees almost invariably exploit by starting at the bottom and working upwards (Heinrich 1975b, 1979a; Pyke 1978a) (Plate 7). In some plants, the lower flowers in the raceme produce more nectar, so that bees forage upwards until low rewards stimulate departure (Pyke 1978b). It makes sense for bees to start at the most rewarding point and depart when rewards become too low to be worthwhile. However, bumblebees continue to forage upwards when the distribution of nectar is artificially reversed so that the topmost flowers are most rewarding (Waddington and Heinrich 1979). Some flowers such as *Linaria vulgaris* provide most nectar at the top of inflorescences, but bumblebees usually forage upwards on these too (Corbet et al. 1981). Rather than relating to the distribution of nectar, upwards foraging may simply be a result of the position which bees generally adopt when foraging on flowers (facing forwards and upwards). Corbet et al. (1981) noted that *B. terrestris* robbed *L. vulgaris*, and that to do this some individuals perched on the inflorescence facing downwards. These individuals tended to start at the top and work downwards. Whatever the distribution of nectar and the direction taken by the bee, a simple foraging rule in which each individual bee always moves in the same direction ensures that they rarely encounter flowers that they have just visited, at least within a single inflorescence. From the point of view of the plant, systematic foraging is also beneficial since it is likely to reduce the frequency of selfing.

Simple systematic search patterns are also possible on other types of inflorescence. Many plants provide rings of open flowers around a central stem (e.g. *Trifolium* spp.). When visiting inflorescences of this sort, bees simply circle around them until they re-encounter the first floret that they landed on. Interestingly, individual bumblebees exhibit a strong tendency to rotate around the flower in a particular direction. Kells and Goulson (2000) studied the direction of rotation of workers of four UK bumblebee species, *B. lapidarius*, *B. terrestris*, *B. lucorum*, and *B. pascuorum*, when foraging on inflorescences of *Onobrychis viciifolia*. Individuals of all four species tended to rotate in the same direction in successive visits to inflorescences (Figure 8.1). Overall, bees rotated around inflorescences in the same direction as on their previous visit on 68.6% of visits (all bee species combined, S.E. = 9.34). Presumably, a bee which exhibited random rotation would receive the same rate of reward, and hence just as much reinforcement of its behaviour, as a bee with a fixed direction of rotation. Ecologically, the direction of rotation would seem to be trivial, since it has no consequence for either the bee or plant. The direction in which a naïve bee turns on the first inflorescence it encounters may be random, but the bee may then simply repeat this behaviour since it has proved to be successful.

For three of the four species, approximately equal numbers of individuals tended to rotate either clockwise or anticlockwise, but intruigingly *B. pascuorum* exhibited a significant tendency to rotate in an anticlockwise direction (Kells and

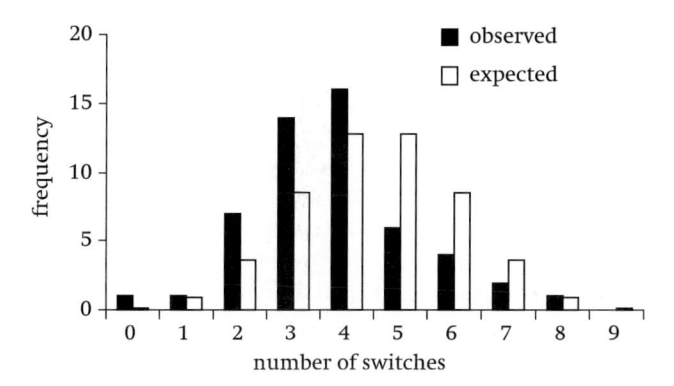

Figure 8.1. The frequency distribution of switches between anticlockwise and clockwise rotations around flowers, compared to that which would be expected if the direction of foraging was random on each visit to an inflorescence. Each bee was observed on ten consecutive inflorescences. Data have been combined for four bumblebee species, *B. lapidarius*, *B. terrestris*, *B. lucorum*, and *B. pascuorum*, when foraging on inflorescences of *Onobrychis viciifolia*. The proportion of switches between clockwise and anticlockwise visits did not differ between the four bee species ($X^2 = 4.86$, d.f. = 3, $p > 0.05$). Overall, bees tended to forage on consecutive inflorescences by rotating in the same direction. The number of switches between directions was significantly lower than the number of times a bee continued to forage by rotating in the same direction ($t = 3.34$, d.f. = 51, $p < 0.001$). From Kells and Goulson (2000).

Goulson 2000). It is tempting to dismiss this as a spurious result, but parallels can be found in higher organisms. In behaviours which require body rotation, children exhibit a tendency to turn in one direction or another, and just as in *B. pascuorum*, most children tend to rotate anticlockwise. This tendency becomes more pronounced with age (Day and Day 1997). Preferred directions are correlated with handedness (Yangzen *et al.* 1996). Similar rotational preferences have been found in other mammals including capuchin monkeys (Westergaard and Suomi 1996) and mice (Nielsen *et al.* 1997), but apparently they do not occur in goats (Ganskopp 1995). The origins and significance of these preferences remain to be explained.

8.1.2 *Non-random choice of patches*

The distribution of foragers among patches depends on the relationships between recruitment rate and patch size, and also how long foragers spend in patches of varying size once they get there. In general, insect foragers preferentially visit large patches (plants with many flowers) (Wilson and Price 1977; Silander and Primack 1978; Schaffer and Schaffer 1979; Augspurger 1980; Davis 1981; Udovic 1981; Thomson *et al.* 1982; Waser 1982b; Schmitt 1983; Bell 1985; Geber 1985; Andersson 1988; Schmid-Hempel and Speiser 1988; Klinkhamer and de Jong 1990; Eckhart 1991; Dreisig 1995; Goulson *et al.* 1998a), although the relationship between recruitment and patch size is often less than proportional (Schmid-Hempel and Speiser 1988; Klinkhamer et al. 1989; Dreisig 1995; Goulson et al. 1998a).

Several studies have found that this combination of higher recruitment to large patches and systematic searching results in a visitation rate per flower which is independent of plant size (i.e. foragers achieve an ideal free distribution) (Heinrich 1976b; Pleasants 1981; Schmitt 1983; Bell 1985; Geber 1985; Schmid-Hempel and Speiser 1988; Thomson 1988; Dreisig 1995; Robertson and Macnair 1995; Kunin 1997). When nectar production rates varied greatly between plants, Dreisig (1995) found that foraging bumblebees achieved an ideal free distribution by preferentially visiting individual *Anchusa officinalis* which had high rates of nectar production, the result of which was that all bees received an approximately equal rate of reward. However, an ideal free distribution was not found in all studies. For example, Klinkhamer and de Jong (1990) found that visits per flower by bumblebees declined with plant size in *Echium vulgare*, while Klinkhamer et al. (1989) describe the reverse in *Cynoglossum officinale*.

It appears that, more often than not, bumblebees achieve an approximately ideal free distribution, but how do they do this? Factors governing recruitment rates to patches have received little attention. Greater recruitment to large patches is presumably at least partly because large patches are more apparent or because they are more likely to be encountered, and does not necessarily imply an active preference by the forager. The general finding that increases in recruitment are less than proportional to increases in patch size is less easily explained. It may be because foragers searching for flowers tend to search in two dimensions (they tend to fly at an approximately uniform height) so that the probability of encountering a patch is a function of its diameter rather than its area (Goulson 1999). Since the number of flowers in a patch is likely to be proportional to its area, this could result in a decelerating relationship between flower number and recruitment. At present, there is insufficient information available to determine whether recruitment patterns are the result of passive encounter rates or active choice by foragers.

8.2 The Marginal Value Theorem

Studies of the response of pollinators to varying patch sizes have found that not only are more foragers attracted to larger patches (see above), but also that they spend longer in them, and visit more flowers while they are there, as one would intuitively expect (Schmitt 1983; Geber 1985; Andersson 1988; Schmid-Hempel and Speiser 1988; Klinkhamer et al. 1989; Klinkhamer and de Jong 1990; Eckhart 1991; Dreisig 1995; Robertson and Macnair 1995; Brody and Mitchell 1997; Goulson 2000b). More interestingly, studies of a diverse range of plant–pollinator systems have also found that the pollinators visit a smaller proportion of the available flowers in larger patches (Beattie 1976; Heinrich 1979a; Zimmerman 1981b; Schmitt 1983; Geber 1985; Morse 1986a; Andersson 1988; Schmid-Hempel and Speiser 1988; Thomson 1988; Klinkhamer et al. 1989; Klinkhamer and de Jong 1990; Pleasants and Zimmerman 1990; Dreisig 1995; Harder and Barrett 1995; Robertson and Macnair 1995; Brody and Mitchell 1997; Ohashi and Yahara 1998; Goulson et al. 1998a) (although Sih and Baltus 1987 found that bumblebees visited a higher proportion of flowers in large patches of Nepeta cataria). The explanation for this pattern is not obvious.

The marginal value theorem is an optimality model for investigating the behaviour of foragers exploiting patchy resources (Charnov 1976). The theorem states that a forager should leave a patch when the rate of food intake in the patch falls to that for the habitat as a whole. This can be used to predict the optimal duration of stay of a forager in a patch, if the shape of the pay-off curve for staying within a patch and the mean travel time between patches are known. Can the marginal value theorem explain why pollinators visit a decreasing proportion of flowers in a patch as patch size increases? Several researchers have applied the marginal value theorem to bees foraging on flowers held in vertical racemes (Pyke 1978c, 1981, 1984; Hodges 1981; Zimmerman 1981c; Best and Bierzychudek 1982; Pleasants 1989). The aim of these studies was to predict when the insect should move to a new inflorescence, and to examine what departure rules might be used to achieve the most efficient strategy. However, this is a special case. Vertical racemes are easy to search systematically (insects typically start at the bottom and work upwards, Heinrich 1975b, 1979a), and usually have a predictable, declining reward in successively higher florets (Pyke 1978b). More commonly, a pollinator has to search amongst loose aggregations of flowers or inflorescences with no clear spatial structuring. Here the search strategy employed by the pollinator will largely determine the shape of the pay-off curve that it gains from visiting a patch of flowers. If we can ascertain the shape of the pay-off curve in different patch sizes, then it will be possible to predict the optimal duration of stay (sometimes known as the 'give up time', Charnov 1976).

The shape of the pay-off curve will depend on whether the forager searches randomly, or has a systematic strategy. As we have seen, bumblebees use systematic search patterns, which include directionality, and turning more frequently when in particularly rewarding or dense patches of flowers. However, it seems probable that a forager will be unable to carry out a systematic search of all the flowers on a large plant without making mistakes, and re-encountering flowers that it has depleted. Thus we would expect the rate of reward acquisition to begin to decline after a period of time spent within the patch. If travel time between patches is short, then an insect should depart soon after this decline begins (Goulson 1999). Two models have been developed applying the marginal value theorem to bumblebees exploiting patches within which flowers were haphazardly arranged (Goulson 1999; Ohashi and Yahara 1999). Both models predict that bumblebees should visit a greater proportion of flowers in small patches if they forage systematically and so are able to avoid flowers that they have just depleted. However, the relevant parameters were not quantified, so only qualitative predictions were possible. No information was available on the proportion of flowers on a patch an insect can visit before it begins to make mistakes (revisit flowers), and how this proportion changes with patch size.

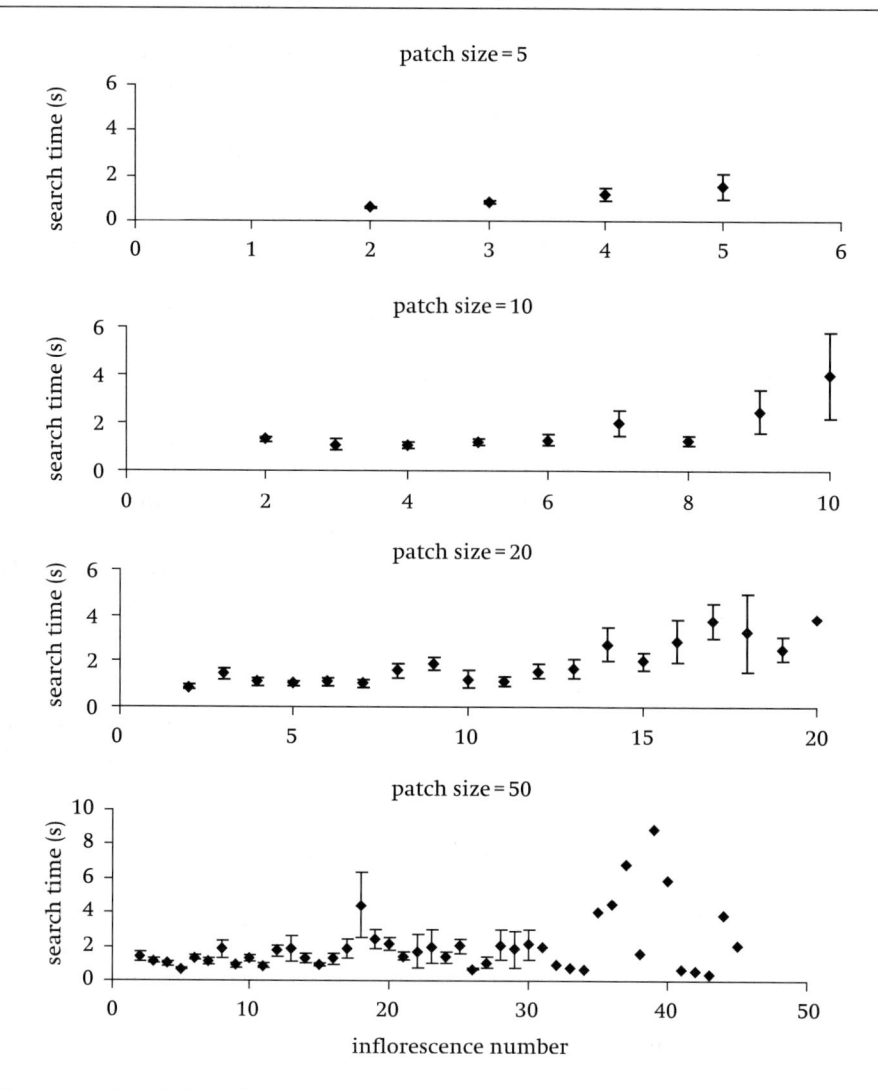

Figure 8.2. Search times for successive inflorescences within patches containing 5, 10, 20, or 50 inflorescences of *T. repens* (±SE). Values shown were calculated by taking the means for five bumblebees that visited each patch, and using these means from five replicate patches to calculate a grand mean and standard error. The bees arrival at the first inflorescence in a patch was deemed to be the time of arrival in the patch, so there is no search time for the first inflorescence. Data from Goulson (2000b).

I recently attempted to quantify the pay-off curve for workers of the bumblebee *B. lapidarius* foraging in artificially created patches of varying size of white clover, *Trifolium repens* (Goulson 2000b). By quantifying travel time between patches, handling time per inflorescence, and search time for each successive inflorescence located, it was possible to construct pay-off curves for different patch sizes, and predict the optimal duration of stay within patches. Search time within patches increased as the proportion of inflorescences visited increased, demonstrating that foraging bumblebees cannot systematically visit all of the flowers within a patch without making mistakes (Figure 8.2). For all four patch sizes that were examined, pay-off curves were very closely described by quadratic equations, with each linear and quadratic term significantly improving the fit of the line (Figure 8.3). Since handling times were not affected by patch size or duration of stay, it is the increase in search time for successive inflorescences that results in the typical pay-off curve with a declining slope (Charnov 1976). The optimum duration of stay in each patch is given by the point of contact between the curve and a tangential straight line plotted through coordinate '—(travel time between patches), 0' (following Charnov 1976). The optimum duration of stay increased with patch size, but was less than proportional so that to achieve a maximal rate of reward per time bees should visit a smaller proportion of inflorescences in larger patches.

For the smallest patch size, the predicted optimum duration of stay was close to the observed value, but as patch size increases, observed and predicted values diverged. However, even in the largest patch size where the discrepancy between observed duration of stay and the predicted optimum was greatest, the bees were still achieving a rate of reward acquisition very close to the optimum due to the shape of the pay-off curve (Figure 8.3). So although bees were apparently behaving in a sub-optimal way in larger patches (assuming that calculation of the pay-off curves is accurate and that the assumptions of the model are met), they are only very slightly sub-optimal.

Both observed and predicted durations of stay within patches result in a declining proportion of inflorescences being visited as patch size increases. It appears that visiting a declining proportion is optimal, but why is it optimal? The answer must lie in the changing patterns in search time. In small patches, use of a systematic search pattern could enable pollinators to visit all of the inflorescences without mistakes, and thus without an increase in search time (the pay-off curve would be a straight line) (Goulson 1999). Similarly, Ohashi and Yahara (1999) suggest that pollinators are able to memorize and avoid the last few flowers that they visited, so that when the number of flowers in the patch is less than or equal to the number that can be memorized, the pollinator should visit every flower in the patch. It seems that, if pollinators can memorize the positions of flowers that they have visited, they can do so for only a very few (<4). Even in patches containing just five inflorescences, search time exhibited a marked increase with the fifth inflorescence taking on average 2.5 times as long to locate as the second (Goulson 2000*b*).

Presumably, searching for the remaining unvisited inflorescences is simpler in a small patch than in a large one. In this respect pollinators visiting flowers represents a rather different situation to that for which the marginal value theorem was originally developed (predators searching for prey) because the flowers remain after they have been visited. By doing so they render locating the remaining unvisited flowers more difficult. The explanation as to why search times overall are longer

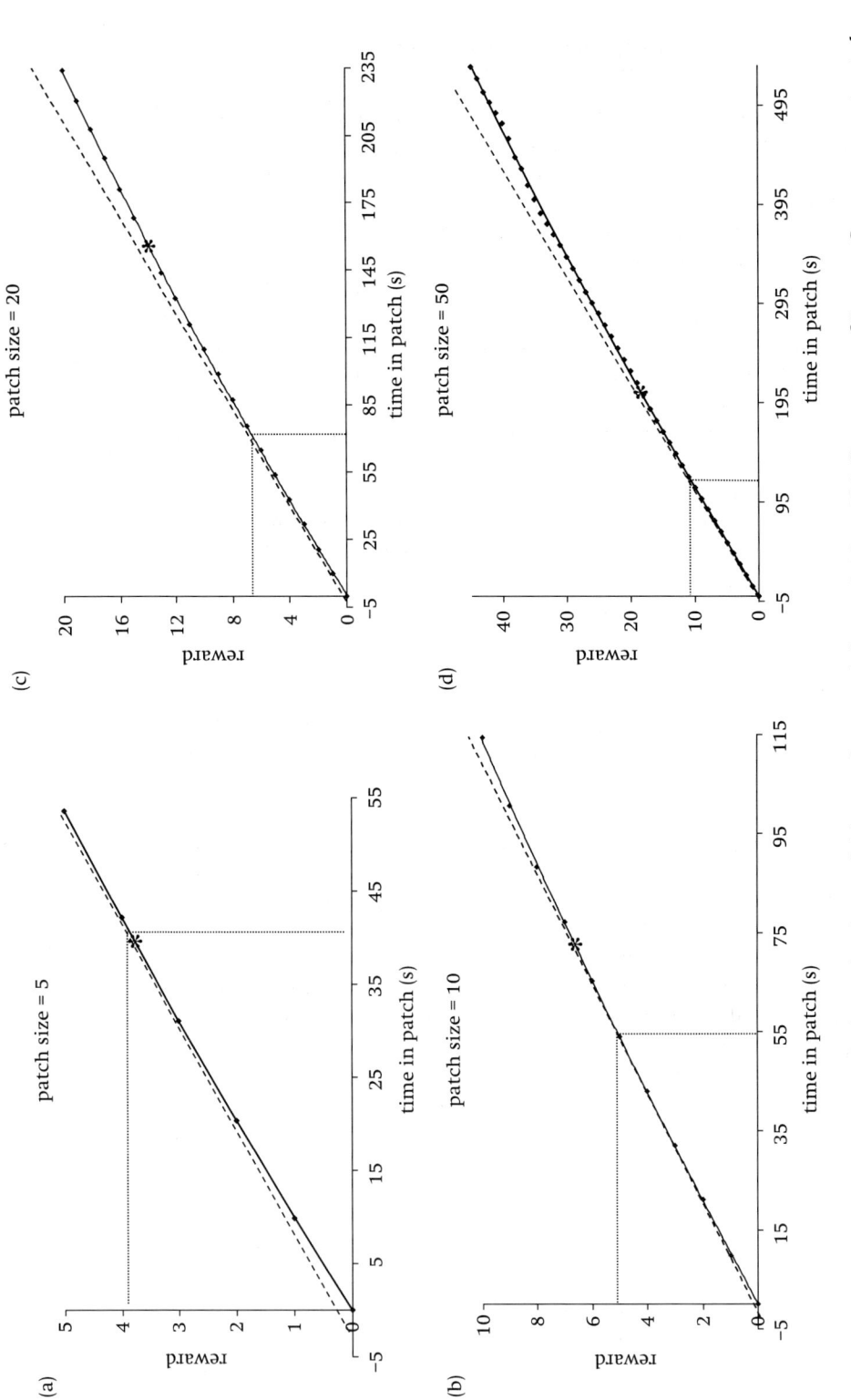

Figure 8.3. Pay-off curves for the bumblebee, *B. lapidarius*, visiting patches containing 5, 10, 20, or 50 inflorescences of *T. repens*. Curves are constructed from measured search times, which increase as the proportion of inflorescences within the patch that have already been visited increases. Handling time is independent of patch size with a mean of 9.79 ± 0.35 s per inflorescence; this value is used for constructing curves. As handling time is independent of patch size, the assumption that reward per inflorescence is equal across patch sizes appears to be valid, since inflorescence handling time is closely correlated to reward received (Harder 1986; Kato 1988). Reward is thus measured as number of inflorescences handled. The mean travel time between patches was 2.29 ± 0.63 s. The optimum duration of stay and number of inflorescences handled in each patch size is marked (dotted lines). * = observed duration of stay within patches. Data from Goulson (2000*b*).

in bigger patches may be illustrated by a simple numerical example. Consider a bee foraging in a patch of 5 inflorescences, that has already visited 3 of them. If it visits the next inflorescence at random, it has a 2/5 chance of locating one of the unvisited ones on its first attempt. However, the simple movement rules of bees render it unlikely that it will visit the inflorescence it has just left, so it actually has a chance of 2/4 of locating an unvisited flower on the first attempt. In contrast, consider a bee in a patch of 50 flowers, of which it has visited 30 (the same proportion). When it departs from the 30th flower it has a 20/49 chance of striking an unvisited flower on its first attempt, a value substantially less than 2/4. Finding the fourth flower of five, and so achieving an 80% visitation rate (as most bees did), is substantially easier than locating the 31st, 32nd, ..., 40th flower in a patch of 50 (and very few bees did so). This argument does not require the pollinator to memorize the positions of flowers that it has recently visited, only that it does not immediately visit the flower that it just departed from.

Another way of considering this is to examine what cues stimulate departure from a patch. It is clear that the size of rewards recently received and the density of flowers influence the probability of departure. In both bumblebees and solitary bees, low rewards promote departure from an inflorescence (Cresswell 1990; Kadmon and Shmida 1992). Similarly in bumblebees and honeybees, low rewards trigger longer flights and so often result in departure from the plant or patch (Heinrich *et al.* 1977; Pyke 1978*a*; Thomson *et al.* 1982; Zimmerman 1983; Plowright and Galen 1985; Kato 1988; Giurfa and Nùñez 1992*a*; Dukas and Real 1993*b*). This has a clear analogy in the triggering of switching between plant species by receipt of low rewards (Chittka *et al.* 1997; Goulson *et al.* 1997*a*; Greggers and Menzel 1993) (see Chapter 9).

There is some disagreement as to the departure rules used by foragers. For some time it was thought that departure from a patch was triggered by the reward from a single flower falling below a threshold (Pyke 1978*a*; Hodges 1981, 1985*a,b*; Best and Bierzychudek 1982; Pleasants 1989). It subsequently became apparent that a simple threshold departure rule was not strictly accurate, at least for bumblebees, but rather that the probability of departure increases with decreasing reward (Cresswell 1990; Dukas and Real 1993*b*). However, any strategy based on only the last visit to a flower seems intuitively likely to be sub-optimal given the high heterogeneity of rewards that is usually found within patches, since it is likely to result in premature departure from highly rewarding patches. In fact, recent studies have demonstrated that both bumblebees and other bee species are able to integrate information over several flower visits (not just the last one) in making decisions about departure from a patch (Hartling and Plowright 1978; Waddington 1980; Cibula and Zimmerman 1987; Kadmon and Shmida 1992; Dukas and Real 1993*b,c*).

Let us return to the data on pay-off curves for *B. lapidarius* feeding on patches of clover. Suppose that a bee departs from a patch if it encounters two inflorescences in a row that it has already visited (and which are thus more-or-less empty). If we assume that the bee is equally likely to encounter any inflorescence (excluding the one that it just left), then it is simple to calculate the probability that this inflorescence has already been visited, and to square this to obtain the probability of this happening twice and the bee departing from the patch. We can thus calculate the probability of a bee departing after one visit, two visits, and so on, and use this to calculate the expected mean number of inflorescences visited per patch for bees using this departure rule. For the patch sizes of 5, 10, 20, and 50 flowers used by Goulson (2000b), we would predict mean numbers of inflorescences visited per patch to be 3.95, 6.06, 9.88, and 18.48. These values are remarkably close to those that were observed (Figure 8.3). Whether this is coincidence is hard to say without explicitly studying the departure rules used, but nonetheless this example illustrates an important point; that a simple departure rule can result in pollinators visiting more inflorescences per patch but a declining proportion of inflorescences per patch, as patch size increases, exactly as is observed in nature.

To summarize, floral resources are patchily distributed. All else being equal, selection will favour foraging strategies that maximize the rate of resource acquisition. Bumblebee workers are less likely to be constrained in their foraging behaviour than most other insects. Bumblebees use systematic searches within flower patches, but these break down as the number of flowers already visited within the patch increases. Thus, the search time increases with duration of stay within a patch. Application of the marginal value theorem to experimentally obtained pay-off curves predicts that bees should visit more flowers in large patches, but should visit a declining proportion of flowers as patch size increases. This is broadly in agreement with a large body of evidence from field studies. A simple departure rule based on two successive encounters with inflorescences that have already been visited closely predicts observed behaviour in the bumblebee *B. lapidarius*.

9
Choice of flower species

Although current plant–pollinator mutualisms represent the result of approximately 100 million years of co-evolution, extreme specialization is unusual (reviewed in Waser *et al.* 1996). There are a small number of plant species which depend on a single or very few pollinator species throughout their range; examples include the Yucca (*Yucca* sp.) (Bogler *et al.* 1995), Figs (*Ficus* sp.) (e.g. Wiebes 1979), various orchids such as *Ophrys speculum* (Orchidaceae) (Nilsson 1992), and a guild of red-flowered plants found in the Fynbos of South Africa which are pollinated by the butterfly *Aeropetes tulbhagia* (Marloth 1895; Johnson and Bond 1994). Examples in which an insect depends exclusively on one plant species for all of its nectar or pollen requirements appear to be even more scarce (Waser *et al.* 1996), and at present include a handful of species of bee (Westrich 1989). Interestingly, three bumblebee species are known that are each almost entirely dependent on one species or genera of flowers; *B. consobrinus* on *Aconitum septentrionale*, *B. derstaeckeri* on *Aconitum* spp., and *B. brodmannicus* on *Cerinthe* spp. (Løken 1973; Rasmont 1988). All three are alpine species with short colony duration, which presumably allows them to specialize.

The vast majority of insects, including most bumblebees, have flexible floral preferences and visit a range of flowers of different plant species according to availability. Similarly, the majority of plants are visited by several or many insect species (Waser *et al.* 1996), although not all may be effective pollinators. Some of the insects that visit flowers exhibit little in the way of specialized adaptations for feeding on nectar or pollen, and are thus only able to exploit simple flowers. For example, the inflorescences of many Apiaceae effectively form a platform upon which a range of polyphagous beetles and flies can graze pollen without requiring specialized morphological adaptations or particular handling skills. However, most flower-visiting insects are specialists in that nectar, pollen, or both represent their major food source, and in that they do possess appropriate morphological adaptations (typically elongated sucking mouthparts and/or hairs or baskets to trap pollen) (e.g. Thorp 1979; Gilbert *et al.* 1981). This group, which includes bumblebees, are able to tackle a broad range of flower species, and are responsible for the pollination of many (perhaps most) insect-pollinated plants.

Bumblebees are usually faced with a choice of flower species. Each will differ in abundance, distribution, the likely rewards that it provides, and the ease with

9.1 Flower Constancy

Insects foraging for nectar or pollen have long been known to exhibit a learned fidelity to flowers of a particular plant species which has previously provided a reward. In doing so, they ignore many other suitable and rewarding flowers which they pass, but of course they also avoid visiting unsuitable flowers. This behaviour was first described by Aristotle in the honeybee in about 350 BC (Grant 1950), and subsequently attracted the attention of Darwin (1876). Naïve bees have innate flower colour preferences, notably for the wavelengths 400–420 nm and 510–520 nm, but their preferences quickly change with experience (Lunau 1990; Gumbert 2000). They exhibit rapid sensory learning, and can use scent, colour, shape, or a combination of all three to identify flower species which previously provided a reward (Koltermann 1969; Menzel and Erber 1978). Foragers can learn to selectively attend to particular cues that are associated with reward, and ignore others that are not (Dukas and Waser 1994). The learning process takes just 3–5 consecutive rewards, and once learned, a preference may persist for minutes, hours, or even for days (Menzel 1967; Heinrich et al. 1977; Dukas and Real 1991; Keasar et al. 1996; Chittka 1998). In honeybees the learned fidelity can be strong, so that 93–98% of all visits in a single foraging bout are to the favoured plant species (Grant 1950; Free 1963). Learned fidelity of this sort became known as flower constancy (a term perhaps first coined by Plateau 1901 and defined by Waser 1986) and has been identified in the foraging regimes of other pollinators, including bumblebees. The preference shown by an individual insect is not fixed, and varies between individual foragers of the same species (Heinrich 1979c; Barth 1985).

Flower constancy is of crucial importance to plant reproductive biology (Levin 1978). From the point of view of the plant, constancy in its pollinators is of great benefit since it minimizes pollen wastage and stigma clogging with pollen from other species. Thus flower constancy influences the outcome of interspecific competition for pollination services (Waser 1982b; Rathcke 1983; Kunin 1993). Flower constancy may also reduce inter-morph pollen transfer in polymorphic flowers and reduce hybridization between related species (Grant 1949, 1952; Jones 1978; Goulson 1994; Goulson and Jerrim 1997). It has been implicated as a contributory factor in sympatric speciation (Free 1963), although current opinion is that flower constancy alone is unlikely to provide sufficient isolation for speciation to occur (Grant 1992, 1993, 1994; Waser 1998; Chittka et al. 1999a).

Flower constancy is also intriguing from a behavioural viewpoint because it seems to represent a suboptimal pattern of foraging which remains to be explained convincingly (Woodward and Laverty 1992). By adopting this strategy, the insects are bypassing rewarding flowers. If they were not flower constant but visited with equal preference all flower species which provided a reward then they could reduce travelling time. This apparent inefficiency is even more striking when, in two-choice experiments, some honeybees remained constant to an artificial flower morph which provides a consistently lower reward than the alternative (Wells and Wells 1983, 1986; Wells et al. 1992).

Studies of flower constancy have continued to focus primarily on Hymenoptera, principally bumblebees and honeybees, but in the last 10 years it has become clear that flower constancy is more widespread. It has recently been identified in butterflies (Lewis 1989; Goulson and Cory 1993; Goulson et al. 1997a) and hoverflies (Syrphidae; Diptera) (Goulson and Wright 1998), and circumstantial evidence from analysis of gut contents in pollen feeding beetles suggests that they may also exhibit constancy (De Los Mozos Pascual and Domingo 1991). It thus seems probable that flower constancy is a general phenomenon amongst foragers which gather nectar and/or pollen. Although flower constancy is now known to be widespread, there is still disagreement as to why it occurs (e.g. Oster and Heinrich 1976; Real 1981; Barth 1985; Waser 1986; Woodward and Laverty 1992; Goulson 2000a). Because bumblebees are docile and easily observed in the field, they have become a popular vehicle for testing the alternative hypotheses that have been put forward.

9.1.1 Explanations for flower constancy

Several explanations for flower constancy have been proposed; perhaps the most favoured theory is based on an idea proposed by Darwin:

That insects should visit the flowers of the same species for as long as they can is of great significance to the plant, as it favors cross fertilization of distinct individuals of the same species; but no one will suppose that insects act in this matter for the good of the plant. The cause probably lies in insects being thus enabled to work quicker; they have just learned how to stand in the best position on the flower, and how far and in what direction to insert their proboscides. Darwin (1895: 419)

This idea has since been elaborated upon and has become known as Darwin's interference hypothesis (Lewis 1986; Waser 1986; Woodward and Laverty 1992). Essentially what Darwin suggested is that insects may be constant because they were quicker at repeating the same task (handling a particular type of flower) than they would be if they switched between different tasks. More recently, this has been interpreted as arguing that constancy is a result of learning and memory constraints; foragers may be limited by their ability to learn, retain and/or retrieve motor skills for handling flowers of several plant species (Proctor and Yeo 1973; Waser 1983, 1986; Lewis 1986, 1989, 1993; Woodward and Laverty 1992). Learning to extract rewards efficiently from within the structure of a flower takes a number of visits to that flower species, resulting in a decline in handling time on successive visits (Laverty 1980; Lewis 1986; Laverty and Plowright 1988; Keasar et al. 1996) (Figure 9.1). Switching between species of flower differing in floral morphology often temporarily increases handling time as Darwin predicted, particularly when the morphology is complex (Heinrich et al. 1977; Lewis 1986; Woodward and

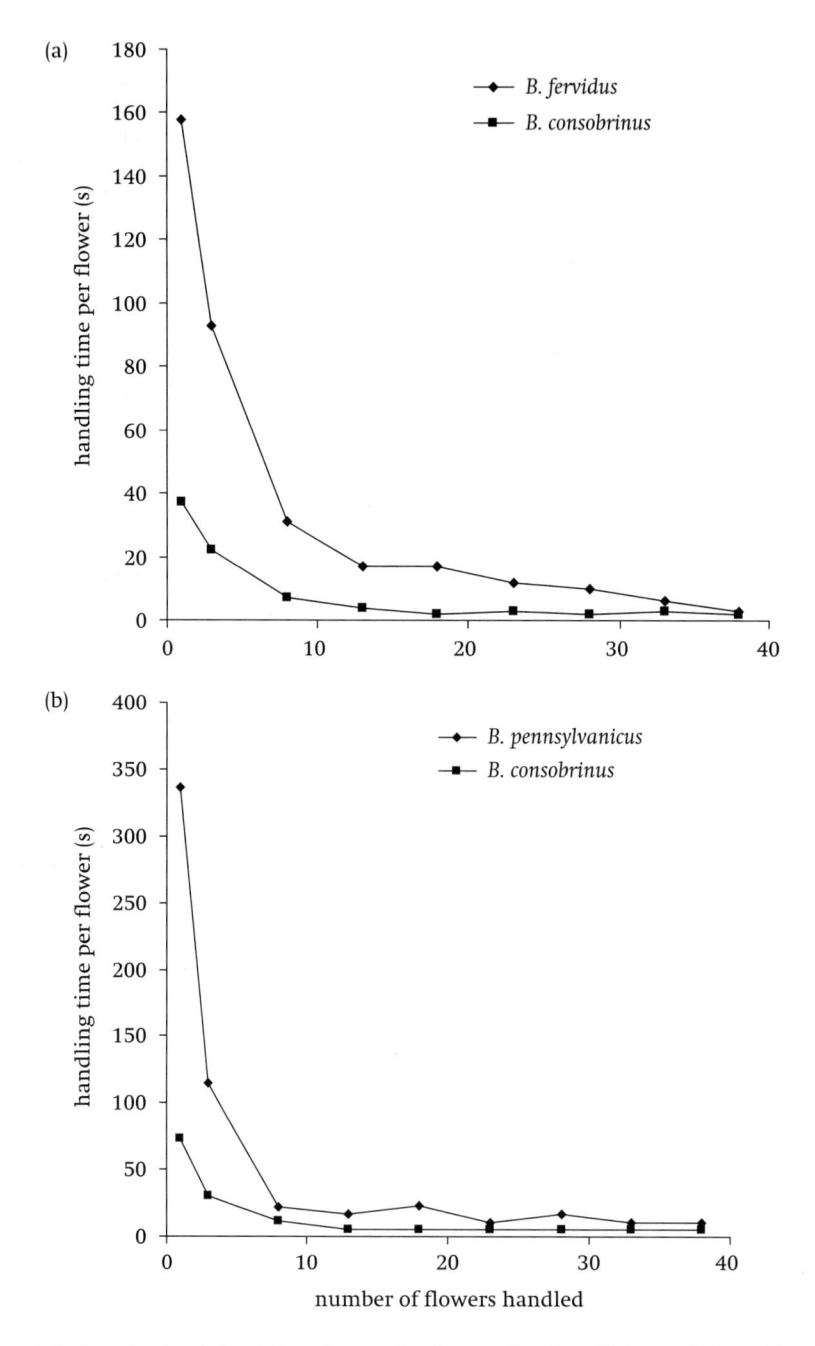

Figure 9.1. Standardized handling times of naïve worker bumblebees visiting flowers of (a) *Aconitum napellus* and (b) *A. variegatum*. Beginning with the first flower visit, means are calculated over five visits. All three bee species show marked improvement as they learn to handle flowers. *B. consobrinus* is the only known example of a specialist bumblebee, feeding primarily on *Aconitum* spp. Even naïve bees of this species are markedly better at handling *Aconitum* flowers than the generalist bee species *B. fervidus* and *B. pennsylvanicus*. From Laverty and Plowright (1988).

Laverty 1992; Chittka and Thomson 1997). It has been argued that memories of handling skills for one flower type are replaced if new skills are learned that is, insects have a limited memory (Lewis 1986). However, considerable research on insect memory has been carried out in recent years and most researchers now agree that memory capacity is not the limiting factor; bees (and probably related insects) appear to have an accurate and large long-term memory (Menzel et al. 1993; Chittka 1998; Menzel 1999). In honeybees and bumblebees, learned handling skills may be retained in long-term memory for weeks even when they are not being used (Chittka 1998; Menzel et al. 1993). For example, B. impatiens trained to locate rewards within a simple maze retained the ability for at least 20 days although there was no reinforcement within this period (Figure 9.2) (Chittka 1998). It appears that bees can learn to suppress associations between sensory inputs and learnt handling skills if they become inappropriate, but that the memories are retained (Chittka 1998). Hence, learned motor (handling) skills are probably not lost as new skills are learned, but there is evidence that errors are likely to be made in retrieving the correct memory in the appropriate context if a bee switches between tasks frequently (Greggers and Menzel 1993; Chittka et al. 1995, 1997; Chittka 1998). After the initial learning process a flower constant forager maintains a low handling time (but requires longer flight times to locate flowers), while a labile forager may incur a penalty of an increased handling time following switches between flower species (but benefits from a higher density of available resources and so a reduced flight time). Thus Darwin's interference hypothesis requires the trade off between handling and flight times to favour constancy.

Attempts to quantify this trade off suggest that this may not be so. Studies of bumblebees and butterflies have found that increases in handling time following switching vary greatly between plant species but are generally too small (0–2 s) to compensate for savings in travelling time (Woodward and Laverty 1992; Laverty 1994a; Goulson et al. 1997b; Gegear and Laverty 1995). Also, if forced to switch between tasks, bumblebees may eventually be able to eliminate interference effects (Dukas 1995), although probably only when foraging on no more than two types of flower with simple structures (Gegear and Laverty 1998). Indeed bees do switch between simple flowers of different species with minimal interference effects (Laverty and Plowright 1988; Chittka and Thomson 1997; Gegear and Laverty 1998). However, switching between three simple flower types or between two complex flower types does induce substantial handling penalties (Gegear and Laverty 1998). It seems that not only do bumblebees become more adept in handling flowers with practice, but that the skills they learn are transferable between flower species; experience with other species of broadly similar flower morphology may actually increase learning rates (Laverty 1994b).

There is an alternative hypothesis that has gained some favour in recent years, that insects might use a search image when looking for flowers (Waser 1986; Goulson 2000a). Tinbergen (1960) introduced the search image as an explanation for prey selection patterns of great tits (Parus major) foraging under natural conditions in woodland. He noted that individual birds tended to collect sequences of

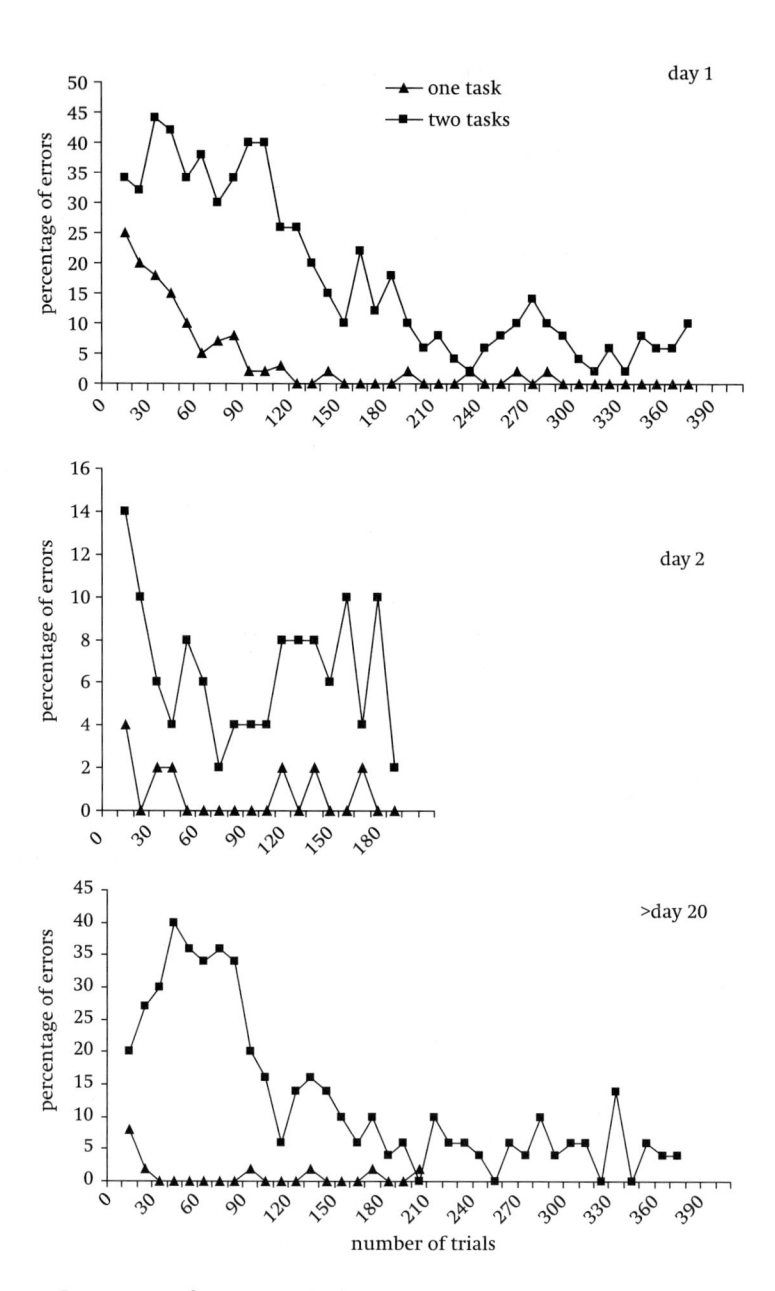

Figure 9.2. Percentage of errors made by the bumblebee, *B. impatiens*, trained to locate rewards within a simple maze. The entrance to the maze was either yellow or blue, and the reward obtained by turning either left or right. Bees trained to a single task (triangles) were provided either with mazes with yellow entrances where food was obtained by turning left, or with blue entrances where food was obtained by turning right. Bees trained to two tasks (squares) experienced both maze types in alternation. Bees experienced 400 trials on day 1, 200 on day 2, and a further 400 after more than 20 days (after Chittka 1998). It is clear that overnight retention of memories is good. However, even after at least 20 days without practice on these artificial flowers, performance was substantially better than that of naïve bees, demonstrating that memories can be retained for long periods.

the same prey species, and that they exhibited positive frequency-dependent selection. Although an intuitively appealing concept, search images have proved difficult to demonstrate convincingly, and it remains unclear how frequently they occur in natural situations (Guilford and Dawkins 1987; Allen 1989). Also, the cognitive mechanisms which give rise to the use of a search image are poorly understood. Almost all definitions of search images specify that they apply to cryptic prey (although Tinbergen himself did not explicitly state this). For example 'as a result of initial chance encounters with cryptic prey, the predator 'learns to see', and selectively attends to those cues that enable it to distinguish the prey from the background' (Lawrence and Allen 1983). This assumption now appears to be valid: experiments using pigeons (*Columba livia*) have found that search image effects are only evident when prey are cryptic (Bond 1983; Bond and Riley 1991; Reid and Shettleworth 1992). Adoption of a search image for a particular prey's visual characteristics enhances its detectability and interferes with incoming perceptual information regarding alternative prey types (Bond 1983). It has recently become apparent that the search image concept has much in common with a phenomenon known to psychologists as selective attention, by which predators learn to detect cryptic prey by selectively attending to particular visual features of the prey which best distinguish them from the background (Langley 1996). Both bumblebees and honeybees are able to use selective attention when distinguishing among flower types (Kosterhalfen *et al.* 1978; Dukas and Waser 1994). Psychological studies of humans and various animals have demonstrated that the brain has a limited capacity for processing information simultaneously that is, it has a limited attention (Blough 1979; Corbetta *et al.* 1990; Eysenck and Keane 1990; Posner and Peterson 1990). An analogous situation has been described in bees; honeybees have a fragile and probably limited short-term memory which is prone to rapid decay and to replacement by new memories (Menzel 1979; Menzel *et al.* 1993; Chittka *et al.* 1999a). Dukas and Ellner (1993) predicted that if predators have a limited attention and prey are cryptic then they should devote all their attention to a single prey type, but that if prey are conspicuous then predators should divide their attention among prey types. Thus, search images may result from both a limited ability to process information simultaneously and from selective attention to cues associated with particular prey types.

There is an obvious flaw in the reasoning so far; plants which are pollinated by animals have evolved brightly coloured flowers to attract the attention of their pollinators. It thus seems implausible to argue that flowers may actually be cryptic, yet, search images are thought to be a mechanism for locating cryptic prey. However, studies of pollinator fidelity (either in the laboratory or field) almost invariably focus on situations where the pollinator is presented with several flower choices at high densities. When viewed against a backdrop of other floral displays (either of the same or different plant species) all of which are vying for the attention of pollinators then any particular flower may be effectively cryptic since it represents a random sample of the background (Endler 1981) (Plate 8). Many flowers which commonly occur together have colours which are extremely similar to insect colour vision systems, and to the human eye (Kevan 1978, 1983; Chittka et al. 1994; Waser et al. 1996). When an insect flies through a meadow containing several flower species, individual flowers appear in the insect's field of view in a very rapid succession; making some conservative assumptions about flight speed, flower density and size, and the insect's visual resolution, Chittka et al. (1999a) estimate that a bee encounters a new flower every 0.14 s. Even if the bee had previously encountered all of the flower species, it seems unlikely that in such a short time period the bee would be able to retrieve memories necessary to recognize the flower, recall the likely rewards and the motor skills required to access them, and then make an economic decision as to whether to visit the flower or not.

The suggestion that search images may be involved in flower constancy is not a new one. Levin (1978) was (to my knowledge) the first to argue that pollinators may develop a search image when foraging for flowers. He proposed that frequency dependent selection by pollinators among colour morphs of the same plant species is best explained by use of a search image (constancy to particular colour morphs when all have identical structure cannot be explained by Darwin's interference hypothesis). More recently, this idea has received additional support.

There is evidence that flower constancy declines as flower density (and thus crypsis) declines (Kunin 1993; Goulson et al. 1997a), in accordance with the predictions of Dukas and Ellner (1993) (although this would also be expected if flower constancy resulted from a trade off between flight time and handling time). Dukas and Real (1993a) demonstrated that bumblebee foraging efficiency is limited in part by their ability to recognize rewarding flower types. Bees made fewer errors in identification when visiting only one rewarding flower type, even when the flower types differed markedly in colour (although Chittka et al. (1999a) argue that the experimental design was flawed). It seems that constraints on recognition (rather than handling) may favour constancy. Several studies have demonstrated that pollinators switch readily between plant species which have similarly coloured flowers (Waser 1986; Kunin 1993; Laverty 1994b; Chittka et al. 1997), even when these flowers have very different structures (Wilson and Stine 1996). Conversely, pollinators rarely switch between flowers of similar structure but of different colour (Wilson and Stine 1996). Darwin's interference hypothesis predicts precisely the reverse, but this is consistent with the hypothesis that search images are used by foragers.

9.1.2 Can flowers be cryptic?

So are flowers really hard to find, despite their bright colours? Recent evidence suggests that they are. Spaethe et al. (2001) demonstrated that the time taken by *B. terrestris* workers to locate artificial flowers against a green background was very strongly correlated with flower size; flowers of 4 mm diameter took approximately 10 times as long to find as flowers of 28 mm diameter (Figure 9.3). This suggests that time taken to locate flowers may be a major component of total foraging time. Any mechanism which improved the efficiency with which flowers were located would be of great benefit. Interestingly, Spaethe et al. (2001) also found that search

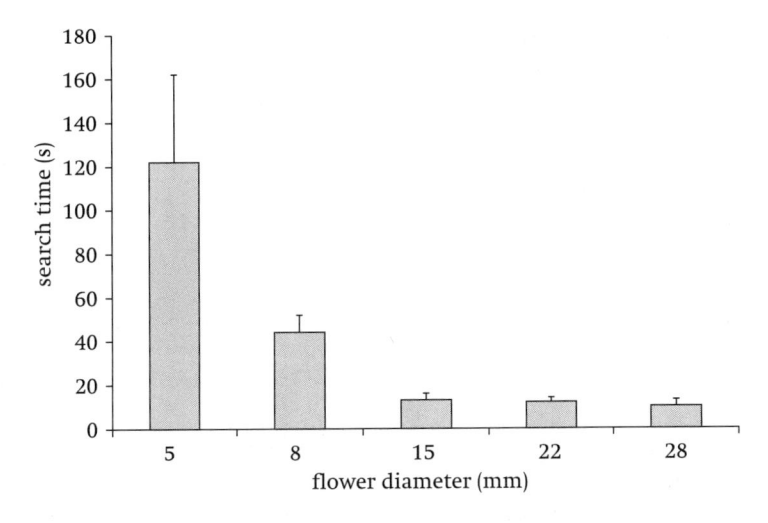

Figure 9.3. Search times for *B. terrestris* when locating artificial blue flowers of varying sizes, viewed against a green background. From Spaethe *et al.* (2001).

times depend greatly on the colour of the flower, suggesting that the foraging efficiency (and thus optimal floral preference) of bees depends not only on floral rewards, flower density and handling times, but also on variation in search times due to flower colour.

These experiments were conducted against a uniform green background, against which the flowers tested were not, strictly speaking, cryptic (they did not resemble a random sample of the background). Nevertheless, small flowers were still hard to find. Flower location is likely to be even more difficult when the desired flower species is viewed against a background of similar coloured flowers of other species. To examine this, I quantified the flight times of wild bumblebees, *Bombus pascuorum*, foraging among grids of flowers of *Lotus corniculatus* (Leguminosae) or *Viccia cracca* (Leguminosae), two species which are favoured by *B. pascuorum* (Goulson 2000a). Flowers of *L. corniculatus* are yellow and those of *V. cracca* are purple. These grids were presented either with or without a background of yellow flowers of species not generally visited by *B. pascuorum*. The background of yellow flowers greatly increased flight times when foraging on the yellow-flowered *L. corniculatus*, but had no effect when foraging on *V. cracca* (Figure 9.4). Bees took on average twice as long to locate *L. corniculatus* flowers when they were presented against a background of other yellow flowers compared to when they were on their own. This is hardly surprising since the apparency of flowers is simply a function of the degree of contrast they make with their background (Lunau *et al.* 1996). Frequently bees were observed to approach to within 1–2 cm of yellow flowers other than *L. corniculatus* but then rejected them after close inspection. Small flowers of different species but with similar colour are likely to be indistinguishable to a bee until it is

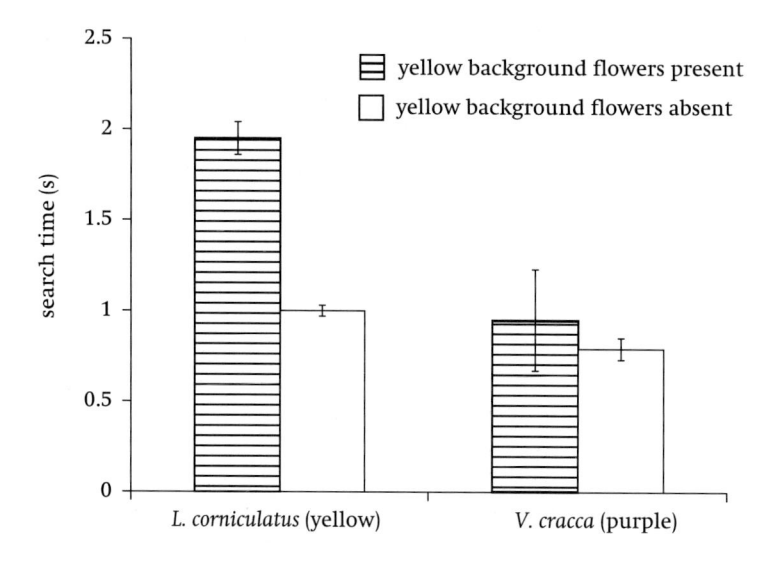

Figure 9.4. Search times (mean (s) ± SE) of *B. pascuorum* when moving between inflorescences of *L. corniculatus* or *V. cracca* arranged in a regular grid with 1 m spacing, with and without a natural background of yellow flowers. Flowers of *L. corniculatus* are yellow and those of *V. cracca* are purple. Times are grand means of individual means from each bee, based on five bees per grid and four replicate grids. When searching for the yellow *L. corniculatus*, search times were much shorter without a background of yellow flowers ($F_{1,12} = 19.6$, $p < 0.001$). From Goulson (2000a).

at very close range. In fact *L. corniculatus* inflorescences are hard for human observers to find when mixed with other yellow flowers (Plate 8). Of course, bee vision is markedly different to our own, and it is likely that some of the flower species present were, to a bee, markedly different in colour or brightness to *L. corniculatus*. However, the results suggest that at least some of the species had a similar spectral reflectance since their presence appeared to increase bee foraging time.

The weight of evidence suggests that bumblebees looking for flowers use search images, that is, they have a limited ability to process visual information from many floral displays simultaneously, and so selectively attend to particular visual features of their preferred flowers. It is likely that other insects also do so, although experimental evidence is lacking. Of course, if bumblebees are using a search image this does not rule out the possibility that Darwin's interference hypothesis may also be valid since the two hypotheses are not mutually exclusive (Wilson and Stine 1996). In fact, the two mechanisms may be synergistic. A perceptual mechanism which renders fidelity a more efficient means of finding flowers (i.e. minimizes search time) could act in conjunction with selection for minimized handling times to promote constancy. Carefully executed experimental tests are required to establish the relative importance of these two processes in promoting flower constancy. To do this successfully, it would probably be necessary to use arrays of artificial flowers in which both colour and structural complexity could be varied independently. Thus it would be possible to manipulate both apparency (difference in flower size and reflectance spectrum compared to the background) and handling time, and examine the conditions which promote constancy.

9.2 Infidelity in Flower Choice

Although flower constancy occurs in a diverse range of insects, it is important to emphasize that constancy is usually far from absolute, and that the terms 'constancy' and 'fidelity' are slightly misleading in this context. Foragers will sample other flowers to keep track of changing rewards over time (sometimes known as minoring in bumblebees), and may also change their preference over time in response to a sequence of low rewards or reduced availability of their preferred flower. Minors may be included as a compromise required to track changing rewards through time (Heinrich 1979c). Bumblebees in particular frequently visit flowers of several species during a single foraging bout, and are markedly less constant than honeybees (Bennett 1883; Brittain and Newton 1933; Grant 1950; Free 1970; Thomson 1981; Waddington 1983a). If the favoured flower ceases to be rewarding bees can rapidly replace learned preferences with new ones (Menzel 1969, 1990; Meineke 1978).

Insects may change their foraging preference in response to rewards received or according to changing frequencies of encounter with different flowers. They appear to follow simple rules. When flowers are scarce, theory predicts that foragers should abandon specialization in favour of generalization (e.g. Levins and MacArthur 1969; Schoener 1969; Colwell 1973; Kunin and Iwasa 1996). Empirical studies have confirmed that, even given the memory constraints previously discussed which favour constancy, this does indeed occur in insects visiting flowers. Bumblebees, honeybees, and hoverflies all abandon constancy when their preferred flower is scarce (Chittka *et al.* 1997; Kunin 1993). Foragers also tend to switch preference following a low or zero reward from the last flower(s) visited. In honeybees, low rewards from individual flowers have been found to promote switching between different coloured artificial feeders in laboratory studies (Greggers and Menzel 1993). Feeding time is known to be an indicator of the reward received in bumblebees and honeybees, and probably also in other insects feeding on nectar (Pyke 1978a; Schmid-Hempel 1984; Bertsch 1987; Greggers and Menzel 1993), enabling examination of the relationship between reward and subsequent behaviour under natural conditions. Both bumblebees and butterflies exhibit higher rates of switching following low feeding times on individual flowers (Chittka *et al.* 1997; Goulson *et al.* 1997b). Switching away from a flower species after receiving a low reward may explain apparently risk-averse behaviour which has been recorded in bumblebees and wasps. In experiments where nectar levels were manipulated, these insects preferred floral types which provided a less variable reward over types which provided a more variable reward with the same mean reward per flower (Real 1981, 1982; Waddington et al. 1981; Harder and Real 1987).

Whatever the mechanism involved in switching, having a flexible preference for particular flowers enables bumblebees to adapt their strategy according to changing spatial and temporal patterns of availability of reward in different flower species.

10
Intraspecific floral choices

Individual flowers exhibit considerable variation in the rate at which they produce rewards within a plant species, and even among flowers on the same plant. This variation may be due to micro-environmental influences, genetic variation, age of the plant, or age of the flower. If foragers can distinguish between more and less rewarding flowers of their preferred species, then they can enhance their foraging success. The time it takes for a bumblebee forager to handle a flower varies greatly according to floral morphology, from as little as 1 s for simple flowers to up to 10 s for complex flowers (e.g. Pyke 1979; Heinrich 1979b; Best and Bierzychudek 1981; Hodges 1981; Osborne 1994; Cresswell 1999). If the flower contains little or no reward then this time is wasted, so there is strong selection pressure on bees to evolve means of choosing the more rewarding flowers.

There is abundant evidence that bumblebees use a variety of cues to indicate which flowers are most likely to provide a high reward.

10.1 Direct Detection of Rewards

Both bumblebees and honeybees are often seen to hover in front of a flower, sometimes briefly touching the corolla, and then depart without probing into the flower structure. These rejected flowers contain, on average, less nectar than flowers which are probed (Heinrich 1979a; Corbet et al. 1984; Wetherwax 1986; Kato 1988; Duffield et al. 1993). Several mechanisms may be in operation. Where the flower structure is open and the anthers are clearly visible, bumblebees are able to directly assess pollen content of open flowers visually (Zimmerman 1982; Cresswell and Robertson 1994). It has been suggested that they may be able to determine the nectar content of some flower species in the same way (Thorp et al. 1975, 1976; Kevan 1976). It has also been proposed that they may be able to assess nectar volumes from the scent of the nectar itself or the scent of fermentation products from yeasts in the nectar (Crane 1975; Heinrich 1979a; Williams et al. 1981). They could plausibly detect nectar volumes from humidity gradients surrounding the flower (Corbet et al. 1979). However, apart from visual detection of pollen availability, no other mechanisms of direct detection of floral rewards have been demonstrated.

10.2 Flower Size

Where direct assessment of the reward contained within a flower is not possible, bees may use other cues to indicate which flowers are most rewarding. Bumblebees generally prefer to visit the largest flowers available (usually measured as corolla width) (e.g. Galen and Newport 1987; Galen 1989; Cresswell and Galen 1991; Eckhart 1991; Ohara and Higashi 1994; Shykoff et al. 1997), although Cresswell and Robertson (1994) found no relationship between size and visitation rate. Higher visitation rates may simply be because large flowers are more apparent, but there is some evidence that flower size is correlated with production of pollen or nectar, so that selection of large flowers may be reinforced by learning (Teuber and Barnes 1979; Brink and Wet 1980; Stanton and Preston 1988; Cresswell and Galen 1991; Duffield et al. 1993).

10.3 Flower Age

Similarly, bees can learn to be selective with regard to the age of the flowers that they choose to visit. Rates of production of nectar may vary with flower age (Boetius 1948; Manning 1956), but there is no general pattern to changes in nectar production with age. In some plants, nectar production declines after the flower opens (Voss et al. 1980) or reaches an early peak and then declines (Carpenter 1976; Bond and Brown 1979; Frost and frost 1981; Bertin 1982; Pleasants and Chaplin 1983; Southwick and Southwick 1983; Cruzan et al. 1988). Conversely in other species nectar production increases with flower age (Pyke 1978b; Brink and Wet 1980; Corbet and Willmer 1980; Best and Bierzychudek 1982; Robertson and Wyatt 1990). In some plant species nectar production is independent of flower age (Bertsch 1983; Pleasants 1983; Marden 1984a; Zimmerman and Pyke 1986).

It has long been known that bumblebees, honeybees, solitary bees, hoverflies, and butterflies are able to discriminate between age classes of flowers using visual cues, and so preferentially select the more rewarding flowers (Müller 1883; Ludwig 1885, 1887; Kugler 1936, 1950; Lex 1954; Jones and Buchmann 1974; Kevan 1978; Thomson et al. 1982; Weiss 1995a). Discrimination among flowers according to their age may be facilitated by clear visual cues given by the plant itself, particularly by colour changes which variously occur in part or all of the flower (Schaal and Leverich 1980; Gori 1983, 1989; Kevan 1983; Delph and Lively 1989; Weiss 1995a). Such changes have been described in 78 families of plant so far (reviewed in Weiss 1995b and Weiss and Lamont 1997). For example, flowers of *Pulmonaria* sp. change from red to blue, enabling bumblebees and flower bees (*Anthophora pilipes*) to select the more rewarding red flowers (Müller 1883; Oberrath et al. 1995). These age-dependent preferences can be flexible; honeybees select 3-day-old capitula of *Carduus acanthoides* in the early morning, and switch to 2-day-old capitula later in the day. This accurately targets the time of maximum nectar production in capitula which is from midway through their second day until early on their third

(Giurfa and Núñez 1992a). Although these colour changes are often triggered by pollination and so benefit the plant by directing pollinators to flowers which are as yet unpollinated (Gori 1983; Weiss 1995b), this is not always so. It is unclear why unpollinated older flowers of some species give clear signals that they are producing little reward (Oberrath et al. 1995).

10.4 Flower Sex

The reproductive success of male flowers (in monoecious and dioecious species) or male-phase flowers (in dichogamous species where male and female function are temporally separated within the same flower) is likely to be more variable than that of female flowers, since a male flower could in theory fertilize numerous female flowers. Thus we may expect male flowers to invest more in attracting pollinators. If males produce more nectar, then we might expect foragers to prefer to visit male flowers, provided that males and females can be distinguished. Higher levels of nectar production in male flowers coupled with pollinator preferences for male flowers have been found in a number of systems (Bell et al. 1984; Devlin and Stephenson 1985; Delph and Lively 1992; Shykoff and Bucheli 1995). For example, in viper's bugloss (*Echium vulgare*), the protandrous flowers produce more nectar and receive higher rates of visitation during their male phase than during their female phase (Klinkhamer and de Jong 1990). However, male flowers are not always more rewarding (reviewed in Willson and Ågren 1989). For example in *Digitalis purpurea*, nectar rewards are higher during the female phase (in this species female-phase flowers are at the bottom of vertical racemes and bumblebees forage upwards, so that this arrangement prevents selfing) (Best and Bierzychudek 1982). Also, foragers differ in their requirements; some gather nectar, some pollen, while others may gather both. Those which are collecting pollen clearly benefit from avoiding female flowers, and appear to be able to do so. For example, honeybees which are collecting nectar prefer inflorescences of *Lavandula stoechas* with a high proportion of female flowers (which produce more nectar than male flowers), while individuals which are collecting both nectar and pollen choose inflorescences with a greater proportion of male flowers (Gonzalez et al. 1995). Similar preferences for pollen or nectar producing flowers according to requirements have been recorded elsewhere, in bumblebees (Alexander 1990; Cresswell and Robertson 1994), honeybees (Kay 1982; Greco et al. 1996), and solitary bees (Eckhart 1991). Clearly bees are able to distinguish between sexes or sexual phases of the flowers of at least some plant species, and are able to learn which provide the greatest reward.

From the point of view of the plant, forager preferences for flowers of a particular sex clearly threaten efficient pollination; if for example male flowers invest more in attracting visitors so that female flowers are ignored then pollination will be poor. Hence, neither sex will benefit if sexual differences are too marked.

10.5 Flower Symmetry

All flowers exhibit either radial or bilateral symmetry, although this symmetry is never perfect if measured carefully enough (Neal *et al.* 1998). Recently it has become clear that insects may use floral symmetry both in floral recognition and in discrimination between more or less rewarding flowers. Honeybees show an innate preference for symmetrical shapes, which can be reinforced by learning (Giurfa *et al.* 1996), although naive bumblebees show no preference for either radial or bilateral symmetry in flowers (West and Laverty 1998). Symmetrical artificial flowers placed in the field attracted more foraging Hymenoptera, Diptera, and Coleoptera than less symmetrical flowers (Møller and Sorci 1998). There is a fascinating parallel between the use of fluctuating asymmetry (small random departures from perfect bilateral symmetry) as an indicator of mate quality in animals (reviews in Møller 1993; Møller and Pomiankowski 1993; Watson and Thornhill 1994; Markow 1995; Møller and Thornhill 1998) and these preference by pollinators which also exert sexual selection pressure, but in this situation upon plants. So why might foragers prefer symmetrical flowers? In mate choice in animals, symmetry is thought to be an indicator of genetic quality, so it makes sense to discriminate. But bees are not mating with flowers, just extracting rewards from them. The answer seems to be that, at least in some systems where pollinators exhibit a preference for symmetrical flowers (e.g. *Epilobium angustifolium*), floral symmetry is a good indicator of floral reward (Møller 1995; Møller and Eriksson 1995). Also, handling times of bumblebees are lower on symmetrical artificial flowers than on asymmetrical flowers (West and Laverty 1998). Thus there are at least two potential benefits of preferentially visiting symmetrical flowers. If these preferences are widespread, then they should exert strong stabilizing selection upon plants for floral symmetry.

However, it is worth noting that not all studies have found pollinator preferences for symmetrical flowers or a positive relationship between floral symmetry and reward, and more studies are needed before any firm conclusion can be reached (Møller and Eriksson 1995). Studies of fluctuating asymmetry in animals have often suffered from a range of methodological flaws which researchers of floral symmetry would do well to learn from (reviews in Palmer 1994, 1996; Markow 1995).

10.6 Other Factors

Most work on discrimination among flowers by pollinators has concentrated on visual cues such as size, shape, or colour, since these are easily recorded. However, many pollinators undoubtedly also use scents produced by flowers as an important source of sensory information, particularly at close range (reviewed in Von Frisch 1967; Williams 1982; Waddington 1983*b*). The use of modern analytical techniques has revealed that many flowers exhibit intraspecific variation in floral scent quality or quantity (Tollsten and Bergstrom 1993; Knudsen 1994; Olesen and Knudsen 1994;

Tollsten and Ovstedal 1994). To my knowledge only two studies have tried to examine whether pollinators discriminate among flowers of the same species using scent. Pellmyr (1986) found that floral scent variation in *Cimicifuga simplex* determined whether bees or butterflies were attracted, while Galen and Newport (1988) found that flowers of *Polemonium viscosum* produce either 'skunky' scented flowers which are preferred by flies, and sweet scented flowers which are preferred by bumblebees. It is possible, perhaps likely, that floral scent variation is far more widespread than is currently appreciated, but that it has been largely overlooked due to our own particular sensory biases.

It is clear that bumblebees can use a broad range of cues to indicate which are the more rewarding flowers of those available. Interestingly, their choosiness can also vary according to levels of energy reserves in the colony. Cartar and Dill (1990) experimentally manipulated the reserves of nectar stored in honey pots within nests, either by draining the pots or filling them with sucrose solution. In colonies with depleted stores of nectar, bees tended to be less selective; they visit smaller inflorescences, they probed flowers at a higher rate, and they tended to fly between inflorescences rather than walk. The net result was that the rate at which they gathered nectar increased. So why do bumblebees not always behave in this way? Presumably there is a cost to this enhanced level of activity. The life span of honeybees appears to be limited by energy expenditure, so that the more active an individual is, the shorter its life expectancy (Wolf and Schmid-Hempel 1989). Similarly, in the bumblebee *B. terrestris*, more active individuals exhibit a weaker encapsulation response (the defence response to parasitoid eggs) (König and Schmid-Hempel 1995). In bumblebees, the foraging behaviour that results in the higher rate of reward also entails higher activity levels (more flight and more rapid probing). Cartar (1992b) found that bumblebees with increased natural wing wear (which presumably accumulates during flight) had elevated levels of mortality, and that artificial wing-clipping also increased mortality. Surprisingly, realistic levels of wing clipping do not appear to increase the metabolic cost of flight in bumblebees, but it may make them more susceptible to predation (Hedenström et al. 2001). Thus, the cost of gaining high rewards may be a shortened life expectancy; when colony reserves are low workers may sacrifice longevity for short-term replenishment of nectar stores.

11
Communication during foraging

The waggle dance of the honeybee has been frequently described and much studied, and is one of the most complex systems in insect communication. In contrast, it has long been assumed that foraging in bumblebees is essentially a solitary endeavour; that workers do not communicate with each other about good sources of forage, so that each individual has to learn for itself which flowers provide reward. Indeed, it has been known for many years that bumblebees (of a range of species) are unable to recruit nestmates to particular places (Jacobs-Jessen 1959; Esch 1967; Kerr 1969). However, it has become apparent that bumblebee foragers do communicate, and that recruitment does occur, but not to specific locations. In an elegantly simple experiment, Dornhaus and Chittka (1999, 2001) demonstrated that on their return to the nest, successful foragers of *B. terrestris* stimulate other workers to forage, and communicate to them the scent of the food source that they have located (Figure 11.1). The returning forager runs around on the surface of the nest in an excited manner, frequently bumping into nest mates and buzzing her wings (very similar behaviour occurs in some stingless bees). This stimulates workers to leave the nest and search for the source of the scent. This communication system is less complex than that of honeybees, for the new recruits do not appear to be given any positional information as to the location of the food source. It would appear that there is some sort of pheromone signal released by the returning forager, for activity in adjacent nests is stimulated unless air flow between them is prevented (Dornhaus and Chittka 2001).

Why do bumblebees not communicate positional information? Dornhaus and Chittka (1999) argue that conveying the location of food sources may be less important to bumblebees than to honeybees; honeybees evolved in tropical ecosystems where they rely heavily on flowering trees, a highly clumped resource which may be several kilometers from the nest and so be difficult to locate. In the temperate habitats in which bumblebees probably evolved, the (mainly herbaceous) plants on which they feed are generally more scattered. There is nothing to be gained in recruiting more workers to a specific small patch that one bee can adequately exploit single-handedly. However, communication as to the types of flowers that are providing rewards will allow the colony to rapidly recruit to feeding on a rewarding plant species when it comes in to flower, and so keep track of the changing seasonal availability of different species.

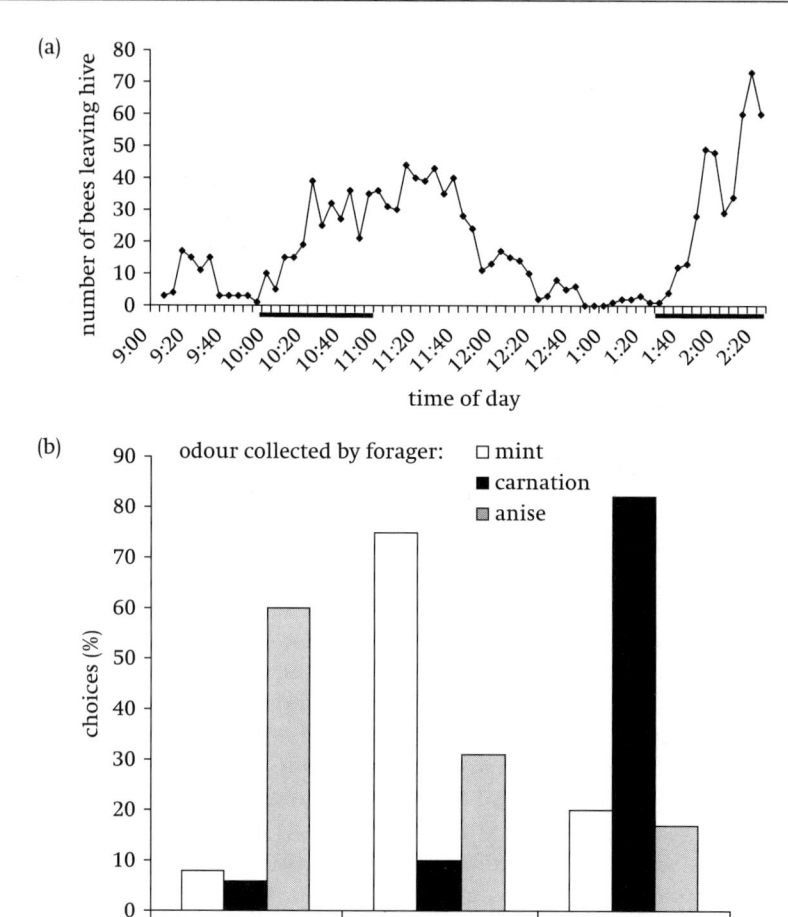

Figure 11.1. Recruitment of *B. terrestris* foragers to a newly discovered food source. (a) The number of bees leaving the nest to forage increases when a successful forager returns. The number leaving increases dramatically when an incoming forager returns with food (horizontal black bars). (b) The newly recruited foragers tend to choose a food source with the same odour as that brought back by the successful forager. From Dornhaus and Chittka (1999).

A second possibility is that bumblebees may forage over shorter distances than honeybees (see Chapter 7), perhaps as a result of their smaller colony size, so rendering communication as to the precise location of forage less important.

There is also a cost to conveying location. Honeybee recruits can take over an hour to decide where to forage when presented with just two returned foragers advertising different locations. They also take a long time to find the food source that is being advertised, for the locational information is not precise (Wenner and

Wells 1990). Thus, in bumblebees it may be that the costs of conveying this information outweigh any gains.

Communication also occurs between bees whilst foraging. Both bumblebees and honeybees can distinguish between rewarding and non-rewarding flowers of the same species without sampling the reward available. When doing so, they are often observed to hover by and 'inspect' each flower, accepting some and rejecting others (Figure 11.2). In some circumstances the bees may be directly assessing the reward levels, or perhaps examining correlates of reward such as flower size and symmetry (Chapter 10). However, there is now clear evidence that perhaps the most important cue used by bees to decide whether to probe or reject a flower are scent-marks left by bees on previous visits (Cameron 1981; Free and Williams 1983; Marden 1984; Kato 1988; Schmitt and Bertsch 1990; Giurfa 1993; Stout et al. 1998; Goulson et al. 1998a). Such marks may increase foraging efficiency by reducing the time spent handling flowers that have recently been emptied by another bee (Kato 1988; Schmitt and Bertsch 1990; Goulson et al. 1998a).

Figure 11.2. A worker of *B. pratorum* inspects a flower of *Symphytum officinale*. The bee hovers close to the flower with its antennae extended. It then either lands and probes for nectar, or rejects the flower. Rejected flowers have less nectar than those that are accepted. The cue used to discriminate between flowers is a scent mark deposited by the feet of bees that visited the flower previously; flowers that have been recently emptied have a stronger scent mark than those that have not been visited for a long while.

Bumblebees, honeybees, and carpenter bees (Hymenoptera, Anthophoridae *Xylocopa* sp.) leave short-lived repellent marks on flowers that they visit and conspecifics use these to discriminate between visited and unvisited flowers (Núñez 1967; Frankie and Vinson 1977; Cameron 1981; Wetherwax 1986; Giurfa and Núñez 1992b; Giurfa 1993; Giurfa et al. 1994; Goulson et al. 1998b; Stout et al. 1998; Williams 1998). When foraging on artificial flowers both bumblebees and honeybeesbees can also leave longer-lived scent-marks that are attractive to themselves and to conspecifics, and thus concentrate subsequent foraging bouts on more rewarding flowers (Ferguson and Free 1979; Kato 1988; Schmitt and Bertsch 1990).

In honeybees, the chemical cue which causes repellency is thought to be secreted from the mandibular glands (Vallet et al. 1991), while Nasanov secretions induce an attractant effect (von Frisch 1923; Free and Williams 1972; Free et al. 1982a,b). A Dufour's gland secretion is probably responsible for carpenter bees (*Xylocopa virginica texana*, Anthophoridae) avoiding recently visited flowers (Frankie and Vinson 1977). In bumblebees both attractant and repellent effects appear to be induced by a chemical cue found on the tarsi, and presumed to be produced from the tarsal glands (Schmitt et al. 1991). The tarsi of queen, worker, and male bumblebees all contain a substantial secretory gland described in detail by Pouvreau (1991).

For *Bombus terrestris* the components of both tarsal glands and the deposited scent marks have been identified and are very similar (Schmitt 1990; Schmitt et al. 1991). Tarsal glands produce primarily straight chain alkanes and alkenes of between 21 and 29 carbon atoms, with compounds with odd numbers of carbons predominating. The alkenes are thought to be mostly (Z)-9 and (Z)-11 configurations (Schmitt 1990; Schmitt et al. 1991). These compounds are common cuticular hydrocarbons found in a broad range of insects (Lockey 1980; Blum 1981, 1987).

More recent analyses of the hydrocarbons on the tarsi of three *Bombus* species, *B. lapidarius*, *B. pascuorum*, and *B. terrestris* have revealed notable differences between the three bee species (Goulson et al. 2000). For example, while tricosane was found in significant quantities in all three species, tricosene was only found in abundance in *B. lapidarius*. Pentacosenes were major constituents of the extracts of *B. lapidarius* and *B. pascuorum*, but were virtually absent in *B. terrestris* (Table 11.1). Species specificity has previously been discovered in the composition of labial gland secretions of male bumblebees (Bergstrom et al. 1981) and in Dufour's gland secretions of bumblebees (Tengö et al. 1991). Oldham et al. (1994) analysed cuticular hydrocarbons in *B. lapidarius*, *B. pascuorum*, and *B. terrestris*, and also compared *B. terrestris terrestris*, and *B. terrestris audax*. Although they did not examine tarsal glands, they concluded that the mix of cuticular hydrocarbons was constant across different body parts, but that species and the two *B. terrestris* subspecies differed in the relative quantities of different compounds. The composition of tarsal extracts described by Goulson et al. (2000) closely follows that for cuticular hybrocarbons found over the rest of the body (Oldham et al. 1994). During foraging, many parts of the bumblebee body may come into contact with the corolla depending upon the shape of the flower, not just the tarsi. Thus it seems probable that scent marks are not exclusively placed by the feet.

Table 11.1. Amounts of each compounds present in tarsal washes (ng/tarsus ± SE) of three bumblebee species, based on four replicate samples per species

Compound	MW	B. terrestris	B. pascuorum	B. lapidarius
Heneicosane	296	12.5 ± 2.41	—	+
Tricosenes	322	9.38 ± 6.63	5.90 ± 0.95	70.5 ± 15.1
Tricosane	324	110 ± 13.4	99.3 ± 1.21	94.8 ± 8.63
Methyl-tricosane	324	—	—	+
Tetracosenes	336	—	12.5 ± 6.03	+
Tetracosane	338	—	+	+
Pentacosenes	350	+	174 ± 12.3	155 ± 11.5
Pentacosane	352	114 ± 17.9	106 ± 5.98	170 ± 6.06
Heptacosenes	378	—	64.5 ± 13.2	+
Heptacosane	380	174.5 ± 28.6	35.5 ± 5.60	+
Nonacosenes	406	102.9 ± 26.1	+	+
Total		514 ± 68.7	491 ± 30.5	490 ± 32.4

MW = molecular weight, += trace. Samples were prepared by cutting the tarsi and approximately 1/2 of the tibia from five individuals of one species and combining them in 0.5 ml of pentane. The samples were analysed with a VG-Analytical 70–250SE mass spectrometer coupled to a Hewlett Packard 5790 gas chromatograph. The column was a BP1 of dimension 25 m × 0.33 mm with a film thickness of 0.25 μm, and the carrier gas was helium. Temperature programming was as follows: 60 °C for 3 min; heating 20 °C/min; 300 °C for 10 min; 280 °C for 12 min. Nonadecane was used as an internal standard to quantify the amounts of compounds present. After Goulson et al. (2000).

In bumblebees, it appears that repellent scent marks are effective among species; interspecific tests between B. terrestris, B. hortorum, B. pascuorum, and B. pratorum reveal that each is repelled by scent marks deposited by the other species (Stout et al. 1998; Goulson et al. 1998b). Also, tarsal extracts from B. terrestris artificially applied to flowers mimic the repellency of natural scent marks, and induce repellency in a range of Bombus species (Stout et al. 1998; Goulson et al. 2000). Even applications of a range of pure synthetic chemical constituents of scent marks (rather than the mixtures that naturally occur) produce the same repellent response. It seems that Bombus species exhibit a generalized response to flowers marked with any of the common hydrocarbons found on the cuticle of conspecifics or heterospecifics. This makes sense, for flowers are commonly visited by a range of Bombus species; the advantage to be gained from detecting empty flowers would be small if only those flowers visited by conspecifics could be detected. Since these compounds are common to most insects, not just Bombus sp., it is possible that bumblebees may be able to detect flowers which have been visited by other insects. Recent studies indicate that Bombus species may be able to detect scent marks deposited by honeybees, and vice versa (Figure 11.3) (Stout and Goulson 2001; but see Williams 1998 for conflicting evidence). This requires further investigation.

The repellent effect of scent marks wanes over time, presumably as the volatiles deposited on the flower evaporate. Thus for example, when visiting *Symphytum officinale*, foraging B. terrestris rejected nearly all flowers that had been visited in the previous 3 min, but by 40 min many flowers were acceptable and the response did

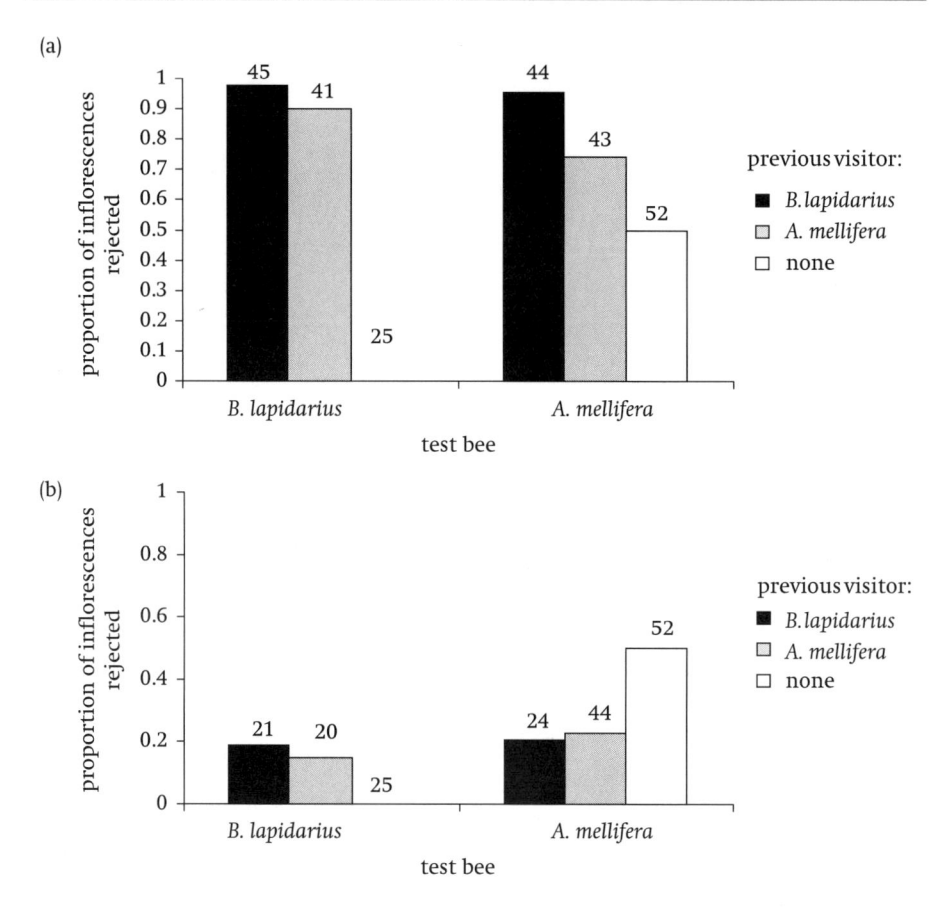

Figure 11.3. The proportion of flowers rejected by *B. lapidarius* and *A. mellifera*, (a) <3 min after the first visitor and (b) 24 h after the first visitor. The frequency of rejection of flowers that had no previous visitors is also shown. Numbers above the bars represent sample sizes. Responses are recorded as acceptance if bees landed and probed for nectar, or as rejection if the bees approached the flower, but departed without landing or feeding. From Stout and Goulson (2001).

not differ significantly from that to unvisited flowers (Figure 11.4(a)) (Stout et al. 1998). This broadly matches the rate of accumulation of nectar in *Symphytum officinale*; after 40–60 min nectar levels have been replenished (Figure 11.4(b)) (Stout et al. 1998). However, different flower species vary greatly in the rate at which they secrete nectar, so a repellent response of 40 min duration would not be appropriate for all flower species. For flowers that replenished nectar more rapidly than *S. officinale*, this would result in bees rejecting many flowers that were full of nectar, and conversely, if the secretion rate were slower, many of the acceptable flowers would contain little nectar. Also, if visitation rates are high or flowers are

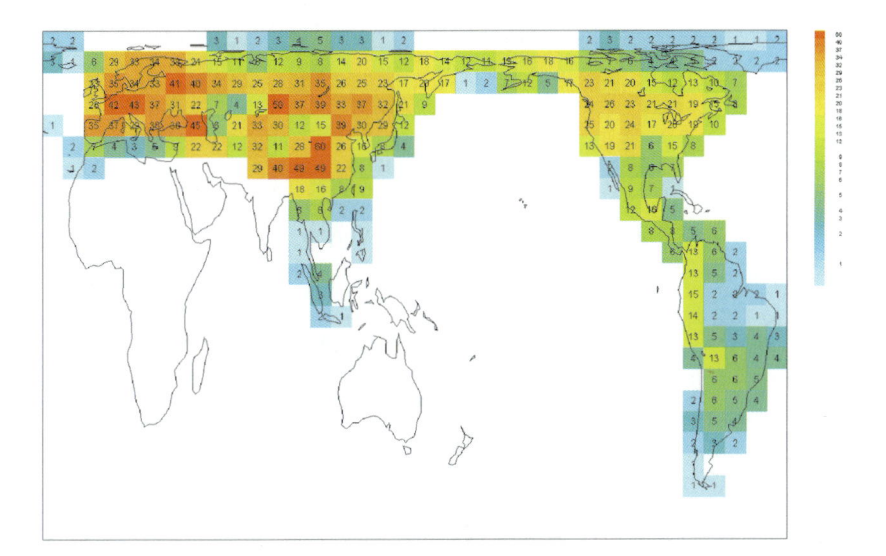

Plate 1. Global distribution of species richness in *Bombus*. The map uses a cylindrical equal-area projection that is orthomorphic at 46° North and South (where bumblebee records are particularly plentiful). Grid cells are of equal area (about 611 000 km^2). The grid covers the known natural range of bumblebees. The colour scale indicates species richness. The greatest species richness occurs in the mountains of central Asia (60 species). After Williams (1994).

Plate 2. Inside a bumblebee nest: a mature nest of *B. terrestris* viewed from above. The wax cover has been removed to reveal the nest structure. Both workers, new queens and males are visible. The queens are conspicuously infested with mites of the genus *Parasitellus*, which preferentially attach themselves to new queens on which they can persist through the winter. Some of the workers are marked with numbered discs used in behavioural studies. Also visible are sealed pupal cells, and hatched pupal cells in use as nectar pots.

Plate 3. Size variation within *B. terrestris*. The queen (bottom) is substantially larger than most workers, which themselves vary greatly in size (7 shown). Larger workers tend to be foragers and the smaller ones work mostly within the nest. The males (right) are about the same size as large workers, but can be distinguished by their longer antennae.

Plate 4. Larvae of the wax moth, *Aphomia sociella* (Lepidoptera: Pyralidae). These are thought to be among the more serious natural enemies of bumblebees, and can destroy nests. The larvae are gregarious, living within and beneath the nest and feeding on wax, pollen, and bee pupae and larvae. The moth larvae spin a dense silken web, which presumably protects them from the adult bees. This photograph is of the inside of a commercial nest box of *B. terrestris*, after removing the (heavily damaged) bee nest itself.

Plate 5. Müllerian mimicry rings in UK bumblebees. Left column, two yellow bands with white tail, from top: *B. Bombus terrestris* worker, *B. Kallobombus soroeensis* worker, *B. Bombus lucorum* ♀, *B. Psithyrus vestalis* ♀. Central column, three yellow bands with white tail, from top: *B. Psithyrus barbutellus* ♀, *B. Megabombus hortorum* ♀, *B. Megabombus ruderatus* ♀, *B. Subterraneobombus subterraneus* ♂, *B. Pyrobombus jonellus* ♀. Right column, black with red tails, from top: *B. Psithyrus rupestris* ♀, *B. Thoracobombus ruderarius* ♀, *B. Melanobombus lapidarius* ♀. Subgenera follow Williams (1998). Bumblebees are often remarkably difficult to identify since many species share very similar colour patterns. This is presumably the result of müllerian mimicry, since colour does not closely correspond to taxonomic divisions. Thus for example, *B. Thoracobombus ruderarius* closely resembles *B. Melanobombus lapidarius*, and is quite different in colour to its close relatives within the subgenus *Thoracobombus*, which are mostly brown/ginger (not shown). Note also that cuckoo bees (subgenus *Psithyrus*) often resemble their hosts: *B. P. rupestris* attacks *B. Melanobombus lapidarius*, *B. P. vestalis* attacks *B. Bombus terrestris*, and *B. P. barbutellus* attacks *B. Megabombus hortorum*.

Plate 6. A *B. terrestris* forager fitted with a transponder for use in harmonic radar studies. Photo provided courtesy of Andrew Martin, Institute of Arable Crops Research, Rothamsted.

Plate 7. *B. hortorum* foraging on a vertical raceme of a foxglove, *Digitalis purpurea*. Newly opened flowers (towards the top) are male phase, and provide less nectar than the female phase flowers at the bottom. Bees exploit inflorescences by starting near the bottom and working upwards until low rewards trigger departure. This benefits the plant by preventing selfing (Best and Bierzychudek 1982).

Plate 8. Can flowers be cryptic? The vegetation shown is typical of temperate semi-natural grassland, in that it contains a diverse array of plants many of which flower at the same time. Flowers of at least ten different species of insect-pollinated plant are visible. Many flowers which commonly occur together have colours which are extremely similar to insect colour vision systems (Kevan 1978, 1983; Chittka *et al.* 1994; Waser *et al.* 1996), and to the human eye. In this example there is a predominance of yellow flowered plants (six species). Flowers of different plant species have different structures and require different handling skills. A great diversity of bumblebees, honeybees, flies and butterflies forage among such flowers for nectar or pollen. Each insect has to make foraging decisions while on the wing as to which flowers to visit. Due to constraints of morphology and experience a particular insect will probably only be able to quickly access just one or two of the available plant species, so that mistakes in flower choice will be costly in terms of time wasted. How apparent is the birds-foot trefoil (*Lotus corniculatus*)? When viewed against a background mosaic of yellow, purple, and green, a small yellow flower is cryptic although it is brightly coloured.

Plate 9. Offering manipulated flowers of *Symphytum officinale* to a foraging bumblebee. Bumblebees are remarkably singleminded when foraging, so that they are not disturbed by close observation or manipulation of the flowers on which they are foraging. In this example, an inflorescence from which the nectar has been artificially removed is offered to a worker of *B. pratorum*, which lands and attempts to feed.

Plate 10. *B. terrestris* robbing nectar from a flower of an *Aquilegia* sp. The curled corolla tube is too deep to allow conventional access by this short-tongued bumblebee species, which instead bites through the corolla near the nectaries to obtain nectar.

Plate 11. An uncropped headland, one of many schemes introduced to European farming to enhance farmland biodiversity. Here the outer 6 m of the field is not sown with crop, and weeds allowed to naturally regenerate. This also provides a buffer between the field and the boundary, reducing pesticide drift into hedgerows. In this example several flowers favourable to bumblebees are visible, notably poppies which are a favoured source of pollen.

Plate 12. Salisbury Plain Military Training Area. This is one of the last surviving strongholds for several rare bumblebee species in southern England, notably *B. humilis* and *B. soroeensis*. The area has escaped the agricultural intensification which has severely reduced biodiversity in much of western Europe. It consists of very large areas of calcareous grassland, managed by low-density grazing of cattle and sheep. The resulting sward, rich in legumes such as *Trifolium, Melilotus, Lotus*, and *Onobrychis*, provides abundant bumblebee forage. The social organization of bumblebees means that they have a very low-effective population size (most individuals do not reproduce), and probably require large areas of suitable habitat such as this to maintain a viable population.

Plate 13. *Bombus hypnorum* worker feeding on *Cotoneaster* sp. This bumblebee species flourishes in gardens and has recently expanded its range in Europe. In 2001, it was recorded for the first time in England, and it now appears to be established in some areas along the south coast. It is not known how it traversed the English Channel. This is one of several bee species to have undergone range expansions as a result of the activities of man.

Plate 14. *B. terrestris* foraging on a *Callistemon* sp. in Tasmania. The natural ranges of these organisms are separated from one another by many thousands of kilometers, but *B. terrestris* readily adopts this and many other native Tasmanian plants as sources of forage. Depending on the match between the morphology of the bee and that of the flower, alien bee species may increase or decrease pollination services for native plants.

Plate 15. *B. terrestris* foraging on *Lupinus arboreus* in Tasmania. *L. arboreus* is a native of California, introduced to New Zealand and Tasmania to stabilize dune systems. It is an obligate outcrosser, adapted for bumblebee pollination. In New Zealand, where bumblebees have been present for over 100 years, this species has become a major environmental weed. In Tasmania, the plant has remained scarce. However, the recent arrival of *B. terrestris* (in 1992) has increased seed set considerably, and it seems likely that the plant may become an important weed in the near future.

Plate 16. Spreading expanses of *Lupinus arboreus* on the coast of SE Tasmania. Introduced *B. terrestris* are now abundant in the area, and have greatly improved pollination service to this non-native plant.

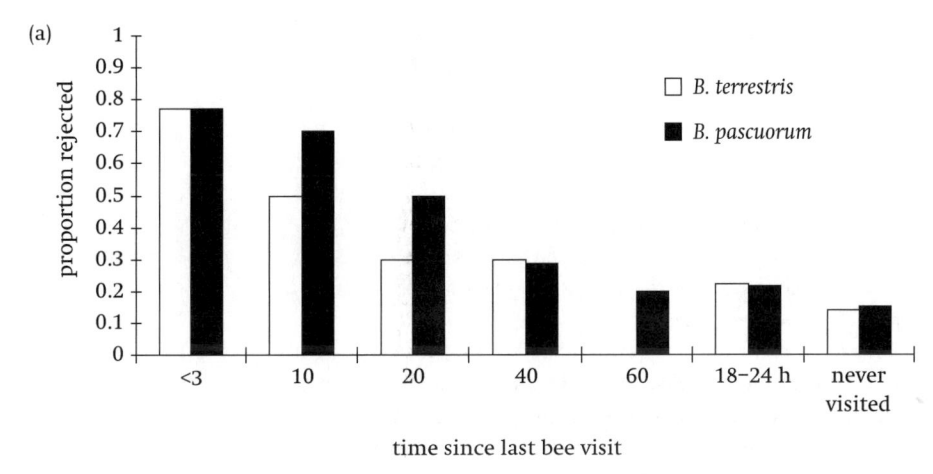

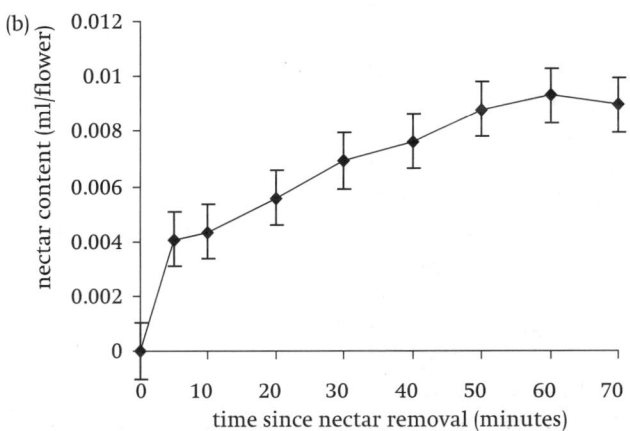

Figure 11.4. (a) The proportion of flowers of *Symphytum officinale* rejected by foraging *B. terrestris* and *B. pascuorum*, according to the time that had elapsed since the previous visit to the flower. (b) The rate of accumulation of nectar in flowers of *S. officinale* after they have been emptied. From Stout *et al.* (1998).

scarce, we would predict that bees should be less choosy and more likely to accept flowers which were visited quite recently.

It seems likely that bees may learn to use an appropriate concentration of scent mark as the threshold for rejection depending on the circumstances (Stout *et al.* 1998). Given that most individual bees are flower constant, they have the opportunity to learn an appropriate threshold concentration of scent mark for their preferred flower species. It is known that bumblebees do sample available floral rewards and modify their behaviour accordingly (Dukas and Real 1993a,b). If bumblebees can estimate the time since the last bumblebee visit according to how strong the scent mark is (as suggested by Schmitt et al. 1991 and Stout et al. 1998), then it would be possible for them to learn what concentration of scent corresponds to an appropriate threshold for acceptance of a flower. There is evidence that this does indeed occur. Williams (1998) found that repellency was of very short duration (about 2 min) when *Bombus* species were foraging on *Borago officinalis*, which has an unusually high rate of nectar secretion. Conversely, Stout and Goulson (2002) found that repellent scent marks deposited by *B. lapidarius* on *Lotus corniculatus* flowers lasted for 24 h; *L. corniculatus* has a low nectar secretion rate and was extremely abundant at the study site so that bees could afford to select the most rewarding flowers.

There is a prominent anomaly in recent studies of scent marking in bumblebees which requires an explanation. Schmitt et al. (1991) found that scent marks were used to mark rewarding flowers, and so had an attractant effect, while more recent studies have only found repellent effects, whether using natural marks, tarsal extracts, or synthetic compounds (Stout et al. 1998; Williams 1998; Goulson et al. 1998b). It has previously been postulated that scent marks might be initially repellent, but as the volatiles evaporate they may become attractants (Stout et al. 1998). However, recent studies have found no evidence for attractant marks when applying dilution series of tarsal extracts to flowers, even when the lowest concentrations contained less than one molecule per flower (Goulson et al. 2000), so it cannot be argued that an attractant response may have been detected at still lower doses.

An alternative possibility is that the more volatile components produce repellency, and the less volatile ones attraction. However, when a range of pure synthetic compounds present in natural extracts were bioassayed, all induced repellency (Goulson et al. 2000). Thus this seems unlikely. It is possible that the changing composition of a scent mark over time as the more volatile components of the natural mixture evaporate could result in attractive marks. However, bumblebees tend to reject flowers of *S. officinale* for about 40 min following a visit, but flowers visited 1, 4, or 24 h previously have acceptance rates equal to flowers that have never been visited (Stout et al. 1998). At no point were flowers that had previously been visited found to be more attractive than controls. Generally, unvisited (and unmarked) flowers receive very high rates of acceptance, so there was little scope for a scent mark to increase attractiveness of flowers (Goulson et al. 1998, 2000; Stout et al. 1998). Overall, it seems unlikely that attractant marks are in operation.

Close examination of the experimental design used by Schmitt et al. (1991) suggests another explanation. Their study used artificial flowers that were either always rewarding (regardless of whether they had been visited or not), or were never rewarding. In this circumstance, bees would inevitably spend longer feeding on the rewarding flowers, so that rewarding flowers would become covered in cuticular hydrocarbons. Given that bees are readily able to learn associations between sensory cues and rewards (reviewed in Menzel and Müller 1996), it is likely that they may have learned to preferentially visit the marked flowers, since these were the rewarding ones. It is less easy to conceive how short-range attractant marks could operate with real flowers that never provide unlimited rewards.

It is possible that attractant marks may be used to indicate plants with unusually high nectar secretion rates; studies to date have not explicitly examined whether differences in reward rates between patches influence how bees interpret scent marks.

Since bumblebees do not forage randomly they rarely encounter inflorescences which they themselves have just visited, so that the evolutionary benefit gained by leaving scent marks is not immediately apparent. Presumably they help in avoiding errors in systematic foraging. In social bees the depositors of scent marks may also benefit through improving the foraging efficiency of siblings. However, bumblebee colonies are rather small (compared to honeybees), so that the majority of beneficiaries of marks left by bumblebees are often probably not siblings. Competition between bee species is known to occur in some communities (Inouye 1978; Pyke 1982), and thus scent-marking may benefit both siblings and probable competitors. Of course, it is possible that the action of scent-marking did not initially evolve as a benefit to the marker or her siblings. Rather, the ability to detect chemicals accidentally deposited on flowers during foraging is more likely to have been the first (and perhaps only) evolutionary step towards a system of scent marking. Indeed, there is no evidence that repellent scent marks are deliberately deposited. They are comprised of alkanes and alkenes which commonly occur in the cuticles of a broad range of insects (Lockey 1980; Blum 1981, 1987), and which are bound to be left behind in tiny amounts if any part of the body comes into contact with flower parts. Thus it is debatable whether this is truly a form of communication, since the signal may be accidental.

It has only recently become apparent that the use of scent marks by bees when choosing which flowers they are going to visit is not confined to honeybees. As yet we do not know how widespread this phenomenon is. Do solitary species, bees not belonging to the Apidae, or other flower visiting insects use scent marks? Since the compounds used are widespread, interspecific interactions between distantly related taxa are possible, perhaps even likely.

12
Competition in bumblebee communities

There are about 250 species of bumblebee, mostly distributed through the temperate, alpine, and arctic regions of the northern hemisphere (Williams 1989a). In most communities, several different species of bumblebee occur sympatrically (Ranta and Vepsäläinen 1981; Williams 1989b). All *Bombus* species occupy a broadly similar niche. They are all large (relative to other bees), hairy, and facultatively endothermic; they exhibit remarkably little morphological variation; they nearly all have an annual cycle and are active at similar, overlapping, times of the year; and they all feed almost exclusively on nectar and pollen throughout their lives. One might expect fierce interspecific competition to shape bumblebee communities (Brian 1954; Heinrich 1976a). How then do many species manage to coexist?

Although *Bombus* species are all superficially similar in shape, they differ markedly in one characteristic; the length of their tongues. Some species, notably *B. hortorum* and *B. ruderatus*, have very long tongues (approximately 14 mm) compared to others such as *B. terrestris* (approximately 8.5 mm). The former also have a noticeably longer head. In combination, this enables these species to reach the nectaries in deep, narrow flowers that exclude access by other bumblebees. As a consequence, *B. hortorum* tends to visit flowers with deeper corollas (mean 8.8 mm) compared to *B. terrestris* (mean 6.3 mm) (Prys-Jones 1982). *Delphinium* provide a familiar garden example of a flower in which the nectar is hidden in a narrow tubular spur beneath the flower, and in the UK *Delphinium* are a favourite with *B. hortorum* but are rarely visited by the other bumblebee species found in gardens, all of which have relatively short tongues. In contrast, species such as *B. terrestris* feed on shallow flowers (e.g. bramble, *Rubus fruticosus*). Thus variation in tongue length between species leads to differences in the floral preferences of bumblebees (Hulkkonen 1928; Stapel 1933; Brian 1957; Hobbs *et al*. 1961; Hobbs 1962; Holm 1966; Macior 1968; Ranta and Lundberg 1980; Harder 1985; Graham and Jones 1996).

Resource partitioning with respect to tongue length is thought to be an important factor in allowing a number of bumblebee species with otherwise very similar biology to coexist (Teräs 1976; Heinrich 1976a; Inouye 1978, 1980a; Pyke 1982; Barrow and Pickard 1984; Harder 1985; Johnson 1986; Graham and Jones 1996). Inouye (1976) noted that sympatric bumblebee species in valleys near Crested Butte, Colorado, differed in mean tongue length by a constant factor of 1.2–1.4, and inferred that this pattern was the result of competition. To test whether this was so, one pair of species was examined in more detail (Inouye 1978). *B. appositus* is a long-tongued species that preferentially foraged on *Delphinium barbeyi*, while *B. flavifrons* has a medium length tongue and preferentially foraged on *Aconitum columbianum*. In three separate experiments, foragers of one species were caught and removed, and the behaviour of the other species recorded. On each occasion the remaining species increased its visitation rate to the plant that it did not normally visit (although this difference was only statistically significant for one experiment). This was interpreted as an example of competitive release, whereby each bee species is restricted to one preferred flower species by intraspecific competition. Subsequent studies at the same site indicated that *B. appositus* chose *D. barbeyi* because they obtained a higher rate of reward than if they visited *A. columbianum*. However, in the absence of their competitor, *B. flavifrons* gained equal rates of rewards on both flower species, again suggesting that in nature they are confined to *A. columbianum* through the effects of competition (Graham and Jones 1996).

In a famous study Pyke (1982) examined the distributions of seven bumblebee species in the same locality. He found that the seven species could each be assigned to one of four groups according to tongue length: long-, medium-, and short-tongued, and a short-tongued species that was also a nectar robber. Each group tended to feed upon different flower species with corolla depths appropriate to their mouthparts, with the nectar robbing species feeding primarily on bird-pollinated flowers that no other bumblebees were able to visit. The bumblebee community at any particular site tended to consist of at most four species, and never more than one from each group, although the actual species differed between sites. This he interpreted as evidence for powerful competition within groups leading to competitive exclusion of all but one species.

It is not immediately obvious why bees with long tongues should generally avoid flowers with shallow corollas. Indeed, one might imagine that long-tongued species would generally be at an advantage since they would be able to feed on both shallow and deep flowers (Ranta and Lundberg 1980). Field observations have demonstrated that long-tongued species can feed on shallow flowers, even though they generally choose not to (Heinrich 1976a; Ranta and Lundberg 1980). However, overall they do tend to visit a greater range of flower species than short-tongued bumblebees (Harder 1985). Yet, there are generally more short-tongued than long-tongued species in any given area, and long-tongued species are generally less abundant (Anasiewicz 1971; Teräs 1976; Anasiewicz and Warakomska 1977; Ranta and Lundberg 1980). It is species with medium and long tongues that have declined most in Europe, while species with unusually short tongues (*B. terrestris*, *B. lucorum*, and *B. lapidarius*) are still widespread and abundant. Kugler (1940) suggested that a long tongue might be a hindrance when feeding on shallow flowers. This explanation was confirmed by recent studies in Canada: Plowright and Plowright (1997) found that bees with long tongues fed more slowly on shallow flowers than bees with shorter tongues. Presumably a long tongue is rather unwieldy in these circumstances. Thus short-tongued bumblebees may exclude longer-tongued species from shallow flowers. This would explain the neat partitioning of floral resources described by Pyke (1982); bees are at their most efficient when feeding on flowers with a corolla depth that matches the length of their tongue.

More recent studies in Europe have failed to find such clear patterns. North and Central European bumblebee communities commonly consist of 6–11 species, with considerable overlap in tongue lengths (Ranta and Vepsäläinen 1981; Ranta et al. 1981). Several short-tongued species are commonly found to coexist. In the UK, six bumblebee species are abundant, widespread, and generally occur together. Yet, four of them have short tongues of very similar length (Williams 1989b; Goulson et al. 1998a). Several studies of local assemblages of bumblebees have failed to find any pattern in the tongue lengths of species in relation to their co-occurrence (Ranta 1982, 1983; Ranta and Tiainen 1982; Williams 1985b, 1988). Ranta and Vepsäläinen (1981) attribute coexistence of species with similar tongue lengths in Europe to spatio-temporal heterogeneity in nest distribution and floral resources (see also Tepedino and Stanton 1981; Ranta 1982, 1983). They argue that the strength and direction of competitive interactions between colonies of different species will fluctuate greatly over the season (as the availability of different flower species varies) and also from nest to nest, since flower distributions are patchy. Thus competition will not drive species to local extinction.

This explanation is plausible enough, but begs the question as to why this does not occur in Colorado too. It could be that floral resources are not limiting in Europe (and are in Colorado), but this seems unlikely. A more promising explanation is that there are niche dimensions other than tongue length that may vary between species, but which have received comparatively little attention. Harder (1985) examined flower choice by bumblebees in Ontario and concluded that although tongue length was an important factor, the relationship between tongue length and flower choice varied over time, and was influenced by numerous factors such as flower abundance and species richness, and also by body size and wing length of individual bees. Within species, differently sized workers tend to feed on different flower species (Cumber 1949a; Heinrich 1976a; Morse 1978b; Inouye 1980a; Barrow and Pickard 1984; Johnson 1986). This may be in part because size relates to tongue length, but is also probably because smaller bees have lower metabolic costs during foraging and so can profitably forage on flowers that provide low rewards per flower (Corbet et al. 1995). It seems likely that the substantial size differences found between bumblebee species also influence foraging preferences. Morse (1977) examined competition between *B. ternarius* and *B. terricola* when feeding on goldenrod, *Solidago canadensis*, in coastal Maine, USA. He found that competition led to the smaller species, *B. ternarius*, being excluded from proximal parts of the inflorescences where the larger florets occur. *B. ternarius* continued to visit the smaller distal florets, so that resources became neatly partitioned according to the size of both the bee and the flower.

Morphological variation between species, such as in size or tongue length, is easily recorded and is likely to influence foraging niche, but is only part of the picture. It is now clear that bumblebee species also differ with respect to their physiology. Teräs (1985) found that in Finland, long-tongued bumblebee species such as B. hortorum tended to visit flower species that had deep corollas, but also preferred those that were sparsely distributed. Similarly, both Sowig (1989) and Carvell (2002) found that B. hortorum, B. humilis and other longer-tongued species tend to visit flowers that occur in small patches, while short-tongued bumblebees including B. terrestris and B. lucorum favoured plants that provided large patches of flowers. Why should some species prefer clustered flowers and others scattered flowers? It seems that these preferences may reflect differences between bumblebee species in their abilities to generate heat internally (Newsholme et al. 1972; Prys-Jones 1986). Bumblebees must attain a high body temperature to take off, and they are able to generate heat in their thorax (Heinrich 1975a). The exact mechanism of thermogenesis is a bone of contention; Heinrich (1979b) maintains that it is produced through shivering, while other researchers claim that heat can be produced through substrate cycling (see Chapter 2).

Fructose bisphosphatase is a key enzyme involved in substrate cycling (if indeed it occurs), and has unusually high activity in the flight muscles of bumblebees (Newsholme et al. 1972; Prys-Jones and Corbet 1991) (Table 12.1). In non-flying bumblebees, the rate of substrate cycling is inversely related to ambient temperature, enabling the bees to maintain an internal temperature that is independent of ambient conditions even when they are not active (Clark et al. 1973; Clark 1976). Although all bumblebees that have been examined have this enzyme, the amount varies greatly between species (Newsholme et al. 1972; Prys-Jones 1986). Bumblebee species with high enzyme activity can more readily generate heat, and thus need to fly less frequently to maintain a high body temperature. While feeding on a flower, the body temperature of a bumblebee will tend to fall. If feeding on large inflorescences, the temperature may fall below the threshold for flight, about 30 °C (Heinrich 1993). B. lapidarius has a relatively high level of fructose bisphosphatase activity, can maintain a high temperature while feeding for long periods on a large inflorescence, and so can take off at any time (Prys-Jones 1986).

Table 12.1. Activity of fructose biphosphatase in the flight muscles of different bee species, and their tendency to visit plants that present massed flowers

Species	Fructose biphosphatase activity (μmol min^{-1} g^{-1} muscle, mean ± S.E.)	Proportion of visits to massed flower arrangements
B. lapidarius	131 ± 7 (8)	0.54 (210)
B. lucorum	80 ± 16 (5)	0.39 (84)
B. pratorum	73 ± 10 (13)	0.19 (177)
B. terrestris	59 ± 13 (7)	0.38 (188)
B. pascuorum	45 ± 6 (20)	0.18 (254)
B. hortorum	23 ± 1 (11)	0.07 (159)

The two variables are strongly correlated: $r = 0.88$, df = 5, $p < 0.02$ (from Prys-Jones and Corbet 1991).

In contrast, *B. hortorum* has a low enzyme activity; if it were to spend a long period feeding on a single inflorescence it would cool and then be unable to take off without a period of shivering of the flight muscles. But because they preferentially forage on scattered flowers, necessitating frequent flights, they do not need high levels of fructose biphosphate activity to keep warm (and they also minimize competition with species such as *B. lapidarius*). There appears to be a clear relationship between the preference of bee species for plants with massed flower arrangements and their fructose bisphosphatase levels (Table 12.1).

Interestingly, despite its greater potential for thermogenesis through substrate cycling, *B. lapidarius* has a higher minimum air temperature threshold for activity than other common European species such as *B. terrestris* and *B. hortorum* (Reinig 1972; Corbet et al. 1993). We do not know what physiological or metabolic factors determine differences in the temperature range over which bumblebees are active (other than size). The latitudinal ranges of bumblebee species vary greatly; for example, *B. distinguendus* is a northern European species while its close relative *B. subterraneus* has a more southerly distribution. Presumably, these species are adapted to activity under different temperature regimes; *B. subterraneus* does have a noticeably more sparse coat. This aspect of niche differentiation in bumblebees has received surprisingly little attention. Williams (1986, 1989a,b) argues that the patterns of abundance of bumblebee species in the UK are best explained by their climatic optima, rather than by competition. Species which are near the edge of their range tend to be less abundant and confined only to the highest quality sites. In the UK, tongue length does not seem to be an important factor in determining which species co-occur.

In addition to thermal factors, bumblebee species differ subtly in may other ways; they use different nest sites (Alford 1975; Svensson et al. 2000), queens emerge at different times (Prys-Jones 1982), and they reach peak worker abundance at different times of year (Goodwin 1995). The successional emergence of queens from hibernation must inevitably reduce interspecific competition, and this may be particularly important at a time when flowers are scarce. Differences in the timing of peak worker foraging may serve the same purpose. For example, *B. pratorum* is one of several ubiquitous short-tongued bumble bee species in the UK, but it differs from its potential competitors by having a very short colony duration. Worker abundance peaks in May or early June and reproductives are produced from May onwards (Alford 1975; Goodwin 1995). In contrast, most of the other UK species do not reach peak abundance until July. Species also differ in their proclivity for collecting pollen versus nectar. For example, *B. lucorum* and *B. terrestris* appear to collect significantly more pollen, and proportionally less nectar, than *B. pascuorum* (Brian 1957).

Factors other than tongue length have received little scrutiny, yet one of the most convincing demonstrations of competition between North American bumblebees strongly suggests that factors other than tongue length are important. Bowers (1985b) experimentally produced sympatric and allopatric populations of *B. flavifrons* and *B. rufocinctus* in subalpine meadows of Utah. In the absence of competition, both species fed on a similar range of flowers. When sympatric,

B. rufocinctus were excluded from their preferred flower species and the body weights of foragers were smaller, indicating that they had received less food during development. No effects of competition were detected in *B. flavifrons*. These two species are indistinguishable with regard to tongue length or size, so their floral choices and the asymmetry of competition between them must be due to other factors. The only obvious difference is in phenology; *B. flavifrons* emerges from hibernation several weeks before *B. rufocinctus*, so that by the time workers of *B. rufocinctus* appear, workers of *B. flavifrons* are already numerous. This may provide *B. flavifrons* with a competitive advantage, but exactly how is not clear.

Tongue length is undoubtedly an important factor in determining niche overlap and community structure in some bumblebee communities, but it is certainly not the only important factor. We as yet have only a sketchy knowledge of the details of the physiology and ecology of most bumblebee species; almost nothing is known about the rarer species. It seems that interspecific competition can be important, but the coexistence of numerous species with short tongues in Europe remains to be adequately explained.

13
Bumblebees as pollinators

Pollination is defined as the transfer of pollen from the anthers of one flower to the stigma of the same or a different flower. In the majority of plants pollination is necessary for seed set. Plants may employ a variety of vectors to transport pollen, including wind, water, birds, and bats, but a significant majority are pollinated by insects. Unlike any other insect group, adult bees feed their offspring on pollen. To gather sufficient resources for its offspring, a bee has to maintain a high work rate (compared to, say, a butterfly, which stops at flowers only to feed itself). Because of this work rate they make excellent pollinators, and a great many plants are adapted for bee pollination. The efficiency of a social lifestyle means that social bees tend to be far more numerous than their solitary counterparts, and throughout much of their range bumblebees are the most abundant native pollinators, both of crops and of wild flowers.

Plants adapted for pollination by bees tend to show a number of characteristics, a 'pollination syndrome'. Those pollinated by bumblebees are often large and brightly coloured (especially blue or purple). They are frequently bilaterally symmetrical (rather than radially), and provide large nectar rewards, often located in a deep spur (Corbet et al. 1991). However, there are a great many exceptions, and the pollination syndrome of a flower can only be taken as an indication of the likely pollinator (Waser et al. 1996).

13.1 Pollination of Crops

A broad variety of crops depend upon insect pollinators. Some, such as alfalfa (*Medicago sativa*) and clovers (*Trifolium* spp.), set no seed unless they are cross-pollinated (pollen is transferred from flowers on one plant to another). Self-fertile crops such as oilseed rape (*Brassica napus*), brown mustard (*Brassica juncea*), and tomato (*Lycopersicon esculentum*) are capable of self-pollination, but insect visits are needed to move pollen from the anthers to the stigma. In oilseed rape, adequate pollination further benefits the grower by ensuring an early and uniform ripening of seeds; otherwise, seed ripening is staggered and some seeds are shed before harvest (Williams et al. 1987). Some crops, notably sunflower (*Helianthus annuus*), are partially self-fertile, and produce better quality seed when cross-pollinated. Even fully self-fertile crops can benefit from cross-pollination through improved quality of the offspring; for example, field beans (*Vicia faba*) will set seed in the absence of pollinators, but the offspring produced will themselves set few or no seed without insect visitors (Stoddard and Bond 1987). In fruits such as strawberry (*Fragaria x ananassa*), melon (*Cucumis melo*), and kiwifruit (*Actinidia deliciosa*), fruit size is related to the number of seeds produced (and hence to the number of ovules fertilized). Adequate pollination ensures maximum fruit size. Remarkably, we are ignorant of the pollination requirements of a great number of crops despite the fundamental and well-appreciated relationship between pollination and yield (Corbet *et al.* 1991). For example in Europe, a region better studied than most, about 250 plant species are grown as crops. Of these, about 150 are thought to be insect pollinated, but for most we do not know which insects pollinate them, or whether yields are being limited by inadequate pollination (Corbet *et al.* 1991; Williams 1995). The current drive to diversify arable production is leading to the introduction of yet more crops, many of which require insect pollination (for example lupin, *Lupinus* spp.), yet whether we have sufficient appropriate insects to pollinate them is unknown.

It is exceedingly hard to estimate the total value of bee pollination (see Gill 1991), but various estimates have been produced and all agree that the contribution made by bees is vast. Estimates for the USA vary from $1.6 billion to $40 billion per year (Martin 1975; Levin 1983; Robinson *et al.* 1989; Southwick and Southwick 1992). Gill (1991) estimated the value to be A$156 million for Australia, while Winston and Scott (1984) put the value for Canada at C$1.2 billion. A comparable estimate for the EC suggests that insect pollination was worth 5 billion ECUs in 1989, of which 4.2 billion was ascribed to honeybees (Borneck and Merle 1989). More than a third of all human food is thought to depend upon insect pollination (McGregor 1976).

13.1.1 Honeybees versus bumblebees

The honeybee, *Apis mellifera*, is overwhelmingly the most widely managed pollinator of crops, and many farmers are entirely unaware that there are other insects that are capable of pollination. The economic value of pollination is often credited entirely to honeybees (Parker *et al.* 1987), and is often used to justify public subsidizing of honey bee keeping. Even the scientific literature is frequently blinkered in this respect (discussed by Richards 1993; Batra 1995). For example, honeybees were promoted for pollination of alfalfa up until the 1980s even though Henslow noted in 1867 that honeybees were incapable of tripping the flowers (Olmstead and Wooten 1987; Robinson *et al.* 1989; Batra 1995). In 1909, it was discovered that other species of bee, notably those belonging to the Megachilidae, did trip the mechanism and provide efficient pollination (Brand and Westgate 1909), but through a combination of inertia and poor advice to farmers it was not until the 1970s that use of Megachilidae for alfalfa pollination became widespread.

There is now growing appreciation that there are alternatives to the honeybee, and that in some situations the alternatives may be better (Westerkamp 1991). Honeybees do have a number of advantages as pollinators: they form vast colonies that can pollinate large areas of crops; there is a substantial body of expertise in the management of these colonies; and they provide honey. However, they also have disadvantages. First, honeybees are fair weather foragers (Willmer et al. 1994). In cold conditions, and when it is raining, they will not forage. In an unpredictable climate such as that of the UK this can be important, particularly when growing crops such as apples that flower early in the year when a spell of poor weather is likely. Second, honeybees are not able to adequately pollinate some crops. They have very short tongues, and so are not keen to visit crops with deep flowers such as red clover (*Trifolium pratense*). In some plants, such the Solanaceae (which includes tomatoes and potatoes) the pollen is presented in poricidal anthers. These are essentially similar to an inverted salt cellar; to obtain the pollen an insect has to shake the anthers (known as buzz pollination). Honeybees are not able to do this, and thus cannot efficiently pollinate these crops (Rick 1950). Lastly, reliance on a single species for pollination of crops is an inherently risky strategy. This was made all too clear during the recent epidemic of the mite *Varroa destructor*, which all but exterminated the honeybee through vast parts of its range. Similarly, the invasion of the USA by Africanized honeybees has greatly reduced the availability of commercial hives for crop pollination (Richards 1993).

In contrast, bumblebees are remarkably hardy and will forage in very cold conditions and even when it is raining (Corbet et al. 1993). In North America bumblebee queens have been seen foraging when the air temperature was below freezing, while in the Scandinavian summer they will forage for 24 h each day. Under the same conditions, bumblebees tend to forage faster than honeybees, and so pollinate more flowers per bee (Poulsen 1973; Free 1993). Thus, they provide a reliable pollination service despite the vagaries of the weather. Because different bumblebee species differ in their tongue lengths, between them they can pollinate a range of crops. For example, short-tongued bumblebee such as *B. terrestris* are important pollinators of oilseed rape, particularly in poor weather when honeybees are inactive (Delbrassinne and Rasmont 1988). Species with medium or long tongues (*B. pascuorum* or *B. hortorum*) are needed to pollinate field beans and red clover (Fussell and Corbet 1991).

Bumblebees are capable of buzz pollination, and make excellent pollinators of Solanaceae such as tomatoes (Van den Eijnde et al. 1991). The anthers of these flowers only release pollen when vibrated, which bumblebees achieve by placing their thorax close to the anthers and contracting their flight muscles at a frequency of about 400 Hz (King 1993). Members of the Ericaceae such as cranberries and blueberries (*Vaccinium* spp.), and also kiwifruit (*Actinidia deliciosa*) also benefit from buzz pollination (Buchmann 1985), and so are more effectively pollinated by bumblebees than by honeybees (Kevan et al. 1984; Mohr and Kevan 1987; Cane and Payne 1988; MacKenzie 1994).

In general, adequate pollination requires an approximate match between the size and shape of the flower and that of the pollinator. For some plants, honeybees are ineffective at pollen transfer (Westerkamp 1991; Wilson and Thomson 1991). Thus for example, on cranberry (*Vaccinium* spp.), alfalfa (*Medicago sativa*), and Delicious apples

(*Pyrus malus*), honeybees gather nectar while making little or no contact with the reproductive structures, and thus are poor pollinators (Gray 1925; Roberts and Struckmeyer 1942; Farrar and Bain 1946; McGregor 1976; Robinson 1979). Similarly, bumblebees have been demonstrated to be better pollinators than honeybees for watermelon (*Citrullus lanatus*), cucumber (*Cucumis sativus*) (Stanghellini *et al.* 1997, 1998), and for apples (Thomson and Goodell 2001). Bumblebees are hairier than honeybees, which may contribute to their efficacy in transferring pollen; for example, when visiting raspberry flowers, bumblebees deposited significantly more pollen on the stigmas than did honeybees (Willmer *et al.* 1994).

In Europe and North America, bumblebees are among the most important wild pollinators of crops (Corbet 1987; Plowright and Laverty 1987; Corbet *et al.* 1991). At least 25 major crops grown within the EC are visited and pollinated by bumblebees, including field beans, red clover, alfalfa, oilseed rape, and various hard and soft fruits (Corbet *et al.* 1991) (Table 13.1). There are almost certainly more crops that benefit from bumblebee pollination, but as noted earlier, the pollination requirements of most crops have not been investigated.

13.1.2 *Approaches to enhancing bumblebee pollination*

There are two alternative approaches to using bumblebees as pollinators; they can be bred for the purpose, and the captive colonies placed in the crop, or the grower can exploit natural populations of bees. The former approach is perhaps best suited to high-value crops grown intensively in glasshouses (Plowright and Laverty 1987). Until recently, pollination of glasshouse tomatoes was carried out by hand using a vibrating wand, no doubt a very tedious job and costly in terms of labour (Cribb 1990). Honeybees have been used for tomato pollination but they provide an erratic yield, and from preference will not visit tomato flowers (Spangler and Moffett 1977; Banda and Paxton 1991). In contrast, bumblebees are highly effective pollinators, and give increased yield compared to honeybees or hand pollination (Banda and Paxton 1991). Some even claim that bumblebee-pollinated fruit taste and smell better than those produced by hand pollination! (Heinrich 1996).

The efficacy of bumblebees as tomato pollinators was discovered in the 1980s in the Netherlands. Several companies began commercial rearing of *B. terrestris*, and within 3 years 95% of tomato growers in the Netherlands had switched to bumblebee pollination. *B. terrestris* is now the standard pollinators for glasshouse tomatoes in Europe; in 1990 over 500 ha of glasshouse tomatoes were pollinated by bumblebees in the Netherlands alone (Van den Eijnde *et al.* 1991). There are also widely used for aubergines and curcubits (e.g. Fisher and Pomeroy 1989). More recently, use of *B. terrestris* for glasshouse pollination has spread to North Africa, Japan, and Korea. The major provider of bumblebees for pollination in Europe is Koppert Biological Systems, who now sell more than 100 000 *B. terrestris* colonies per year to growers. Japan alone imports about 40 000 colonies (Asada and Ono 2000). Colonies are contained within a shoe-box sized artificial nest box, and are readily delivered by courier. They are simply placed within the glasshouse and the colony entrance

Table 13.1 Crops known to benefit from bumblebee pollination

Crop	Need for pollination	Other probable pollinators
Actinidiaceae		
Actinidia deliciosa, kiwifruit	***	H
Brassicaceae		
Brassica napus, rape	*	H, S
Brassica campestris, turnip rape	**	H, S
Asteraceae		
Helianthus annuus, sunflower	***	H, S
Ericaceae		
Vaccinium macrocarpon, cranberry	***	H
Vaccinium angustifolium, lowbush blueberry	***	H
Vaccinium ashei, rabbiteye blueberry	***	H
Vaccinium corymbosum, highbush blueberry	***	H
Grossularidaceae		
Ribes grossularia, gooseberry	*	H
Ribes spp., currants	**	H
Malvaceae		
Gossypium spp., cotton	*	H, S
Fabaceae		
Phaseolus multifloris, runner bean	**	H
Phaseolus lunatus, lima bean	*	H
Vicia faba, field or broad bean	**	H, S
Vicia villosa, vetch	**	H, S
Medicago sativa, lucerne or alfalfa	***	H, S
Melilotus spp., sweet clover	***	H, S
Trifolium spp., clovers	***	H, S
Glycine max, soya bean	*	H
Lupinus spp., lupins	**	—
Rosaceae		
Prunus avium, sweet cherry	***	H
Prunus cerasus, sour cherry	***	H
Prunus communis, pear	***	H
Prunus domestica, plum	**	H
Pyrus malus, apple	***	H, S
Rubus fruticosus, blackberry	**	H, S
Rubus ideaus, raspberry	*	H
Rutaceae		
Citrus spp., orange, lemon etc.	*	H, S
Solanaceae		
Solanum melongena, aubergine	*	H
Lycopersicon esculentum, tomato	*	H, S
Capsicum spp., pepper	*	H
Cucurbitaceae		
Cucumis melo, muskmelon	**	H
Cucumis sativus, cucumber	***	H
Citrullus lanatus, watermelon	***	H
Cucurbita spp. squash, pumpkin, and gourd	***	H, S

H = honeybees, S = solitary bees. *** = insect pollination essential, ** = insect pollination necessary for a good yield, * = insect pollination improves yield to some degree. Data derived primarily from Corbet *et al.* 1991, Delaplane and Mayer 2000.

opened. The workers quickly acclimatize themselves to their new surroundings and within a matter of minutes begin their pollination duties.

North American growers were quick to realize the value of bumblebees for tomato pollination, but import of *B. terrestris* to Canada and the USA was banned. In the early 1990s commercial rearing of the native *B. impatiens* was developed. This has proved to be similarly successful for pollination of glasshouse crops, notably tomato, muskmelons, and sweet peppers (Fisher and Pomeroy 1989; Kevan *et al.* 1990; Meisels and Chiasson 1997). Recent studies by Morandin *et al.* (2001) suggest that 7–15 colonies of *B. impatiens* per hectare (equivalent to about 2000 bee trips per hectare per day) are sufficient for tomato pollination in glasshouses.

As yet, rearing of bumblebee colonies is expensive, and it has been argued that for most field crops it is uneconomical (Plowright and Laverty 1987). However, trials have taken place in various crops in Europe, North America, and New Zealand (Van Heemert *et al.* 1990; Ptácek 1991; Whidden 1996; Stubbs and Drummond 2001) demonstrated that only 5 colonies of *B. impatiens* per hectare of lowbush blueberry produced yields equal to using 7.5 honeybee colonies per hectare, despite presumably having far fewer workers per colony. Costs of hire of bumblebee and honeybee colonies vary from year to year, but were similar at the time of their study, suggesting that use of bumblebees may be more economical than honeybees. Lowbush blueberry blooms too early in the season for wild bumblebee populations to be adequate (only queens are present). However, for most field crops, exploiting natural populations of bumblebees is likely to be the best option. This is an approach that has been championed particularly in Europe (Corbet *et al.* 1991).

As discussed in Chapter 14, modern farming practices have led to a decline in the abundance of bumblebees both in Europe and North America (Peters 1972; Williams 1982, 1986; Rasmont 1988, 1995; Kosior 1995; Banaszak 1996; Buchmann and Nabhan 1996; Westrich 1996; Westrich *et al.* 1998). Boyle and Philogène (1983) counted only 5 bumblebees in a 3-year census of orchard pollinators in Ontario. Bumblebees are abundant in other parts of Ontario, but are thought to have been driven from the fruit-growing regions by intensive use of pesticides. These crops now rely solely in pollination by honeybees. Similarly, native populations of bumblebees are rarely adequate to pollinate cranberries in North America (Marucci and Moulter 1977; Winston and Graf 1982; Kevan *et al.* 1990). Cranberry farmers are forced to rent honeybees colonies to effect pollination (Robinson *et al.* 1989), but, as with tomatoes, honeybees do not favour cranberry flowers and from preference will forage elsewhere (Marucci and Moulter 1977; Kevan *et al.* 1983). Even when they do visit cranberries they provide a far less effective pollination service than bumblebees (MacKenzie 1994). If field sizes are very large then there may simply not be enough bumblebees to go around (Fussell *et al.* 1991). Farms with large field sizes necessarily have a low proportion of hedgerows or other field margins, and since these are the places that provide nest sites and floral resources for bees when crops are not flowering, then farms with large fields will have relatively few bumblebees (regardless of the pesticide regime adopted). Yield of crops may be limited if there are insufficient bees to visit all of the flowers. For example, in fields exceeding 12 ha in size the yield of field beans was reduced through inadequate pollination by long-tongued bumblebees (Free and Williams 1976). Similarly, Clifford and Anderson (1980) estimated that if field sizes exceeded 5 ha then yield of red clover in New Zealand declined through a shortage of bumblebees.

At present, the area of entomophilous crops in the EC and USA is increasing, and some researchers have predicted that we will soon be facing a serious shortage of both wild and managed bees (Borneck and Merle 1989; Torchio 1990). If pollination is inadequate then farmers may be tempted to switch to growing crops that do not require insect pollination (Osborne et al. 1991). For example, red clover is now rarely grown for seed production in Europe because yields are poor, probably because of a lack of appropriate pollinators. Ironically, most seed is imported from New Zealand where long-tongued bumblebees (originally from the UK) are the main pollinators (Osborne et al. 1991). The introduction of novel crops may also be limited by pollinator availability. A diversity of new crops have been introduced in Europe in recent years, as yet grown only on a small scale. Many are insect pollinated; for example, lupins (*Lupinus* spp.), borage (*Borago officinalis*), camelina (*Camelina sativa*), cosmea (*Cosmea maritima*), cuphea (*Cuphea* spp.), and niger (*Guizotia abyssinica*) (Corbet et al. 1991). The potential of these crops may never be realized if yields are limited by a paucity of suitable insects to pollinate them.

There are ways in which farmers can encourage natural populations of bumblebees. Schemes such as uncropped field margins and conservation headlands were not designed specifically to increase numbers of wild bees, but probably do so. Appropriate management of uncropped areas to encourage wild pollinators may prove to be a cost-effective means of maximizing crop yield (Prescott and Allen 1986). Depending on the crops that they grow, farmers may wish to encourage particular species. For example if they grow field beans in the UK then they require healthy populations of *B. pascuorum* and *B. hortorum*. Field beans are robbed by *B. terrestris* and *B. lucorum*, which gain access to the nectar by biting through the rear of the flower, and by doing so do not come into contact with reproductive parts of the flower. To encourage long-tongued species but discourage nectar robbers, the farmer might sow wildflower strips containing deep flowers such as white deadnettle (*Lamium album*) and red clover (*T. pratense*) (Fussell and Corbet 1992a). Of course the crops themselves provide vast areas of forage, but only for short periods. However, planting a succession of crops that flowered at different times could greatly enhance pollinator abundance while simultaneously maximizing yields.

In addition to providing extra floral resources, there has been interest in providing artificial nests sites to encourage queens to nest close to target crops (Fye and Medler 1954; Hobbs et al. 1960, 1962; Wojtowski and Majewski 1964; Hobbs 1967b; Palmer 1968; Donavan and Weir 1978; Barron et al. 2000). We have little idea whether nest sites are generally in short supply, but it seems likely that they may be in areas with intensive farming regimes. In New Zealand, red clover is grown for seed on a large scale, and provision of artificial nesting boxes for bumblebees has been shown to increase yields (Donavan and Wier 1978; MacFarlane et al. 1983).

Artificial nests placed in intensively managed agroecosystems in New Zealand had a very low take-up rate (2%), compared to those placed in less disturbed sites with a higher availability of flowers (Barron et al. 2000). Planting food sources for foraging queens is thought to encourage them to nest nearby (Teräs 1985; Williams 1989b), and in combination with provision of nest sites may be a good strategy to enhance pollinator availability for pollination of crops later in the year (Woodward 1990). Nest boxes also appear to be more successful when left in place for a number of years (Barron et al. 2000). Previously occupied boxes are more likely to be re-occupied, perhaps because queens return to their maternal site to found their nest (Donovan and Wier 1978; Pomeroy 1981). Alternatively, they may search for suitable nest sites using olfactory cues to locate sites which have shown themselves to be suitable for bumblebee nest development.

Occupancy of artificial nest boxes appears to be much lower in the UK than in New Zealand (Fussell and Corbet 1992c). This may be because natural nest sites are scarcer in New Zealand, due to the limited number of small burrowing mammals, or because bumblebee populations are higher due to a paucity of natural enemies in New Zealand. Provision of nest boxes has not to my knowledge been adopted as an economically viable practice in any country other than New Zealand, and even there, their use is not widespread.

Management of farmland with the specific aim of enhancing wild bee populations is in its infancy, and at present is largely based on educated guesswork. Large scale experimental trials are urgently needed to establish which methods are most cost-effective, and must take in to account the costs of lost crop area and establishment and management of bee resources, versus the financial benefits gained through improved yields. Enhancing populations of wild bees is likely to be most successful if it is carried out at a landscape scale, which would require cooperation and coordination at a regional level (Richards 1993).

One area of environmental concern relating to the use of bumblebees for pollination is their introduction to areas to which they are not native. Four species of bumblebee were introduced to New Zealand in 1885 for the pollination of red clover, and this led to an immediate and substantial increase in yield of seed (Hopkins 1914). Bumblebee populations there remain high, probably in part because they are free from most of their natural enemies (Donovan and Wier 1978). The efficacy of bumblebees as pollinators of glasshouse tomatoes is probably the reason for the recent arrival of *B. terrestris* in Tasmania (they were probably smuggled into the country from New Zealand). Interestingly, the original introduction of *B. terrestris* to New Zealand was misguided since this species has a short tongue and shows little interest in red clover (the other three species that were introduced have longer tongues and are effective pollinators of clover). However, *B. terrestris* has become valued for pollination of alfalfa (Gurr 1955). *B. ruderatus* has since been introduced to Chile for clover pollination. *B. terrestris* has also been introduced to Chile, and applications have been lodged to introduce it to mainland Australia, South Africa, and Argentina, all motivated by the desire to use it for tomato pollination. Bumblebees are so effective as tomato pollinators that tomato growers in regions where bumblebees are not available suffer a considerable economic disadvantage on the world market. The merits and pitfalls of introducing bumblebees beyond their natural range are discussed in Chapter 15.

13.2 Pollination of Wild Flowers

Because of their ability to remain active at low temperatures, bumblebees are reliable pollinators in unpredictable climates. They also have large foraging ranges, compared to smaller solitary species, and thus are better able to pollinate plants which exist as small, fragmented populations, a situation which generally prevails in Europe (Gathmann et al. 1994; Steffan-Dewenter and Tscharntke 1999). Many wild flowers in the temperate, arctic, and alpine zones of the northern hemisphere are pollinated mainly or entirely by bumblebees, and sometimes by particular species of bumblebee. For example, high-altitude populations of Polemonium viscosum possess a suit of adaptive features that have coevolved with their bumblebee pollinators (Galen 1989). Unfortunately the pollination requirements of the vast majority of wild flower species have never been studied. For most we can only make an educated guess based on the pollination syndrome of the flower, and this approach is not particularly reliable. Some plant families are thought to be very largely dependent on bees for pollination. These include the Boraginaceae, Ericaceae, Iridaceae, Lamiaceae, Malvaceae, Orchidaceae, Fabaceae, Scrophulariaceae, Solanaceae, and Violaceae (Corbet et al. 1991).

The decline in bumblebee abundance must have resulted in reduced pollination services for some plants. The consequences of this depend on whether the plant species in question are limited in their seed set by pollination, and if so, whether their populations are limited by recruitment of seedlings. The relative importance of pollen versus resource limitation in determining seed set remains contentious (e.g. Bierzychudek 1981; Stevenson 1981; Wilson et al. 1994), but pollen is certainly limiting in some species, including ones pollinated by bumblebees (Galen 1985; Snow 1989; Zimmerman and Aide 1989; Primack and Hall 1990; Johnston 1991). A recent review suggests that pollen limitation may be common (Burd 1994). This issue is complex, for even if pollen is limiting in any particular year, fruit production may ultimately be resource limited. For example, Lathyrus vernus is exclusively pollinated by bumblebees in Sweden. Supplementing pollen increased seed set, indicating pollen-limitation of seed set (Ehrlén 1992). However, plants paid for this in the subsequent year, for plants that had received supplementary pollen became markedly smaller and produced fewer flowers (Ehrlén and Eriksson 1995). Overall pollen supplementation did not affect their lifetime reproductive success.

These issues aside, it seems intuitively likely that a reduced pollination service will adversely affect some plant populations, and given the large number of plants that are probably pollinated by bumblebees these effects are likely to be widespread. A decline in pollination services can have more subtle effects than reduced seed output; it may also lead to reduced outcrossing and thus to inbreeding. For example, Phyteuma *nigrum* is an endangered plant in the Netherlands which exists mainly as small, isolated populations. These fail to attract adequate numbers of bees and receive little or no outcross pollen from other populations (Kwak *et al.* 1991a,b). Both reduced seed set and increased inbreeding may lead to declines in the abundance of plant species, which can be very detrimental when plants are already scarce and threatened directly by the same changes in land use that threaten the bees (Senft 1990; Jennersten *et al.* 1992; Laverty 1992; Oostermeijer *et al.* 1992; Kwak *et al.* 1996; Young *et al.* 1996; Fischer and Matthies 1997; Steffan-Dewenter and Tscharntke 1997). Unfortunately, for most wild flowers we do not know their pollination requirements, let alone whether they are pollinator-limited, so it is impossible to predict which species are most at risk. It is likely that many rare plants are receiving a less reliable pollination service than they once did, but this will generally go unnoticed since no-one is studying them (Corbet *et al.* 1991). If perennial plants fail to set seed it may be many years before effects are seen. Alteration of the relative reproductive success of plant species according to their pollination system may lead to profound changes in plant community structure, and in turn this will have knock-on effects for the associated animal community. Rare habitats such as Mediterranean garigue and Atlantic heathland are dominated by bee-pollinated plants, and so may be particularly susceptible to changes in bee abundance (Osborne *et al.* 1991). However, very few long-term studies are carried out in any habitats, and it will be exceedingly difficult to separate effects of pollinator abundance from those of other changes in the environment, such as climate change.

13.2.1 *Nectar robbing*

The relationship between plants and their pollinators is mutualistic since both plant and pollinator benefit from the association. However, mutualistic relationships are susceptible to cheating; if one partner evolves the ability to obtain the reward from its mutualist without providing anything in return then it will flourish (at least until the partner evolves counter-measures) (Boucher *et al.* 1982). In the case of insects and flowers, there is probably little direct pressure on insects to minimize the pollination service that they provide, since carrying a few pollen grains between flowers is not a costly activity; the fitness of an insect is likely to be largely independent of whether it provides an adequate pollination service to the flowers that it visits. However, if they are able, insects will readily gather rewards from flowers without effecting pollination.

A great many (perhaps the majority) of insect visits to flowers do not result in pollination. This commonly happens if there is a mismatch between the morphology of the insect and that of the flower (because the flower is adapted for pollination by a different insect species). For example, if the insect is small, it may be able to enter a flower and gather nectar without contacting either the stamens or the stigma. Ants are common 'nectar thieves' of this sort. For example, in Colorado the ant *Formica neorubfibarbus gelida* commonly takes nectar from flowers that are adapted for bumblebee visitation (Galen 1983). Insects may also extract nectar from polypetalous flowers by pushing in between the petals at the base of the flower corollas and by-passing the reproductive structures of the flower ('base foragers').

Both bumblebees and honeybees sometimes forage in this way (Free and Williams 1973). Finally, some animals make holes in sympetalous flower corollas to allow direct access to the nectaries ('nectar robbers') (Inouye 1980b). Nectar robbers are either primary robbers (individuals which actually make holes in the flower corolla by piercing or biting) or secondary robbers (individuals which use the holes made by primary nectar robbers). If flowers have previously been robbed, primary nectar robbers may re-use holes and act as secondary robbers.

Bumblebees are common nectar robbers of many flower species in both Europe and North America, and have been recorded robbing over 300 different plant species (Lovell 1918; Inouye 1983) (Plate 10). For example, in the UK, *Bombus terrestris* and *B. lucorum* are common primary robbers whilst *B. lapidarius, B. pratorum,* and *B. pascuorum* sometimes secondarily rob (Free 1962). All of these robbing species have relatively short tongue lengths (with the exception of *B. pascuorum* which has an intermediate tongue length) and are thus unable to reach nectar in flowers with a deep corolla by foraging legitimately. The species with the longest tongue that is found in the UK, *B. hortorum*, is rarely seen to rob nectar from flowers (Brian 1957), and in general long-tongued bumblebees show no interest in robbing flowers even when they are unable to handle them legitimately (Inouye 1983). Some of the nectar robbing species have adaptations for the purpose; *B. mastrucatus* and *B. occidentalis* both have mandibles with distinct teeth, unlike most bumblebees (Løken 1949; see also Figure 13.1).

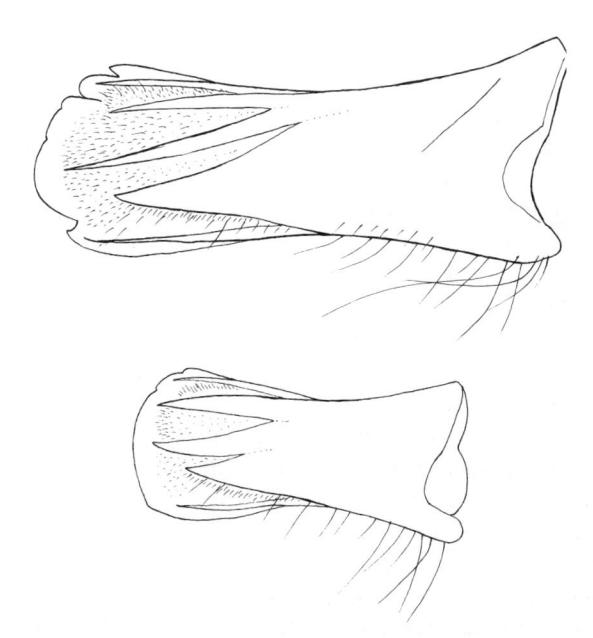

Figure 13.1. Mandibles of *B. terrestris* (top), a nectar robbing species, and those of *B. pascuorum* (bottom), a species that does not rob flowers (although it sometimes secondarily robs flowers using holes made by *B. terrestris*). Note the teeth on the mandible of *B. terrestris*.

Intuitively, nectar robbing is a process that we would expect to be costly to the plant. Darwin (1872) wrote that 'all plants must suffer in some degree when bees obtain their nectar in a felonious manner by biting holes through the corolla'. There is concern that nectar robbing by bumblebees may reduce the yields of some crops. For example in Europe *B. terrestris* and *B. lucorum* commonly rob field beans, *Vicia faba* (Poulsen 1973). In British Columbia, *B. occidentalis* robs nectar from highbush blueberries, *Vaccinium corymbosum*. In this region honeybee hives are stationed close to the crop to provide pollination, but the honeybees preferentially behave as secondary nectar robbers where holes have been provided by bumblebees (Eaton and Stewart 1969). Similarly, it has been argued that nectar robbing of wildflowers by *B. terrestris* in areas where the bee is not native (e.g. in New Zealand and Tasmania) may reduce seed set of some plants and so adversely alter the composition of native plant communities (Stout and Goulson 2000). However, there is little hard evidence for such affects.

The impact of nectar-robbing on plant fecundity has been assessed in various tropical and temperate plant species (reviewed in Maloof and Inouye 2000). Nectar robbers do sometimes have a detrimental effect on seed set in the plants they visit. Robbers reduce the amount of reward available to pollinators which may result in decreased visitation rates by pollinators (McDade and Kinsman 1980) and a reduction in seed set (Roubik 1982b; Roubik *et al.* 1985; Irwin and Brody 1999). Robbers can also damage floral tissues and thus prevent seed production (Galen 1983). Surprisingly, however, nectar robbery has often been found to have no adverse effects on plant fecundity (Newton and Hill 1983; Kwak 1988; Scott *et al.* 1993; Arizmendi *et al.* 1995; Morris 1996). In some instances this is because the 'nectar robbers' are still effective pollinators, despite their unconventional means of accessing the nectaries (Koeman-Kwak 1973; Higashi *et al.* 1988). For example, Koeman-Kwak (1973) found that nectar robbing *B. terrestris*, *B. lucorum*, and *B. jonellus* still transferred pollen between flowers of *Pedicularis palustris* (the use of the term nectar-robber is clearly misleading in such examples, and Higashi *et al.* (1988) suggest the use of the term 'robber-like pollinators'). Similarly, Palmer-Jones *et al.* (1966) found that red clover, *Trifolium pratense*, set more seed in the presence of nectar robbing *B. terrestris* than when no bumblebees were present (although it set far more when long-tongued bumblebees were present). The same is true of the bean *Phaseolus coccineus* (Kendall and Smith 1976). Some nectar robbers may be pollinators because in addition to robbing nectar, they also collect pollen in the conventional manner (Kwak 1988; Scott *et al.* 1993; Morris 1996). For example, when foraging on *Linaria vulgaris*, *B. terrestris* take nectar from the rear of the flowers by robbing, but some individuals then visit the front of the flowers to gather pollen (Stout *et al.* 2000). Similar observations have been made by Meidell (1944) and Macior (1966). Finally, nectar robbing may have no impact on fecundity if pollinators are present in sufficient abundance and are not deterred by robbers (Newton and Hill 1983; Arizmendi *et al.* 1995). For example, Stout *et al.* (2000) found that although flowers of *L. vulgaris* were very frequently robbed by *B. terrestris*, they were still visited with adequate frequency by their main pollinator, *B. pascuorum*, so that pollen did not limit seed set.

It has even been suggested that some plants may actually benefit from the activity of nectar robbers since legitimate foragers are forced to visit more flowers per foraging bout and to make more long-distance flights hence increasing genetic variability through outcrossing (Zimmerman and Cook 1985; Cushman and Beattie 1991; Maloof and Inouye 2000). Although experimental evidence for this hypothesis is largely lacking, Zimmerman and Cook (1985) did induce a greater frequency of long-distance movement of pollinators by artificially robbing flowers. Most recently, Maloof (2000, 2001) demonstrated that pollinating *B. appositus* moved further between flowers when visiting patches of *Corydalis caseana* that had been previously robbed by *B. occidentalis* than in unrobbed patches. However, if plants actually gained fitness through having lower rewards in their flowers, it is hard to explain why they would have evolved to produce higher rewards in the first place.

Overall, in a recent review Maloof and Inouye (2000) found that of 18 studies of the effects of nectar robbers on plant fecundity, 6 had a negative effect, 6 had no effect, and 6 actually increased seed set. Evidently it is not possible to generalize on the effects nectar robbers have on plants in different systems. Many factors, including the breeding biology of the plant, the foraging strategy of the robber and the abundance and efficiency of the pollinator affect the impact that nectar robbing has on the plant–pollinator system.

# 14
Conservation

The available evidence suggests that many bumblebee species have declined dramatically in recent decades, both in the UK, in continental Europe and in North America (Peters 1972; Williams 1982, 1986; Rasmont 1988, 1995; Kosior 1995; Banaszak 1996; Buchmann and Nabhan 1996; Westrich 1996; Westrich et al. 1998). Unfortunately detailed information on the abundance and distribution of most species is not available, and so it is difficult to accurately estimate the extent of this decline. The most comprehensive records available are from the UK, where detailed surveys of the distribution of bumblebee species have been carried out under the Bumblebee Distribution Maps Scheme (Alford 1980). Between 1970 and 1974, data was collected on the bumblebee fauna of 2317 10-km grid squares, comprising most of the British Isles. This can be compared with a considerable body of 'pre 1960' records (see Williams 1982). This comparison, which is already 30 years out of date, revealed a dramatic decline in the distributions of many bumblebee species (Williams 1982). More recent data, albeit for a restricted group of species, is available from the studies of the Bumblebee Working Group (see Edwards 1998, 1999, 2000, 2001). This suggests that the declines documented by Williams (1982) have continued.

The UK is accredited with 25 species of *Bombus*, of which 6 species are cuckoo bees (subgenus *Psithyrus*). One of these bumblebee species, *B. pomorum*, has not been recorded since about 1864 and was only ever known from a few specimens. The extinction of this species should perhaps not be regarded as too serious a loss, since it may never have been a long-term resident of the UK. The second species to become extinct was *B. cullumanus*, a chalk-grassland species of southern England that was probably always local and scarce; this species was last recorded in about 1941 (BMNH collection). More recently, declines appear to have accelerated, particularly in the agricultural lowlands of the south. Post-1960 populations of *B. subterraneus* were scattered across southern England from Cornwall to Kent and East Anglia. However, recent searches in its known haunts have failed to find any specimens, and it is likely that this species may have become the third British bumblebee to become extinct. *B. ruderatus*, another species that once occurred throughout England is now exceedingly scarce and confined to one or two sites in eastern England. *B. distinguendus* has disappeared entirely from England and is only known from the far north of mainland Scotland, the Orkney Islands and the Hebrides. A further six species, *B. muscorum*, *B. humilis*, *B. soroeensis*, *B. ruderarius*,

B. monticolla, and *B. sylvarum* have all but disappeared from most of their range. The future of these species in the UK is precarious.

These declines may not be immediately apparent to the casual observer because some bumblebee species remain abundant. In the UK, six species are widespread and numerous, particularly in gardens, so that it is easy to get the impression that bumblebees are faring well. Many rare and declining species are similar in appearance to the more common ones, so that their absence is easily overlooked. The reality is that more than half of the UK *Bombus* species are extinct already or could face extinction in the UK within the next few decades. If the UK is representative of the situation elsewhere, and the available evidence suggests that it is, then bumblebees are facing a crisis.

14.1 Causes of Declining Bumblebee Numbers

14.1.1 *Declines in floral diversity*

Most researchers are convinced that declines in numbers of bumblebees are linked to the intensification of farming practices (Williams 1986; Osborne and Corbet 1994). In Europe this process has been underway for 250 years, but accelerated during the latter half of the twentieth century.

The habitat preferences of most bumblebee species have not been quantified in detail. Interestingly, the data we have suggest that most bumblebee species are generally not strongly associated with particular habitats; for example, prior to its probable extinction in Britain, *B. subterraneus* occurred in habitats as diverse as shingle, saltmarshes, sand dunes, and calcareous and neutral unimproved meadows (Sladen 1912; Alford 1975; Williams 1988). Other good habitats for bumblebees include heathland, wet meadows and fens (Osborne *et al.* 1991). These habitats are characterized by a high abundance and diversity of flowers. The area of all of these habitats has declined markedly in Europe.

In the UK the Second World War led to a drive for self-sufficiency. The main thrust of the 1947 Agriculture Act was to increase farming productivity by improving yields on farmed land and by bringing unfarmed areas into production. This approach was subsequently adopted by much of Europe under the Common Agricultural Policy. Permanent unimproved grassland was once highly valued for grazing and hay production. The development of cheap artificial fertilizers and new fast-growing grass varieties meant that farmers could improve productivity by ploughing up ancient grasslands, and this they were encouraged to do (Stapledon 1935; Waller 1962). Hay meadows gave way to monocultures of grasses, notably rye grass, *Lolium perenne*, which are directly grazed or cut for silage. Between 1932 and 1984 over 90% of unimproved lowland grassland was lost in the UK (Fuller 1987).

Development grants were also introduced to grub out hedgerows, to plough and re-seed pasture and to drain marshy areas. This lead to a steady decline in the area of unfarmed land and of unimproved and semi-improved farmland. In the drive for increased production many farmers took to ploughing right up to field boundaries, and to cutting hedges very low to the ground (Marshall and Smith 1987). With the loss of hedgerows and unimproved herb-rich grassland (including neutral grasslands, wet meadows and calcareous downland) we have inevitably lost botanical diversity. The process has been further accelerated by increasing use of herbicides, which directly impact on flowers, and by increasing use of fertilizers which allow a few rapid growing plant species to outcompete and exclude slower growing species. When there is no buffer zone between the crop and the hedge, pesticides and fertilizers can penetrate into the hedge bottom, degrading the flora. In combination, changes in farming practices have resulted in the decline or loss of many plant species that were formerly common. These declines in floral diversity have been well documented both in the UK (Williams 1982; Greaves and Marshall 1987; Muir and Muir 1987; Corbet et al. 1991) and elsewhere in Europe (Ingelög 1988; Høiland 1993).

Bees are, of course, entirely dependent on flowers, since they feed more-or-less exclusively on pollen and nectar. Studies in Poland, Finland and the UK have all demonstrated a direct correlation between the floral diversity of an area and the number of bee species (Banaszak 1983; Kells et al. 2001; Bäckman and Tiainen 2002). Loss of floral diversity is widely considered to be the major cause of loss of bee diversity in agricultural landscapes (Banaszak 1983, 1992; Gathmann et al. 1994; O'Toole 1994). On farmland, the crops themselves may provide an abundance of food during their brief flowering periods. Leguminous crops (notably clovers, *Trifolium* spp.) used to be an important part of crop rotations in much of Europe, and these are highly preferred food sources, particularly for long-tongued bumblebees. Since the introduction of cheap artificial fertilizers, rotations involving legumes have been almost entirely abandoned, and it has been argued that this is one of the primary factors driving the decline of long-tongued bumblebees (Rasmont 1988; Rasmont and Mersch 1988; Edwards 1999, 2000). For bumblebees to thrive, they require a continuous succession of flowers from April to July, and crops alone are unlikely to provide this. Bumblebees do not store large quantities of honey in the way that honeybees do, and they store little pollen, so they are vulnerable to discontinuities in the food supply (Shelly et al. 1991; Williams and Christian 1991). The nest establishment phase in spring when the queen has to singlehandedly gather sufficient forage to feed her first batch of offspring may be the time when availability of flowers is most vital, but few crops flower this early (Bohart and Knowelton 1953; Alford 1975). Thus unless farms contain areas of wildflowers, they will not support bumblebees.

Uncropped areas of farmland, such as hedgerows, roadside verges, shelterbelts and borders of streams and ponds can provide flowers throughout the season, and tend to support far greater number of foraging bumblebees than cultivated areas (Banaszak 1983; Barrow 1983; Mand et al. 2002). However, these areas will only be adequate if there are enough of them, and if they have not been degraded by drift of herbicides and fertilizers. Even where flower-rich field margins and road verges remain, they are often regularly cut so that most of the flowers are destroyed.

When uncropped areas are scarce or in poor condition, there will be less food available for bees, and there may be gaps in the succession of flowering plants during which bumblebee colonies will starve and die. Bumblebee colonies frequently die out without producing reproductives. For example of eighty nests of *B. pascuorum* followed by Cumber (1953) only 23 produced any new queens (a further nine produced only males). Extinction of colonies can occur for a variety of reasons, but is more likely when floral resources are scarce (Bowers 1985a). And in turn, if bees decline, then the plants that they pollinate set less seed, so that there is even less food for the bees (Corbet 1987; Osborne *et al.* 1991; Rathke and Jules 1993; Osborne and Corbet 1994). This kind of positive feedback has been rather dramatically described as an 'extinction vortex', in which mutually-dependent species drive each other to extinction. We do not as yet know whether this process is really occurring, but it is clear that farmland provides less food for bees than it used to.

14.1.2 *Loss of nest and hibernation sites*

In addition to reducing the availability of food, modern farming practices are likely to have had other impacts on bees. Bumblebees need suitable nesting sites, the precise requirements for which vary between species. The carder bees such as *B. pascuorum* tend to nest in dense grassy tussocks, while other species such as *B. terrestris* nest underground in cavities. Both groups often use abandoned rodent nests. Studies of solitary bees show that ground-nesting species have declined disproportionately in Europe, suggesting that a lack of undisturbed nest sites may be a major factor driving declines in bee numbers (Westrich 1989). Certainly the loss of hedgerows and of unimproved pastures is likely to have reduced availability of nest sites for both above and below-ground nesting bumblebee species (Banaszak 1983; von Hagen 1994). Those species that nest above ground frequently have their nests destroyed by farm machinery, particularly by cutting for hay or silage. The scarcity of weeds and field-margin flowers on modern intensive farms means that there are less seeds for voles and mice to eat, and lower populations of these mammals will lead to fewer nest sites for both below and above-ground nesting bumblebee species.

Bumblebees also need suitable hibernation sites where young queens can remain undisturbed through the autumn and winter (although for most of the less common species we have no idea where these sites are). For the more common bumblebee species for which some data are available, it seems that hibernation sites are quite different from the sites used for nesting; most species hibernate in soil on north-west facing slopes or in the shade of trees. It is likely that such sites are not as easy to find as they once were.

14.1.3 *Pesticides*

The second half of the twentieth century saw the widespread introduction of organic insecticides, compounds that were initially developed during the Second World War. Little is known as to how much effect these compounds have on wild bees in natural situations. Pesticide risk assessments are routinely carried out for honeybees, but the results from these are probably not directly applicable to bumblebees because they have different floral preferences, and are active at different times of the day (Thompson and Hunt 1999). For example, pyrethroids are commonly applied to flowering oilseed rape in the early morning or evening, to avoid honeybees. Pyrethroids are repellent to most insects, so that sprayed crops are avoided by honeybees. However, spraying in the early morning or evening is likely to result in direct contact with foraging bumblebees since these are precisely the times when bumblebees are most active. This problem is exacerbated by the higher toxicity of pyrethroids at low temperatures (Inglesfield 1989).

Stimulated by the growing use of bumblebees in glasshouses for crop pollination, laboratory and field bioassays appropriate to bumblebees have been developed (van der Steen 1994, 2001), but these are not widely used so that few toxicological data are available (reviewed in Thompson 2001). Almost all tests conducted so far have been on *B. terrestris*. From these studies it seems that toxicity to *B. terrestris* and honeybees tends to be similar.

There are three possible routes of exposure for bumblebees to agrochemicals; through direct contact with sprays (such as when sprays are applied to flowering crops or drift onto flowering weeds where bees are foraging); through contact with contaminated foliage; and through uptake of chemicals in nectar. The latter is most likely with systemic insecticides. Tests with dimethoate and carbofuran suggest that they are selectively transported into the nectar, where they can reach high concentrations (Davis and Shuel 1988). Given the large volume of nectar consumed by bumblebees, this could prove to be the most important route of exposure.

When colonies are large it is likely that they can support some loss of workers. However, in the spring when queens are foraging, and subsequently when nests are small and contain just a few workers, mortality may have a more significant effect (Thompson 2001). Thus spring applications of pesticides are of particular concern.

Despite risk assessments, widespread poisoning of honeybees has been reported in fields of oilseed rape in the UK and elsewhere (Free and Ferguson 1986). Such effects are readily noticed in domestic hives where dead bees are ejected and form piles by the nest. It seems probable that pesticides have similar effects on bumblebees, but they are unlikely to be noticed in most situations. Probably for this reason there are few records of mortality in wild bumblebees caused by pesticides. In Canada, the use of the insecticide fenitrothion in forests led to a decline in yield of nearby *Vaccinium* crops due to a reduction in abundance of bumblebee pollinators (Ernst *et al.* 1989). In the UK, bumblebee deaths have been reported following applications of dimethoate and of alphacypermethrin to flowering oilseed rape, and of lambda cyhalothrin to field beans (Thompson and Hunt 1999; Thompson 2001). Most insecticides are broadly toxic against both honeybees and bumblebees (reviewed in Thompson and Hunt 1999), and their inappropriate use will inevitably lead to bee mortality.

14.1.4 Effects of habitat fragmentation

That changes in farming practices are largely to blame for the loss of many bumblebee species is beyond doubt. In the UK there is now a 'central impoverished region' in which only the six most common *Bombus* species are regularly found (Williams 1982, 1986). This region closely corresponds with the 'predominantly planned countryside' of Rackham (1976), and consists almost entirely of intensively farmed arable land and improved pasture, with some urban areas. The rare UK species now persist in isolated and peripheral areas, notably in south-west England, in south and west Wales, and in remote regions of Scotland, areas that have been less affected by the drives for increased agricultural productivity. Some of the strongest remaining bumblebee communities are in military training areas such as Salisbury Plain and Castlemartin Range in Pembrokeshire (Edwards 1998, 1999; Carvell 2000, 2002). These areas are still farmed, but the grasslands have not been improved and traditional grazing regimes have been retained. Other strongholds are areas of coastal marsh (e.g. Dungeness in Kent and the Thames corridor, both in south east England), areas which are particularly unsuited to agricultural activities. In mainland Europe, similar patterns are evident; bee declines have been far greater in the agricultural lowlands of western Europe than in Mediterranean and mountain regions where agricultural practices are generally less intensive (Rasmont 1995).

Of course, bumblebees are not the only wildlife to have declined in the last century. Many butterflies have decreased in abundance, particularly those associated with chalk and limestone grassland. For example the number of known British populations of the adonis blue butterfly, *Lysandra bellargus*, approximately halved every 12 years between 1950 and 1980 (Thomas 1983). Similarly, birds such as the skylark, lapwing and grey partridge have suffered catastrophic declines; in the last 20 years of the twentieth century they went from being among the most abundant farmland birds in the UK to being relative rarities. And there are many other examples.

It seems that bumblebees have been particularly hard hit as a group. It is interesting to speculate as to why this might be so. The population structure of social insects is rather different to that of most insects (reviewed in Chapman and Bourke 2001). A healthy population of butterflies can persist on quite a small area of land for many decades. For example a south-facing downland of just a few hectares near Folkestone in south-east England has supported a population of many thousands of adonis blue butterflies (*Lysandra bellargus*) for at least 100 years, even though this population is very isolated (most surviving adonis blue populations are in Dorset and Wiltshire, more than 200 km to the west). Contrast this with a bumblebee. A single colony will forage over a radius of perhaps 1 km (>3 km^2), sometimes much more. Most nature reserves are very small, and probably only provide sufficient forage for perhaps one or two colonies. Yet, a colony is, essentially, just one breeding pair (since queens of most species are monogamous, Estoup *et al.* 1995; Schmid-Hempel and Schmid-Hempel 2000). So a species that may appear to be abundant, in terms of workers, may have a very small effective population size. To maintain a viable population in the long term, it is presumably necessary to provide enough habitat for dozens of colonies. Edwards (1999) suggested that a healthy bumblebee population requires at least 10 km² of suitable habitat (this is an educated guess since we have very poor data on the nest densities or foraging ranges of bumblebees). No surviving populations of *B. sylvarum* or *B. distinguendus* in the UK are known from areas smaller than this. It is easy to see why bumblebee species with specialized habitat requirements have become rare. Nature reserves have preserved some fine examples of natural and semi-natural habitats, but in densely populated countries such as the UK most of these reserves are tiny fragments of the original area, often of just a few hectares. Very few are large enough to support a viable population of bumblebees. Thus only a few generalist species, those able to eke out a living in an impoverished agricultural landscape, have survived in most regions.

14.2 Population Structure

Most insects found in semi-natural ecosystems are now thought to exist as metapopulations, groups of variably isolated populations persisting on fragments of suitable habitat. Small, isolated populations are vulnerable to extinction, either due to inbreeding, habitat degradation, or simply due to stochastic events (Hanski and Gilpin 1991). Effective population size in social insects is likely to be several orders of magnitude lower than the observed population size, so they are particularly prone to these effects (Chapman and Bourke 2001). For the metapopulation to persist as a whole, local extinctions must be balanced by colonization events, whereby patches of suitable habitat that are unoccupied are invaded by dispersing individuals. We do not know whether this model is fully applicable to bumblebees, and as we have seen, the minimum patch size able to support a population for any length of time is probably large. Nonetheless, dispersal is likely to be a key feature influencing the population dynamics of bumblebees. In general, organisms with low dispersal are more prone to inbreeding and local extinction, and less able to colonize suitable unoccupied habitats. Unfortunately very little is known about the dispersal abilities of bumblebees (Mikkola 1978). The dispersal phase of bumblebees is most probably young queens, which have on rare occasions been observed to travel considerable distances after hibernation (dispersal of males has received very little attention). For example, in Scandinavia, bumblebee queens have been recorded moving in streams along the coastline in spring; Mikkola (1984) recorded up to 900 queens of *B. lucorum* passing through a coastal strip 150 m wide within one hour (although it is not clear where they were coming from or going to). On a rather smaller scale, Bowers (1985*a*) found movements of up to 1 km of nest-searching queens in the spring using mark-recapture, and this is almost certainly an underestimate since this method is not appropriate for detecting rare long-distance movements.

Another source of information on the dispersal abilities of bumblebees comes from monitoring of their spread when introduced to area where they are not native. When bumblebees were first introduced to New Zealand, they spread by up to 140 km per year (Hopkins 1914), suggesting a potential for very rapid movement. However, we cannot be certain that their dispersal was not artificially aided. They also successfully colonized islands up to 30 km off shore (Macfarlane and Griffin 1990), but conversely they are absent from islands at distances ranging from 16–55 km from the mainland (MacFarlane and Gurr 1995). In Tasmania, to which *B. terrestris* was introduced in 1992, spread has been much slower, at about 10 km per year (Stout and Goulson 2000).

It seems that sea barriers of over 10 km wide are sufficient to restrict gene flow and allow the development of subspecies. Subspecies of *B. terrestris* and *B. lucorum* have developed where populations are separated by straits of 10 km (Elbe/Corsica), 12 km (Italy/Sardinia), 16 km (Spain/North Africa) and 32 km (Great Britain/Europe) (Rasmont 1983). In Japan, incipient speciation is evident in populations of *B. diversus* separated by straits of 19 and 43 km width (Ito 1987). Bumblebee queens (notably *B. terrestris*) are occasionally caught at lightships at sea, up to 30 km offshore.

Dispersal of small organisms such as insects is exceedingly hard to observe directly under normal circumstances, and can provide us with little information on the frequency or magnitude of movements. An alternative approach is to use neutral genetic markers to quantify patterns of relatedness within and between populations. If all populations within a region are genetically homogenous, the implication is that movement of individuals between populations is frequent. Populations can only diverge (via genetic drift of founder effects) if gene flow is negligible. During the 1970s and 1980s studies using allozymes were popular in this respect; these are polymorphic enzymes that can be separated using gel electrophoresis. However, it seems that *Bombus* exhibit unusually little variation in their allozymes, rendering this approach of little value (Pamilo *et al.* 1984; Scholl *et al.* 1990; Owen *et al.* 1992; Estoup *et al.* 1996). More recently the development of an array of molecular techniques have made it possible to directly assess variation in DNA sequences between individuals, and this approach has revealed considerable variation within bumblebee species. Mitochondrial DNA sequences and microsatellites have both proved to be variable and informative (Estoup *et al.* 1996; Pirounakis *et al.* 1998; Widmer and Schmid-Hempel 1999). Studies to date have focused largely on two European species, *B. terrestris* and *B. pascuorum*. In *B. terrestris*, there appears to be little population substructuring within mainland Europe, suggesting that movements are frequent and that there are no substantial isolating barriers between populations (Scholl *et al.* 1990; Estoup *et al.* 1996). However, populations on various Mediterranean islands and Tenerife (Canary Islands) were distinct, suggesting that substantial bodies of water do provide a more-or-less complete barrier to movement (Estoup *et al.* 1996; Widmer *et al.* 1998). In *B. pascuorum*, populations throughout most of mainland Europe are similar, but differ markedly from those found south of the alps in Italy, suggesting that substantial mountain ranges can also act as barriers to dispersal (Pirounakis *et al.* 1998; Widmer and Schmid-Hempel 1999). There were also small differences between populations in Scandinavia and those in the body of Europe. Widmer and Schmid-Hempel (1999) conclude that

B. pascuorum probably invaded Europe from two refugia following the last ice-age, with one population coming to occupy most of Europe from Spain to Sweden, and the other remaining trapped in Italy.

Overall it seems that bumblebees are capable of long-distance movements provided that there are not clear barriers to movement such as sea or mountains. However, as yet molecular methods have only been applied to widespread and abundant species, and it is perhaps not surprising that these species exhibit little population substructuring. It would be exceedingly interesting to apply molecular tools to some of the rarer bees with more restricted ranges, particularly those found only in discrete habitat islands. Theoretical considerations suggest that social Hymenoptera may be particularly prone to inbreeding since they have a low effective population size, and males (being haploid) contribute half of the usual genetic variation to the next generation (Chapman and Bourke 2001). Inbreeding is likely to reduce fitness directly, and may lead to production of diploid (sterile) males which will lower colony fitness. No studies have been conducted on the level of inbreeding found in fragmented bumblebee populations.

14.3 Why are Some Bumblebee Species Still Abundant?

Some bumblebee species appear to have been largely unaffected by changes in the management of the countryside. In the UK, six species are widespread and common (*B. terrestris*, *B. lucorum*, *B. lapidarius*, *B. pratorum*, *B. hortorum*, and *B. pascuorum*). Some species have actually expanded their range in recent years. Both *B. lapidarius* and *B. terrestris* have expanded their range northwards in Scotland in the last 10 years (MacDonald 2001), while *B. hypnorum* has very recently colonised the UK from mainland Europe (Goulson and Williams 2001). We have no historical data on abundance for comparison, but it seems probable that abundance of these common species has declined in heavily farmed areas. Nonetheless, these species are still found in a broad range of habitats.

Williams (1986, 1988) proposes an explanation, the 'marginal mosaic model', based on some earlier general models of species distribution and abundance (Andrewartha and Birch 1954; Brown 1984). He points out that each bumblebee species occupies a particular climatic range. Within the centre of this range the species should be able to forage most profitably, and persist in a range of habitats including those that are not ideal. However, towards the edge of their range, each species will only be able to survive in the very best habitats, and even here it would be expected to be less abundant than those species near the centre of the range. If the quality of a habitat declines, it would thus be species near the edge of their climatic range that become extinct first (and the extinction process may be hastened by competition with species that are near the centre of their range that are better adapted to local conditions). This argument does appear to explain the pattern of species loss in the UK, at least in part. Species at the northern edge of their range (e.g. *B. subterraneus*) or the southern edge (e.g. *B. distinguendus*) have been affected most by declining habitat quality. In contrast, species such as *B. terrestris*, which is near the centre of its range, remains abundant even in poor quality habitats. However, this is clearly not the whole story. Species such as *B. terrestris* appear to be able to occupy vast ranges, and are abundant throughout. Some species have declined much more than others even within the heart of their range. For example, *B. subterraneus* is now thought to be very rare throughout western Europe, but apparently increased near the northern edge of its range in Finland in the 1970s (Pekkarinen *et al.* 1981).

In general, we know rather little about the ecology of many of our rarer bumblebee species. There may be other important differences between species that render some more sensitive than others to the effects of habitat degradation and fragmentation. Walther-Hellwig and Frankl (2000) provide (admittedly weak) evidence that bumblebees may differ greatly in foraging range, with species such as *B. terrestris* and *B. lapidarius* foraging further afield than 'doorstep foragers' such as *B. pascuorum*, *B. sylvarum*, *B. ruderarius*, and *B. muscorum*. It is perhaps significant that the former two species remain ubiquitous in much of the UK, while three of the four doorstep foragers have declined greatly. Species with greater foraging range may be able to persist when the density of floral resources is lower.

It seems likely that some species are more generalist than others in terms of their floral preferences. *B. terrestris* is the most polylectic bumblebee known, and is also the most abundant bumblebee throughout much of Europe. It is short-tongued, but by nectar robbing it is also able to access some flowers with deep corollas. It has been recorded feeding on many hundreds of flower species both within and outside of its natural range (Free and Butler 1959; Proctor *et al.* 1996; Semmens 1996a,b; Ne'eman *et al.* 2000). However, it is unclear whether the large number of plant species which it has been recorded visiting are the cause of its ecological success, or simply an artefact of its abundance. There have been far more studies of *B. terrestris* than of other bumblebee species, so that it is inevitable that it has been recorded feeding on a lot of plants. In contrast, a small number of bumblebee species are known to be tightly associated with just one plant species; for example *B. consobrinus* with *Aconitum septentrionale* (Løken 1973; Rasmont 1988), and inevitably such species are rare compared to *B. terrestris*. In general, rare bumblebees are little studied, and so the recorded list of plant species that they are known to visit is small. Whether there is a general association between rarity and specialization remains to be established.

14.4 Consequences of Declining Bumblebee Numbers

The plight of our bumblebee fauna deserves particular attention because loss of bee species will have knock-on effects for other wildlife. As already mentioned, a large number of wild plants are pollinated predominantly or exclusively by bumblebees, sometimes by particular species of bumblebee (Corbet *et al.* 1991;

Osborne et al. 1991; Kwak et al. 1991a,b; Rathke and Jules 1993). Reduced bumblebee numbers will result in reductions in seed set, which has obvious implications, and can also lead to reduced outcrossing and thus to inbreeding. These processes can be very detrimental when plants are already scarce and threatened directly by the same changes in land use that threaten the bees (Senft 1990; Jennersten et al. 1992; Laverty 1992; Oostermeijer et al. 1992; Kwak et al. 1996; Young et al. 1996; Fischer and Matthies 1997; Steffan-Dewenter and Tscharntke 1997). It seems probable that reductions in the abundance and species richness of bumblebees may lead to widespread changes in plant communities (Corbet et al. 1991). And of course these changes will have further knock-on effects for associated herbivores and other animals dependent on plant resources.

Bumblebees also directly support a diverse array of parasites, commensals, and parasitoids, organisms that feed on bumblebees adults, immature stages, or on detritus in the nest. Over 100 species of insects and mites have been discovered living in bumblebee nests, and many are found nowhere else (Alford 1975). Because of their vital role in supporting a diverse range of other organisms, bumblebees can be regarded as keystone species (Kevan 1991; Corbet 1995). If our bumblebees disappear then much else will go with them.

Aside from the implications for conservation, there are good financial reasons for conserving bumblebees. The yields of many field, fruit and seed crops are greatly enhanced by bumblebee visitation (Corbet et al. 1991; Free 1993; Osborne and Williams 1996; Carreck and Williams 1998). For example, field beans are largely dependent on pollination by longer tongued species such as *B. pascuorum* and *B. hortorum*, and without them, yields are poor (Free and Williams 1976). Bumblebees are acknowledged to be more reliable pollinators than honeybees, particularly because they will continue foraging even when it is cold and wet. In a poor spring, bumblebee queens (and perhaps also solitary bees such as *Osmia rufa*) may be the only insects that remain active enough to pollinate early-flowering crops such as hard fruits. Reliance on honeybee pollination is also risky since if a disease or parasite epidemic removes this one species, and no alternatives are available, then crops will fail. During the outbreak of the mite *Varroa* in the 1990s, that decimated both managed and wild honeybees, many crops became dependent on bumblebees for pollination. Thus, even those with no interest in conserving biodiversity for its own sake should be concerned, for there are direct economic costs to the decline in wild bee abundance.

14.5 Conservation Strategies

It is all very well to bemoan the demise of our native bumblebee fauna, but what exactly can be done about it? If, as I have argued, bumblebee conservation requires vast patches of suitable habitat, then surely it will never be possible to meet their needs within the overcrowded confines of much of Europe, and we must resign ourselves to the disappearance of all but the hardiest species? Perhaps not; there is a glimmer of hope. In Europe the drive for increased farming productivity slowed during the latter decades of the twentieth century with the realization that we were producing more food than was needed. Government policies have changed to accommodate this, and many of the subsidies that were provided to bring more land into production, for example by grubbing out hedgerows, were removed. There is now an emphasis on combining the goals of agriculture and conservation (Firbank et al. 1991; Dennis and Fry 1992; Saunders et al. 1992), and in Europe subsidies are currently available to remove land from arable production. This change in policy has lead to some confusion, there being a time in the 1980s when farmers in the UK could receive subsidies to remove hedgerows and then further subsidies to replant them. However, these anomalies have largely been resolved. The primary aim of these schemes when they were first introduced was to reduce agricultural production, but there is growing emphasis on using land that is taken out of production to encourage farmland biodiversity. In the UK, farmers can now choose to adopt any of a range of schemes which aim to reduce yields and increase farmland wildlife, subsidized by the Countryside Stewardship scheme and more recently the Arable Stewardship scheme (MAFF 1998, 1999). Similar schemes are in operation elsewhere. Options include new hedge-planting, repair of existing hedgerows, conservation headlands (field margins that are not treated with fertilizers or pesticides), beetlebanks (strips of tussock-forming grasses planted across fields), uncropped field margins (either allowed to regenerate naturally or sown with wildflower seed mixtures), and set-aside, whereby the land is left fallow for variable periods of time (Kaule and Krebs 1989; Firbank et al. 1993; Marshall et al. 1994; Sotherton 1995; Kleijn et al. 1998).

As yet little is known as to the relative value of these various forms of management for wildlife, and they are likely to differ between faunal or floral groups. Indeed the objectives of the schemes are often rather vague (Webster and Felton 1993). However, there is no doubt that broadly the schemes do benefit wildlife. For example, hedgerows and beetlebanks provide overwintering sites for beetles, and so boost the overall populations on farmland (Dennis and Fry 1992). They also provide a home for small mammals and nesting sites for birds (Boatman and Wilson 1988; Boatman 1992; Aebischer et al. 1994). Conservation headlands have been shown to increase abundance of farmland butterflies (Dover et al. 1990; Dover 1992; Feber et al. 1996), and to provide nectar for hoverflies (Syrphidae) (Cowgill et al. 1993). All of these schemes increase the abundance and diversity of flowers that are available. For example studies of uncropped field margins (6 m wide field margins that are not sown with crops or treated with agrochemicals) have found that they support approximately 6 times as many flowering plant species and 10 times as many flowers as equivalent cropped field margins (Figure 14.1) (Kells et al. 2001). Any form of management that increases floral resources and reduces the area of crop is likely to benefit bumblebees (Dramstad and Fry 1995), particularly since the reproductive output of bumblebee colonies is directly linked to food availability (Sutcliffe and Plowright 1988, 1990). The uncropped field margins attracted approximately 10 times as many foraging bumblebees as cropped field margins (Kells et al. 2001).

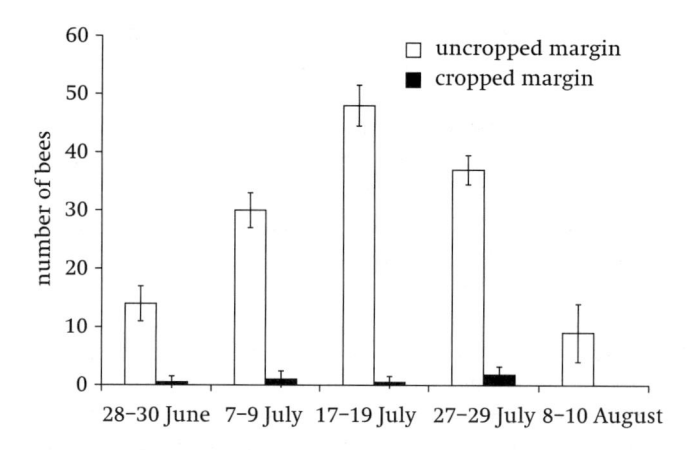

Figure 14.1. Comparison of the number of bees (*Bombus* + *Apis*) in 6 m wide uncropped margins of arable fields and in cropped margins. Counts were made in 1999 on five replicate farms in central England (from Kells *et al.* 2001). Numbers shown are means per 50 × 0.5 m transect (±SE).

Individual bumblebee species have different preferences which can be explained at least in part by their tongue lengths. For example species with medium tongues (e.g. *B. pascuorum*) or long tongues (e.g. *B. hortorum*) tend to visit deep flowers that other bees cannot reach into. They are particularly fond of labiates such as white deadnettle (*Lamium album*) and vetches (*Vicia* spp.). Short-tongued species such as *B. terrestris*, *B. lucorum*, *B. lapidarius* and *B. pratorum* are more generalized, but have notable favourites including dandelion (*Taraxacum officinale*), bramble (*Rubus fruticosus*), and thistles (*Cirsium* spp.). *B. lapidarius* has a particular preference for knapweeds (*Centaurea* spp.) (Fussell and Corbet 1991). In general, it seems that perennial and biennial plants are more favoured by bumblebees than annuals, probably because, as a very broad generalization, they tend to produce more nectar (Fussell and Corbet 1992a; Dramstad and Fry 1995) (Figure 14.2). Longer-tongued bumblebees in particular strongly favour perennials (Parrish and Bazzaz 1979; Williams 1985b; Fussell and Corbet 1991; Saville 1993).

The bumblebees that have declined most in the UK are all medium- or long-tongued species. Farm management schemes such as conservation headlands and annual rotation set-aside which are ploughed up every year will clearly favour annuals (Vieting 1988). This management is appropriate for conserving rare arable weeds (Firbank *et al.* 1993) but is of less value to most other organisms and particularly to bumblebees than long-term schemes where communities of perennial plants are allowed to develop (Osborne and Corbet 1994; Corbet 1995). Long-term set-aside (at least 5 years) has been available in the UK since 1993. There is also a Habitat Scheme, introduced in 1994, that enabled farmers to set aside land for a minimum of 20 years. However, past management may mean that perennial plants are poorly represented in the seedbank. Thus, it may be some time before

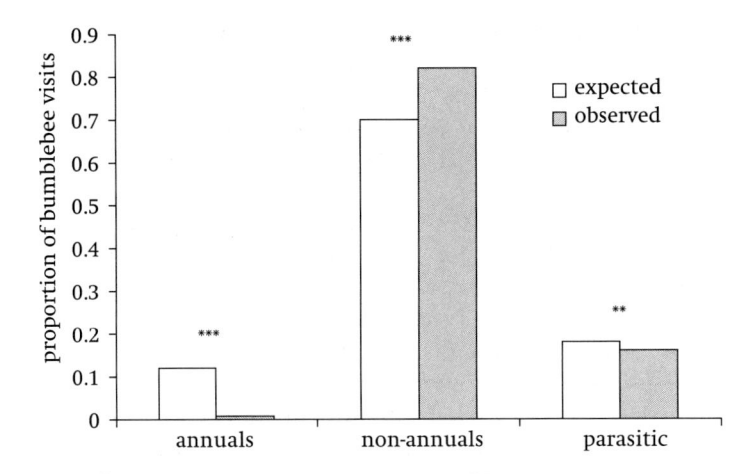

Figure 14.2. Foraging preferences of foraging bumblebees (data pooled for 7 Norwegian species) with respect to life-history strategies of plants. Bumblebees exhibit a preference for non-annuals (biennials and perennials), and also for parasitic plant species. $\chi^2 = 58$ for annuals, 14.4 for non-annuals and 3.64 for parasitic plants (after Dramstad and Fry (1995); reproduced with permission of Elsevier Science).

these areas develop a mid-successional community dominated by perennials that favours long-tongued bumblebees (Corbet 1995). This sort of plant community is also preferred by honeybees and butterflies (Feber 1993; Saville 1993; Feber et al. 1994; Smith et al. 1994). It is probably no coincidence that the two bumblebee species that have survived in reasonable abundance on arable farmland in the UK, B. terrestris and B. lapidarius, both have very short tongues in comparison with other bumblebees.

At present, only one long-tongued and one medium-tongued species remain common in the UK (B. hortorum and B. pascuorum, respectively). There are probably a number of wild flowers with deep corollas that are more-or-less dependent on these species for pollination. Corbet (2000) makes a strong argument for giving conservation of the longer-tongued bee/deep-corolla pollinator 'compartment' a high priority in Europe, since the knock-on effects of losing our last longer-tongued species would probably be severe.

Natural regeneration of a diverse community of perennials may be hampered if soil fertility is high, as a result of previous use of fertilizers. In these circumstances regeneration will consist predominantly of rank grasses and weeds such as Urtica dioica. Where seed banks are impoverished or soil fertility high it may be sensible to accelerate succession by sowing wildflower mixtures, although unfortunately such mixes can be expensive (Smith and Macdonald 1989; Smith et al. 1993, 1994; Corbet 1995). Frequent mowing may also be necessary in the first few years to suppress growth of annual weeds (Smith and MacDonald 1992).

In addition to setting aside entire fields, some farmers now sow flower strips adjacent to their crops, using either wildflower mixes or pure stands of particular flowers. Some plants, such as *Phacelia tanacetifolia* and *Borago officinalis* are quick to flower and particularly attractive to some bumblebees (notably *B. terrestris*), and are well suited to this approach (Patten *et al*. 1993). A wildflower seed mixture comprising mainly annual species has been developed in Germany, aimed primarily at encouraging honeybees (Engels *et al*. 1994). Most of our information on the favoured plants of bumblebees refers only to the six most common UK species (Fussell and Corbet 1992a). If we are to encourage other species, including those that are most threatened, to recolonize farmland than efforts should be made to identify and cater for their requirements.

It has often been suggested that the most critical time of the year for bumblebee colonies is the spring, when the queen has to singlehandedly gather pollen and nectar to rear her offspring (Alford 1975). Provision of forage specifically for queens may thus provide great benefits. Suitable plant species for early-emerging bumblebee species include *L. album* and *Salix* spp. The latter is not infrequent in hedgerows, and it would be a simple matter for farmers to plant more (plants generate readily from cuttings pushed in to the ground). However, many of the rarer bumblebee species emerge later in the year, and for most of them we do not have good data on the forage requirements of nest-founding queens.

Some of the new farm management schemes will also provide new nesting and overwintering sites for bumblebees. In fact as long ago as 1943, Skovgaard argued for the protection of uncultivated refuges in agricultural land to provide for bumblebee nest sites. Tussocky grass favoured for nesting by above-ground nesting bumblebees such as *B. pascuorum* is provided by long-term set aside, permanent uncropped field margins and by beetlebanks. Replanting of hedgerows and repair of damaged hedgerows provides more sites for species that nest underground in holes. In contrast, annual rotation set-aside and conservation headlands will not provide suitable nest sites since the vegetation is not sufficiently dense for above-ground nesting species, and subterranean nesting species rarely use newly tilled soil.

There is a long way to go before farmland in the intensive agricultural regions of Europe such as central and southern UK will support anything like the diversity of wildlife that it did 60 years ago, and this may be an unrealistic target. But at least it is beginning to move in the right direction. Further hope is provided by the move to organic farming. Demand for organic produce has rocketed in the UK in recent years, and far outstrips the supply. At present, much has to be imported. Farmers are understandably reluctant to switch to organic production because they face a four-year transition period during which they cannot market their produce as organic. Despite this, the area of land under organic regimes is steadily climbing and it seems certain that this too will aid wildlife. Rich bumblebee communities including rare species such as *B. sylvarum* have been identified on organic farms in Pembrokeshire in south-west Wales (Edwards 1999). Apart from the obvious avoidance of use of pesticides, organic farms are favourable for bees because they depend heavily on rotations involving legumes such as clover to maintain soil fertility. These are a favoured food source of bumblebees. Red clover (*T. pratense*) in particular is thought to be an important food source for longer-tongued species. Brian (1951) found that pollen from red clover made up 74% of larval food in *B. hortorum* nests in Scotland, and studies in Finland, Sweden and Denmark all suggest the importance of red clover for this bumblebee species (Skovgaard 1936; Teräs 1985; Jennersten *et al.* 1988). Bumblebee species richness in Finnish farmland was recently found to be strongly correlated with abundance of zigzag clover, *Trifolium medium* (Bäckman and Tiainen 2002). It seems that Fabaceae in general are the major source of pollen for most bumblebee species (Edwards 2001). Rasmont (1988) argues that the decline of several long-tongued bumblebees in France and Belgium is largely attributable to a decline in the area of leguminous fodder crops once grown to feed horses. A shift back towards use of legumes could greatly benefit bumblebees. Organic livestock farms may in the long-term provide excellent habitat for bumblebees. Some of the best remaining habitats in the UK are unimproved grasslands maintained by cattle or by grazing of sheep in the winter only (with the sheep moved to higher ground in the summer). Essentially, all that seems to be required is a consistent regime of moderate or rotational grazing without use of artificial fertilizers, which is exactly how many organic farms are managed. Of course, it may be many decades before land that has previously been improved will once again develop high levels of floristic diversity, but this does provide some hope for the future.

There is evidence that gardens are a particular stronghold for some species of bumblebee. Young nests of *B. terrestris* placed in suburban gardens in southern England grew more quickly and attained a larger size than nests placed in arable farmland (Goulson *et al.* 2002) (Figure 14.3). The foragers returning to them were also carrying pollen from a greater variety of plants. Interestingly, these artificial nests were far more frequently attacked by the wax moth, *Aphomia sociella*, when placed in gardens than when in farmland. Since this moth only occurs in bumblebee nests, then if it is more common in gardens it is reasonable to presume that bumblebee nests occur at a higher density in gardens than in farmland. That gardens provide a favourable habitat for several bumblebee species is perhaps not surprising given the density, variety and continuity of flowers that an extensive area of gardens provides (Corbet *et al.* 2001). However, there is undoubtedly room for improvement. Artificial selection has sometimes resulted in modern flower varieties which provide little or no reward, or which are inaccessible to insects. For example, double flowered varieties of *Lotus corniculatus*, normally a plant favoured by bumblebees, provide no nectar (Comba *et al.* 1999; Corbet *et al.* 2001). Some exotic plants provide rewards that are inaccessible to native species. For example, *Salvia splendens*, a native of the Neotropics where it probably pollinated by hummingbirds, provides high levels of nectar but when grown in the UK is not visited by bumblebees or other day-flying insects because the corolla is too deep (Corbet *et al.* 2001). If we are to maximize the utility of gardens for native pollinators, then we should encourage gardeners to choose their plants appropriately. Promoting the cultivation of flowers with deep corollas may provide a means of conserving the longer-tongued bees that have declined most.

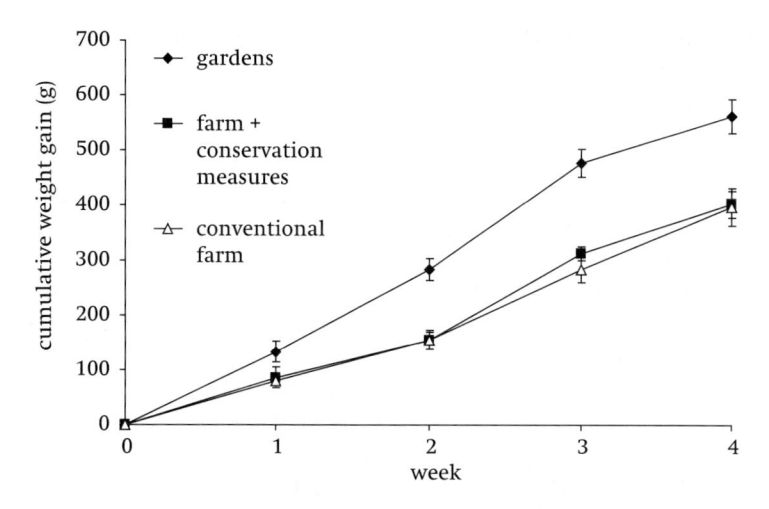

Figure 14.3. Growth rates of nests of *B. terrestris* placed in one of three different habitats: suburban gardens; conventional farmland, or farmland with substantial areas of set-aside. The experiment was conducted in southern UK in 2000. There were 10 nests per treatment (after Goulson *et al.* 2002).

14.1.1 *Summary of farmland management recommendations*

Arable farms are inherently unsuitable for bumblebees and much other wildlife because of the large area given over to crops. However, if sufficient uncropped areas are present, healthy populations of the more common bumblebee species can be expected, and these may contribute substantially to pollination of entomophilous crops (unfortunately we do not as yet know what 'sufficient' is). Conservation headlands, uncropped headlands and set aside on an annual rotation are of some value, but longer-term schemes are to be preferred. Healthy hedgerows with permanent wide margins and long-term set-aside are particularly valuable, perhaps with initial seeding of appropriate wildflower mixes if the seed bank is inadequate. Cutting of grassy areas between April and September should be minimized. Growing entomophilous crops will provide brief periods of plenty for bees. The benefit to the bees and to the farmer can be maximized if several crops are grown that flower at different times. Organic regimes are also very likely to be beneficial, although this has not been quantified.

Unimproved pasture is a very valuable habitat for bumblebee species. Short-term leys using grass monocultures are of no value, and the move from the former to the latter may be the principal cause of the decline in bumblebee numbers. Of course, many farmers do not have any unimproved pasture. However, semi-improved pasture can support a broad range of species including some rarities. Where pasture is to be reseeded, inclusion of some bumblebee flowers such as red clover, *T. pratense*, or bird's foot trefoil, *L. corniculatus*, will provide great benefits. Heavy grazing during the spring and summer is detrimental since most flowers are eaten. Excellent bumblebee habitat can be provided by grazing only during the autumn and winter months, but this is clearly not practical for most farmers. Without grazing, grassland becomes dominated by coarse grasses and loses its value for bumblebees (Carvell 2002). An alternative is rotational grazing through the spring and summer so that there are always some parts of the farm providing forage for bees. In general, grazing by cattle seems to be more favourable to bumblebees than grazing by sheep (Carvell 2002), and of course low summer grazing densities are preferable to high.

Hay meadows can be excellent bumblebee habitats, but cutting during May–August can be disastrous since it both removes all flowers and destroys the nests of any surface nesting species. Leaving field margins and corners uncut will allow some nests to survive and provide some continuity of flowers for bees to forage on.

15
Bumblebees abroad; effects of introduced bees on native ecosystems

The devastating impacts which some exotic organisms have wreaked on native ecosystems are all too familiar, and surely ought to have taught us a lesson as to the perils of allowing release of alien species. The introduction of Nile perch to Lake Victoria, of cane toads, prickly pear, rabbits, foxes, and cats amongst numerous others to Australia, and of water hyacinth, *Eichhornia crassipes*, to waterways throughout the old world tropics, are perhaps some of the best known examples, but they represent only the tip of the iceberg. Australia alone had 24 introduced mammal species, 26 birds, 6 reptiles, 1 amphibian, 31 fish, more than 200 known invertebrates, and no less than 2700 non-native plants at the last count (Alexander 1996; reviewed in Low 1999). These problems are certainly not confined to the Antipodes, although many of the most dramatic examples are to be found there. A strong case can be made that exotic species represent the biggest threat to global biodiversity after habitat loss (Pimm *et al.* 1995; Low 1999).

The threat posed by exotic species is now widely appreciated, and many countries have rigorous measures in place to prevent further introductions. Yet, we seem to have a blind spot with regard to bees. Bumblebees and a range of other bee species continue to be deliberately released in parts of the world to which they are not native. Of course, bees are widely perceived to be beneficial for their role in the pollination of crops and wildflowers. Because of the economic benefits they can provide there appears to be reluctance to regard bees as potentially damaging in environments to which they are not native.

The natural range of bumblebees is largely confined to the temperate northern hemisphere and the mountains of Central and South America (Williams 1994). Various *Bombus* species have been deliberately introduced to new countries to enhance crop pollination. The earliest deliberate and successful introduction specifically for pollination was of bumblebees to New Zealand. In 1885 and again in 1906, 93 and 143 queens, respectively, were caught in the UK and released in New Zealand with the intention of improving seed set of red clover, *Trifolium pratense* (Hopkins 1914; MacFarlane and Griffin 1990). Four species became established, *B. hortorum*, *B. terrestris*, *B. subterraneus*, and *B. ruderatus*. That these introductions were not well thought through is clear from the introduction of *B. terrestris*, which is not effective as a pollinator of red clover but acts as a nectar robber (Gurr 1957). All four species have survived to this day; *B. hortorum* and *B. subterraneous* have restricted distributions within New Zealand, while *B. terrestris* and *B. ruderatus* have become ubiquitous (MacFarlane and Gurr 1995). It is interesting that of these species, *B. subterraneous* is now probably extinct in the UK while *B. ruderatus* is exceedingly scarce.

B. terrestris spread into Israel in the 1960s (Dafni and Shmida 1996), perhaps as a result of the presence of introduced weeds. During the 1990s *B. terrestris* also became established in the wild in Japan following escapes from commercial colonies used for pollination in glasshouses (Dafni 1998; Goka 1998). In 1992, *B. terrestris* arrived in Hobart, Tasmania, perhaps accidentally transported in cargo, and has since spread out to occupy about one quarter of the island (Semmens 1996a; Buttermore 1997; Stout and Goulson 2000). Recently, *B. terrestris* was introduced to Chile. This is the second UK species to arrive in Chile, for *B. ruderatus* was previously introduced in 1982 and 1983 for pollination of red clover (Arretz and MacFarlane 1986). *B. ruderatus* had spread to Argentina by 1993 (Abrahamovich *et al.* 2001). The spread of *B. terrestris* is likely to continue. Because of its efficacy as a pollinator of glasshouse tomatoes, within the last two years applications have been lodged for deliberate release in mainland Australia, South Africa and Argentina.

Interestingly, the most recently recorded range expansion in bumblebees is in *B. hypnorum*, a species not used commercially. This native of mainland Europe was recorded for the first time in the UK in 2001 (Goulson and Williams 2001), and by 2002 appeared to be established on the south coast in the Southampton area (D.G. pers. obs.). Its seems likely that a queen or nest was accidentally transported to the UK by man, for Southampton is a major port but is approximately 150 km from the continental mainland (almost certainly beyond the natural flight range of any bumblebee).

Bumblebees are not the only bees to have been redistributed around the globe by man. The honeybee is thought to be native to Africa, western Asia, and southeast Europe (Michener 1974), although its association with man is so ancient that it is hard to be certain of its origins. It has certainly been domesticated for at least 4000 years (Crane 1990a). Because of its economic value, the honeybee has been introduced to more-or-less every country in the world, and has achieved a global distribution (being absent only from the Antarctic). It is now amongst the most widespread and abundant insects on earth. The alfalfa leafcutter bee *Megachile rotundata* (Fabr.) (Megachilidae) a native of Eurasia, has been introduced to North America, Australia and New Zealand for alfalfa pollination (Bohart 1972; Donavan 1975; Woodward 1996). At least six other Megachilidae have been introduced to the USA for pollination of various crops (Batra 1979; Parker 1981; Cooper 1984; Torchio 1987; Stubbs *et al.* 1994; Mangum and Brooks 1997; Frankie *et al.* 1998). The alkali bee, *Nomia melanderi* (Cockerell) (Halictidae), a native of North America, was introduced to New Zealand in 1971 for pollination of alfalfa and has become established at restricted sites (Donovan 1975, 1979).

So why should these introductions be a cause for concern? There are a number of possible undesirable effects of exotic bumblebees, including:

1. Competition with native flower visitors for floral resources.
2. Competition with native organisms for nest sites.

3. Transmission of parasites or pathogens to native organisms.
4. Changes in seed set of native plants (either increases or decreases).
5. Pollination of exotic weeds.

I will discuss each of these in turn. Far more studies have been carried out on impacts of honeybees, and many of the effects are likely to be similar, so I have also included a summary of this work.

15.1 Competition with Native Organisms for Floral Resources

For there to be the potential for competition to occur between organisms, the niches that they occupy must overlap. The diet of all bee species consists more-or-less exclusively of pollen and nectar collected from flowers (occasionally supplemented by honeydew, plant sap and waxes, and water) (Michener 1974). The two bee species which have proved to be most adaptable in colonizing new habitats, *A. mellifera* and *B. terrestris*, have done so largely because they are generalists. *A. mellifera* usually visits a hundred or more different species of plant within any one geographic region (Pellet 1976; O'Neal and Waller 1984; Wills *et al.* 1990; Roubik 1991; Butz Huryn 1997; Coffey and Breen 1997), and in total has been recorded visiting nearly 40 000 different species (Crane 1990*b*). *B. terrestris* is similarly polylectic. It has been recorded visiting 66 native plants of 21 families in Tasmania (Hingston and McQuillan 1998) and 419 introduced and native plants in New Zealand (McFarlane 1976).

A diverse range of different organisms collect pollen and/or nectar from flowers, including birds, bats, mammals, and insects. Of the insects, the main groups are the bees and wasps (Hymenoptera), butterflies and moths (Lepidoptera), beetles (Coleoptera), and flies (Diptera). Only the bees feed exclusively on floral resources during all stages of their life cycle, but many of the other groups are floral specialists as adults. The wide distribution and polylectic diet of *B. terrestris* and honeybees means that potentially they might compete with many thousands of different native species. It seems reasonable to predict that introduced bees are most likely to compete with native bee species (rather than other native organisms), since these are likely to be most similar in terms of their ecological niche. Studies of niche overlap in terms of flowers visited have all concluded that both honeybees and bumblebees overlap substantially with native bees and with other flower visitors such as nectivorous birds (Donovan 1980; Roubik 1982*a*; Roubik *et al.* 1986; Menezes Pedro and Camargo 1991; Thorp *et al.* 1994; Wilms *et al.* 1996; Wilms and Wiechers 1997; Hingston and McQuillan 1998; Goulson *et al.* in press). Thus there is the potential for competition.

Some potential competitors manage to coexist by exploiting shared resources at different times of the year or day. Honeybees and bumblebees differ from many other flower visitors in having a prolonged flight season; honeybees remain active for all of the year in warmer climates, while bumblebees commonly forage throughout the spring and summer in the temperate climates where they naturally occur. Thus, in terms of the time of year at which they are active, they overlap with almost all other flower visitors with which they co-occur. They also tend to feed throughout the day, beginning before and ending after most native organisms, so that both are feeding together through the middle of the day. Thus, they share resources with native organisms, and are exploiting them at the same time.

Demonstration of niche overlap does not prove that competition is occurring. In fact, it is notoriously difficult to provide unambiguous evidence of competition, particularly in mobile organisms. Because of this there is no clear agreement as to whether non-native bees have had a significant negative impact upon native pollinator populations (for reviews of the impacts of honeybees which draw different conclusions compare Robertson et al. 1989, Buchmann and Nabhan 1996, and Sugden et al. 1996 with Butz Huryn 1997). The majority of studies to date have been carried out in the neotropics, stimulated by the recent arrival and spread of Africanized honeybees, and in Australia, where awareness of the possible impacts of introduced species is unusually high. Australia also has a large native bee fauna of over 1500 species (Cardale 1993) that is arguably the most distinctive in the world (Michener 1965). Considerable circumstantial evidence has accumulated suggesting that introduced bees do impact upon native pollinator through effects on their foraging, but no unequivocal evidence has been found of competitive exclusion at the population level.

15.1.1 *Effects on foraging of native organisms*

Hingston and McQuillan (1999) examined interactions between bumblebees and native bees in Tasmania and concluded that native bees were deterred from foraging by the presence of bumblebees, perhaps because bumblebees depressed availability of floral resources (rather than because of direct interference competition). Honeybees, or the joint action of honeybees and bumblebees, have been shown to depress availability of nectar and pollen (Paton 1990, 1996; Wills et al. 1990; Horskins and Turner 1999), which may explain why other flower visitors then choose to forage elsewhere. Honeybees commonly deter other bee species from foraging on the richest sources of forage (Wratt 1968; Eickwort and Ginsberg 1980; Roubik 1978, 1980, 1996a; Wilms and Wiechers 1997) (although in at least one instance the converse had been reported, Menke 1954). Native organisms are often displaced to less profitable forage (Holmes 1964; Schaffer et al. 1979, 1983; Ginsberg 1983). For example in Panama, the presence of Africanized honeybees effectively eliminated peaks of foraging activity of Meliponine bees because these native species were prevented from visiting their preferred sources of forage; as a result the rate at which pollen was accrued in the nest was lower (Roubik et al. 1986). Displacement of native organisms has been attributed to the larger size of honeybee when compared to the majority of bee species (Roubik 1980), but is not necessarily size related. For example, the presence of honeybees has been found to deter foraging by hummingbirds (Schaffer et al. 1983). Similarly, in a year when honeybees were naturally scarce, native bumblebees in Colorado were found to expand their diet breadth to include flowers usually visited mainly by honeybees (Pleasants 1981).

Both bumblebees and honeybees begin foraging earlier in the morning than many native bee species (Corbet *et al.* 1993; Dafni and Shmida 1996; Horskins and Turner 1999). Honeybees are able to achieve this due to their large size (compared to most bees) and also due to heat retention within their large nests (Roubik 1989). Bumblebees are able to begin foraging earlier still due to their great size, hairy body, and perhaps also due to an ability to produce heat internally by an exothermic metabolic reaction in their flight muscles (Newsholme *et al.* 1972, but see Heinrich 1993 for a conflicting opinion). It has been argued that depletion of nectar before native bees begin to forage may result in a significant asymmetry in competition in favour of these introduced species (Matthews 1984; Hopper 1987; Anderson 1989; Dafni and Shmida 1996; Schwarz and Hurst 1997). In a site in Tasmania, much of the available nectar was found to be removed by the combined action of honeybees and bumblebees before 10 am, by which time native bees had not begun to forage (D.G. pers. obs.).

Asymmetries in competition may also occur because of the ability of honeybees and bumblebees to communicate the availability and/or location of valuable food sources with nest mates, so improving foraging efficiency (von Frisch 1967; Dornhaus and Chittka 1999) (the majority of bee species are solitary, and each individual must discover the best places to forage by trial and error). Thus, social species are collectively able to locate new resources more quickly, which again may enable them to gather the bulk of the resources before solitary species arrive (Roubik 1980, 1981; Schwarz and Hurst 1997). Honeybees and bumblebees also appear to be unusual in the distances over which they are capable of foraging. Honeybees are known to forage over 10 km from their nest, on occasion up to 20 km (Seeley 1985a; Schwarz and Hurst 1997), and *B. terrestris* up to at least 4 km (Goulson and Stout 2001). Little is known of the foraging range of most other bee species, but those estimates that are available suggest that they are generally lower. For example, *Melipona fasciata* travels up to 2.4 km and Trigonini over 1 km (Roubik *et al.* 1986). Solitary bee species are generally thought to travel only a few hundred metres at most (Schwarz and Hurst 1997).

Asymmetries in competition may not be stable, since the relative competitive abilities of bee species are likely to vary during the day according to temperature and resource availability, and are likely to vary spatially according to the types of flowers available (Corbet *et al.* 1995). Bumblebees and honeybees are large compared to most of the native species with which they might compete; *B. terrestris* weighs 109–315 mg (Prys-Jones 1982), and *A. mellifera* workers 98 ± 2.8 mg (Corbet *et al.* 1995). They also have longer tongues than many native species, particularly in Australia where most native species are short tongued (Armstrong 1979; Goulson *et al.* in press) (although in comparison with other bumblebees, *B. terrestris* is relatively short tongued). Large bees are at a competitive advantage in cool conditions because of their ability to maintain a body temperature considerably higher than the ambient air temperature (Newsholme *et al.* 1972). They can thus forage earlier and later in the day than most smaller bees, and during cooler weather. Bees with longer tongues can also extract nectar from deeper flowers. However, large bees are not always at an advantage. The energetic cost of foraging is approximately proportional to weight (Heinrich 1979b). Thus large bees burn energy faster. As nectar resources decline, the marginal rate of return will be reached more quickly by large bees. Also long tongues are inefficient at handling shallow flowers (Plowright and Plowright 1997). Thus, large bees are likely to be at a competitive advantage early in the day and during cool weather, and they will be favoured by the presence of deep flowers that provide them with a resource that other bees cannot access. But small bees with short tongues can forage profitably on shallow flowers even when rewards per flower are below the minimum threshold for large bees; at these times honeybees and bumblebees may survive by using honey stores. Thus the relative competitive abilities of different bee species are not consistent, and the strength of competition is likely to vary with time.

Although in general honeybees and bumblebees are able to forage at cooler temperatures than native bees, there may be occasional exceptions. For example the Australian native *Exoneura xanthoclypeata* is adapted for foraging in cool conditions (Tierney 1994). It has been argued that this species is specialized for foraging on (naturally) uncontested resources early in the day, and that this species may be particularly susceptible to competition with exotic bees which forage at the same time (Schwarz and Hurst 1997).

The outcome of interactions between exotic and native flower visitors depends upon whether floral resources are limiting. Because floral resources are usually produced continuously during the life of a flower (although often at a variable rate), they are rarely completely used up, but as they become more scarce, foraging efficiency will decline. Resource availability is likely to vary greatly during the year as different plant species come in to flower (Carpenter 1978). When an abundant or large plant flowers, it may provide a nectar flush. Competition is unlikely to occur during such periods (Tepedino and Stanton 1981).

Overall, it seems probable that depression of resources by introduced bees is likely to have negative effects on native bee species. To determine whether these effects are largely trivial (such as forcing native bees to modify their foraging preferences) or profound (resulting in competitive exclusion), population-level studies are necessary.

15.1.2 *Evidence for population-level changes in native organisms*

The only way to test unequivocally whether floral resources are limiting and competition is in operation is to conduct experiments in which the abundance of the introduced bee species is artificially manipulated, and the population size of native species is then monitored. If populations are significantly higher in the absence of the introduced bee, then competition is occurring. Although in principle a simple procedure, such experiments have proved to be exceedingly hard to accomplish. Honeybees and bumblebees are highly mobile, foraging many kilometers from their nests (Seeley 1985a; Goulson and Stout 2001). Thus, excluding them from an area is difficult. Within and between season variation is likely to be large, so such experiments need to be well replicated, with replicates situated many kilometers apart, and conducted over several years. No such study has been carried out.

An alternative approach which is far easier but provides more equivocal data is to correlate patterns of diversity of native bees with abundance of exotic bees, without manipulating their distribution. A comparison of native flower visitors in Tasmania in areas colonized by *B. terrestris* with areas outside of the current range of the exotic bee found no evidence for competition, but concluded that this may by due to an overriding abundance of honeybees at all sites (Goulson et al. in press). Aizen and Feinsinger (1994) found that fragmentation of forests in Argentina resulted in a decline in native flower visitors and an increase in honeybee populations. Similarly, Kato et al. (1999) studied oceanic islands in the north west Pacific, and found that indigenous bees were rare or absent on islands where honeybees are numerous, which they concluded was evidence for competitive exclusion. On Mt Carmel in Israel, Dafni and Shmida (1996) reported declines in abundance of medium- and large-sized native bees (and also of honeybees) following arrival of *B. terrestris* in 1978. However, such studies can be criticized on the grounds that the relationship between exotic bee abundance and declining native bee populations (if found) need not be causative (Butz Huryn 1997). Increasing honeybee populations are often associated with increased environmental disturbance by man, which may explain declines in native bees.

Some researchers have attempted to manipulate numbers of honeybees, either enhancing populations in experimental plots by placing hives within them, or conversely by removing hives from experimental plots in areas where hives have traditionally been placed. Areas without hives usually still have some honeybees, since there are likely to be some feral nests, and also because honeybees can forage over great distances. Replicates of the treatment without hives need to be sited many kilometers from replicates with hives to ensure that bees do not travel between the two, so many studies have been carried out without adequate replication (e.g. Sugden and Pyke 1991). Despite these limitations, some interesting results have been obtained. Wenner and Thorp (1994) found that removal of feral nests and hives from part of Santa Cruz Island in California resulted in marked increases in numbers of native bees and other flower-visiting insects. Addition of honeybee hives caused the Australian nectivorous bird *Phylidonyris novaehollandiae* to expand its home foraging range and to avoid parts of inflorescences favoured by honeybees (Paton 1993), but a comparison of areas with and without hives found no difference in the density of this bird species (Paton 1995). Roubik (1978) found a decrease in abundance of native insects when he placed hives of the Africanized honeybee in forests in French Guiana. This approach has never been attempted with bumblebees.

Finding that increasing the abundance of alien bees decreases abundance of native organisms is in itself not good evidence for competition. Measures of native organism abundance are generally made by recording them on flowers. In the presence of alien bees, they may simply be foraging elsewhere. Few studies have attempted to directly measure reproductive success of native flower visitors while manipulating abundance of introduced bees. This is unfortunate, since effects of competition on reproduction are likely to result in reduced population sizes. The studies that have been carried out have found no clear effects. Roubik (1982a, 1983) found no consistent detrimental effects on brood size, honey stores, or pollen stores in nests of two Meliponine bee species in Panama when Africanized honeybee hives were placed nearby for 30 days. Monitoring of numbers of native bee species using light traps over many years since the arrival of Africanized bee has not revealed any clear declines in abundance (Wolda and Roubik 1986; Roubik 1991). Roubik (1996a) describes the introduction of Africanized honeybees to the neotropics as a vast experiment, but it is an experiment without replicates or controls, so interpreting the results is difficult. Sugden and Pyke (1991) and Schwarz et al. (1991, 1992a,b) failed to find clear evidence for a link between abundance of honeybees and reproductive success of anthophorid bees belonging to the genus *Exoneura* in Australia in experiments in which they greatly enhanced honeybee numbers at experimental sites. However, the native species that they studied are themselves abundant generalists, visiting a broad range of flowers (Schwarz and Hurst 1997; Goulson et al. in press). As such they are the species least likely to be affected by competition.

The majority of bee species are specialized; in a review of data for 960 solitary bee species, Schemske (1983) found that 64% gathered pollen from only one plant family, often only on one genus. For example, some Australian halictine bees have only been recorded on flowers of *Wahlenbergia* sp. (Michener 1965). Very little is known about such species, and no studies have been carried out to determine whether they are adversely affected by exotic bees (Schwarz and Hurst 1997). Also, the Australian studies of Sugden and Pyke (1991) and Schwarz et al. (1991, 1992a,b) were carried out in flower-rich heathlands; floral resources are more likely to be limiting in arid regions of Australia (Schwarz and Hurst 1997), and these areas often contain the highest native bee diversity (Michener 1979; O'Toole and Raw 1991). The *Exoneura* species studied in Australia had coexisted with honeybees for 180 years, so it is not surprising that they are not greatly affected by competition with this species. If there are species that are excluded by competition with exotic bees, there is no point looking for them in places where these bees are abundant.

Overall, there is no indisputable evidence that introduced bees have had a substantial impact via competition with native species. Given the difficulties involved in carrying out rigorous manipulative experiments (and the rather small number of attempts to do so) this should certainly not be interpreted as the absence of competition. The abundance of exotic bees, the high levels of niche overlap and evidence of resource depression and displacement of native pollinators, all point to the likelihood that competition is occurring. But we do not know whether such competition results (or resulted) in competitive exclusion. The best way to test for such competition is to carry out replicated experiments in which exotic bee numbers are manipulated and native pollinator numbers and reproductive success monitored over long periods. Ideally, such studies should target native species that are not generalists, and areas where floral resources are not abundant.

15.2 Competition for Nest Sites

B. terrestris and *B. subterraneous* generally nest in existing cavities below ground. *B. terrestris* often uses abandoned rodent holes, and spaces beneath man-made structures such as garden sheds (Alford 1975; Donovan and Weir 1978), while rather less is known of the nesting preferences of *B. subterraneous*. To my knowledge, there have been no studies to determine which native organisms also use these cavities. Numerous organisms including diverse arthropods and small mammals might be expected to come in to contact with subterranean bee nests, but little is known of the outcome of such interactions.

B. hortorum and *B. ruderatus* both usually nest just above the ground surface under dense vegetation, and it seems unlikely that they would have a serious impact on native organisms through competition for these sites. Overall, Donovan (1980) considered it unlikely that bumblebees compete with native bee species for nest sites in New Zealand.

15.3 Transmission of Parasites or Pathogens to Native Organisms

A great deal is known about the pathogens and parasites of honeybees, and to a lesser extent bumblebees, since these species are of economic importance. Bees and their nests support a diverse microflora and fauna including both pathogenic, commensal and mutualistic organisms (Goerzen 1991; Gilliam and Taber 1991; Gilliam 1997). Many pathogens are likely to have been transported to new regions with their hosts, particularly where introductions were made many years ago when awareness of bee natural enemies was low. Thus for example, the honeybee disease chalkbrood, caused by the fungus *Ascosphaera apis*; foulbrood, caused by the bacteria *Paenibacillus larvae*; the microsporidian *Nosema apis*; and the mite *Varroa destructor* now occur throughout much of the world. Similarly, bumblebees in New Zealand are host to a parasitic nematode and three mite species, all of which are thought to have come from the UK with the original introduction of bees (Donovan 1980).

During more recent deliberate introductions of exotic bees, such as that of *N. melanderi* to New Zealand, some care has been taken to eliminate pathogens or parasites before bees were released (Donovan 1979). However, this is not always the case, and some parasites and pathogens can be hard to detect. Recent studies in Japan have demonstrated that *B. terrestris* imported from the Netherlands are frequently infested with the tracheal mite *Locustacarus buchneri* (Goka et al. 2001). Although this mite also occurs in Japan, the European race is genetically distinct. In addition to importing *B. terrestris*, queens of a Japanese bumblebee, *B. ignitus*, are currently sent to the Netherlands and the established nests re-imported back to Japan. The re-imported nests have been found to be infected with the European race of the mite. In laboratory studies, these mite are able to infest various

Japanese bumblebee species. Exposure of hosts to novel strains of mite can have dramatic consequences, as demonstrated by the recent spread of *V. destructor*.

It is hard to exaggerate our ignorance of the natural enemies of most bee species, particularly their pathogens. We do not know what species infect them, or what the host ranges of these pathogens are. Thus very little is known of the susceptibility of native organisms to the parasites and pathogens that have been introduced with exotic bees. In a survey of natural enemies of native and introduced bees in New Zealand, Donovan (1980) concluded that no enemies of introduced bees were attacking native bees, but that the converse was true. A chalcidoid parasite of native bees was found to attack *M. rotundata* and, rarely, *B. terrestris*. One fungus, *Bettsia alvei*, which is a pathogen of honeybee hives elsewhere in the world was recorded infecting a native bee in New Zealand, but it is not known whether the fungus is also native to New Zealand. Indeed the natural geographic range of bee pathogens is almost wholly unknown. Some bee pathogens have a broad host range; for example chalkbrood (*A. apis*), is also known to infect *A. cerana* (Gilliam *et al.* 1993) and the distantly related *Xylocopa californica* (Gilliam *et al.* 1994). The related chalkbrood fungus *Ascosphaera aggregata* is commonly found infecting *M. rotundata*; in Canada, where *M. rotundata* is an exotic species, this fungus also infects the native bees *Megachile pugnata* (Goerzen *et al.* 1992) and *Megachile relativa* (Goerzen *et al.* 1990).

It seems likely that these few recorded instances of exotic bee pathogens infecting native species are just the tip of the iceberg, since so few studies have been carried out. As to whether these pathogens have had, or are having, a significant impact on native species, we do not know; if the introduction of a new pathogen were to lead to an epizootic in native insects, it would almost certainly go unnoticed. In other better known organisms, exotic pathogens have had disastrous impacts; for example the introduction of several crayfish species from North America has led to elimination of the native species *Astacus astacus* and *Austropotamobius pallipes* from large portions of Europe. The native species have little resistance to the exotic fungal pathogen *Aphanomyces astaci* which is carried by the introduced crayfish (Butler and Stein 1985; Lilley *et al.* 1997; Vennerstrom *et al.* 1998). Studies of the incidence and identity of pathogen and parasite infestations of wild populations of native bees are urgently needed. In the meantime, legislation to enforce strict quarantine of bees prior to transportation would seem to be necessary.

15.4 Effects on Pollination of Native Flora

Recently, concerns have been expressed that exotic bees may reduce pollination of native plants, or alter the population structure of these plants by mediating different patterns of pollen transfer to those brought about by native pollinators (Butz Huryn 1997; Gross and Mackay 1998). Efficient pollination requires a match between the morphology of the flower and that of the pollinator (reviews in

Ramsey 1988; Burd 1994). If there is a mis-match, then floral rewards may be gathered without efficient transfer of pollen, a process known as floral parasitism (McDade and Kinsman 1980). Specialized obligate relationship between plants and pollinators do exist (reviewed in Goulson 1999) but are the exception (Waser et al. 1996). Most flowers are visited by a range of pollinator species, each of which will provide a different quality of pollinator service.

The efficiency of honeybees as pollinators of native plants in Australia and North America was reviewed by Butz Huryn (1997). She concluded that honeybees provide an effective pollination service to the majority of the flower species that they visit, although they do act as floral parasites when visiting a small number of plant species such as *Grevillea X gaudichaudii* in Australia (Taylor and Whelan 1988) and *Impatiens capensis* and *Vaccinium ashei* in North America (Cane and Payne 1988, 1990; Wilson and Thomson 1991). Similar results have been found for honeybees visiting Jamaican flora (Percival 1974). That honeybees are effective pollinators of many plants, even ones with which they did not coevolve is not surprising. After all, they have been used for centuries to pollinate a broad range of crops. Thus, pollination of the native Australian *Banksia ornata* was increased by the presence of honeybee hives (Paton 1995), and honeybees have proved to be as effective as native bees in pollinating wild cashews, *Anacardium occidentale* in South America (Freitas and Paxton 1998). However, their presence may result in reduced seed set of some native plants. Roubik (1996b) reported declining seed set in the neotropical plant *Mimosa pudica* when honeybees were the dominant visitors, compared to sites where native bees were the more abundant, while Aizen and Feinsinger (1994) found reduced pollination of a range of Argentinian plant species in areas where forests were fragmented and honeybees more abundant. Gross and Mackay (1998) demonstrated that honeybees were poor pollinators of the Australian native *Melastoma affine*, so that when honeybees were the last visitors to a flower, seed set was reduced. As Roubik (1996b) points out, if native pollinators are lost (be it through competition with exotic bees, habitat loss, or use of pesticides) then we cannot expect honeybees to provide an adequate replacement pollination service for all wild plants and crops.

No studies have yet been reported of the effects of exotic bumblebees on the seed set of native plants. *B. terrestris* has the potential to disrupt pollinator services in a different way. This bee species is known to rob plants. When the structure of the flower renders the nectaries inaccessible, *B. terrestris* (and some other bee species) may use their powerful mandibles to bite through the base of the corolla (Inouye 1980b, 1983). In this way they inevitably act as floral parasites, removing nectar without effecting pollination. In Tasmania they rob some bird pollinated plants in this way (Goulson et al. in press). The effects of this behaviour are hard to predict. Clearly it could result in reduced seed set if the lowered floral resources render the flowers less attractive to pollinators (Darwin 1876). In some instances, robbers have been found to reduce the amount of reward available which may result in decreased visitation rates by pollinators (McDade and Kinsman 1980) and a reduction in seed set (Roubik 1982b; Roubik et al. 1985; Irwin and Brody 1999). However, robbing does not always result in adverse effects on seed set (see Chapter 13).

A second possible detrimental effect of exotic bees is that rather than reduce seed set of native flowers, they may alter the population structure by effecting a different pattern of pollen transport to native pollinators. There is some evidence to support this. In South Australia, Paton (1990, 1993) found that honeybees extracted more nectar and pollen from a range of flower species than did birds, the primary native pollinators. However, honeybees moved between plants far less than did birds, and so were less effective in cross-pollinating. This seems to be a general pattern, for several other studies have reported that inter-plant movement by both bumblebees and honeybees is lower than that of other visitors (McGregor et al. 1959; Heinrich and Raven 1972; Silander and Primack 1978). Of course other pollinators often also move small distances; in terms of maximizing foraging efficiency it makes obvious sense to do so (Waddington 1983b), and it has been argued that honeybees are not unusual in this respect (Butz Huryn 1997). However, this is not true. Workers of social bees are unusual in that they are not constrained in their foraging behaviour by the need to find mates, locate oviposition sites or guard a territory; they are single-minded in their task. In contrast, for example, butterflies intersperse visits to flowers with long patrolling flights in which they search for mates (males) or oviposition sites (females) (Goulson et al. 1997a,b). Thus honeybees, bumblebees, and other social bees do tend to engage in fewer long flights than other species (Schmitt 1980; Waser 1982a). The most obvious possible effect of exotic social bees in this respect is increased self-pollination, which could also result in reduced seed set if the plant is self-infertile. Reduced inter-patch pollen movement could result in reproductive fragmentation of plant populations. However, rare long-distance pollen flow is exceedingly hard to quantify, and there is at present no data available against which to test whether exotic bees have had a significant impact on the genetic structure of native plant populations.

Clearly, it is not possible to generalize as to the effects that exotic bees will have on seed set of native flowers. For some species they will provide effective pollination, for others they will not. Where native pollinators have declined for other reasons, for example as a result of habitat loss and fragmentation, exotic bees may provide a valuable replacement pollinator service of native flowers. Where exotic bees are floral parasites, the effect will depend on whether rates of parasitism are sufficient to deter native pollinators. Any change in seed set (including increases) of plant species within a community could lead to long-term ecological change, but such effects would be exceedingly hard to detect amongst the much larger environmental changes that are currently taking place.

15.5 Pollination of Exotic Weeds

As we have seen, both honeybees and bumblebees visit a broad range of flowers. They also appear to prefer to visit exotic flowers (Telleria 1993; Thorp et al. 1994). For example, in Ontario, 75% of pollen collected by honeybees was from introduced plants (Stimec et al. 1997). Across a range of sites in Tasmania, overall 72.6% of flower visits by honeybees and 83.5% of visits by *B. terrestris* were to introduced weeds (Goulson *et al.* in press). Indeed it has been argued that the distribution of *B. terrestris* in Tasmania is largely limited to areas where European weeds are abundant, since some of these plants provide a protected resource in the form of nectar presented in deep corollas, which the shorter-tongued native bees cannot access (Goulson *et al.* in press). In New Zealand, *B. terrestris* has been recorded visiting 400 exotic plants but only 19 native species (MacFarlane 1976). The three other introduced *Bombus* species also visit mainly introduced plants (Donovan 1980). In the highlands of New Zealand, honeybees rely almost exclusively on introduced plants for pollen during most of the season (Pearson and Braiden 1990). These preferences presumably occur because the bees tend to gain more rewards by visiting flowers with which they are co-adapted.

So do visits by exotic bees improve seed set of weeds? By virtue of their abundance and foraging preferences, they often make up a very large proportion of insect visits to weeds. For example, in a site dominated by European weeds in Tasmania, honeybees and bumblebees were the major flower visitors and comprised 98% of all insect visits to the problematic weed creeping thistle, *Cirsium arvense* (unpublished data). In North America, honeybees increase seed set of the yellow star thistle, *Centaurea solstitialis* (Barthell *et al.* 1994) and are the main pollinators of the major weed purple loosestrife, *Lythrum salicaria* (Mal *et al.* 1992). Donovan (1980) reports that bumblebees are major pollinators of introduced weeds in New Zealand. It thus seems obvious and inevitable that exotic bees will prove to be important pollinators of various weeds (Sugden *et al.* 1996).

Remarkably, this view has been challenged. It is hard to agree with the conclusions of Butz Huryn and Moller (1995) that 'Although honey bees may be important pollinators of some weeds, they probably do not contribute substantially to weed problems'. Butz Huryn (1997) argues that most weeds do not rely on insect pollination, either because they are anemophilous, self-pollinating, apomictic or primarily reproduce vegetatively. This is undoubtedly true of some weed species. For example, of the 33 worst environmental weeds in New Zealand (Williams and Timmins 1990), nine fall into one of these categories (Butz Huryn and Moller 1995). However, 16 require pollination and are visited by honeybees, and one is pollinated more or less exclusively by them (the barberry shrub, *Berberis darwinii*). Eight more are listed as having unknown pollination mechanisms (Butz Huryn and Moller 1995). This group includes the tree lupin, *Lupinus arboreus*, and broom, *Cytisus scoparius*, which are self-incompatible and rely on pollination by bumblebees (Stout 2000; Stout *et al.* 2002). It also includes gorse, *Ulex europeaus*, which is thought to depend on honeybee pollination, and in which seed set is greatly reduced by a lack of pollinators in the Chatham Islands where honeybees and bumblebees are absent (McFarlane *et al.* 1992). Thus, at least four major weeds in New Zealand are pollinated primarily by exotic bees.

L. arboreus is currently a minor weed in Tasmania. However, seed set in areas recently colonized by *B. terrestris* has increased, and it is likely that *L. arboreus* may become as problematic in Tasmania as it is in New Zealand now that it has an effective pollinator (Stout *et al.* 2002). Its zygomorphic flowers have to be forced apart to expose the stamens and stigma; only a large, powerful bee is able to do this, and no such bees are native to Tasmania. *L. arboreus* is only one of many weeds in Tasmania, New Zealand, and southern Australia that originated in the temperate northern hemisphere and are coadapted for pollination by bumblebees.

Demonstrating that exotic bees increase seed set of weeds is not sufficient in itself to conclusively show that the action of the bees will increase the weed population (Butz Huryn 1997). No long-term studies of weed population dynamics in relation to the presence or absence of exotic bees have been carried out. Since most weed species are short-lived and dependent on high reproductive rates, it seems probable that seed production is a crucial factor in determining their abundance. Key factor analysis of the life history could reveal whether seed set is directly related to population size.

At present, Australia alone has 2700 exotic weed species, and the costs of control and loss of yields due to these weeds costs an estimated AU$3 billion per year (Commonwealth of Australia 1997). The environmental costs are less easy to quantify but are certainly large. The majority of these 2700 exotic weeds are at present scarce and of trivial ecological and economic importance. The recent arrival of the bumblebee may awake some of these 'sleeper' weeds, particularly if they are adapted for bumblebee pollination. Positive feedback between abundance of weeds and abundance of bumblebees is probable, since an increase in weed populations will encourage more bumblebees, and vice versa. If even one new major weed occurs in Australia due to the presence of bumblebees, the economic and environmental costs could be substantial.

15.6 Conclusions

Both *A. mellifera* and *B. terrestris* are now abundant over large areas where they naturally did not occur. They are both polylectic, and thus use resources utilized by a broad range of native species. It seems almost certain that abundant and widespread exotic organisms which single-handedly utilize a large proportion of the available floral resources do impact on local flower-visiting fauna. Consider, for example, the Tasmania native bee community. One hundred and eighty years ago this presumably consisted of a large number of small, solitary, and sub-social species. Over 100 species have recently been recorded, and many more probably exist. Nowadays, by far the most abundant flower-visiting insects at almost every site is the honeybee, often outnumbering all other flower visiting insects by a factor of 10 or more (D.G. pers. obs.). In the southeast, the second most abundant flower visitor is usually the bumblebee *B. terrestris*. The majority of floral resources are gathered by these bees, often during the morning before native bees have become active (D.G. pers. obs.). It is hard to conceive how the introduction of these exotic species and their associated pathogens could not have substantially altered the diversity and abundance of native bees. Unfortunately, we will never know what the abundance and diversity of the Tasmanian bee fauna was like before the introduction of the honeybee.

Of course, the same applies to most other regions such as North America where the honeybee has now been established for nearly 400 years. It is quite possible that some, perhaps many, native bee species were driven to extinction by the introduction of this numerically dominant species or by exotic pathogens that arrived with it. Even if it were practical, or considered desirable to eradicate honeybees from certain areas, it would be too late for such species.

Similarly, the introduction of exotic bees must increase seed set and hence weediness of some exotic plants, particularly when, as in the case of the bumblebee in Australia, many of the weeds were introduced from the same geographic region and are co-adapted with the introduced bee.

It must be remembered that introduced bees provide substantial benefits to man in terms of pollination of crops, and in the case of the honeybee in providing honey. Ideally, these quantifiable benefits should be weighed against the likely costs. In areas where weeds pollinated by exotic bees are a serious threat, and/or where native communities of flora and fauna are particularly valued, it may be that the benefits provided by these species are outweighed by the costs. Clearly, further research, particularly rigorous manipulative experiments, are needed to determine how much the introduced bees contribute to weed problems and whether they do substantially impact upon native pollinator communities. Also, further investigation of the potential of native bees to provide adequate crop pollination is needed. A ban on the import of *B. terrestris* to North America led to the swift development of *B. impatiens* as an alternative pollinator for tomatoes. In most parts of the world there are probably native bee species that could be exploited. For example, there are native Australian bee species such as carpenter bees (*Xylocopa* sp.) that are able to pollinate tomatoes, but adequate means of rearing these bees for glasshouse use have not yet been developed (Hogendoorn *et al.* 2000).

The precautionary principle argues that in the meantime we should prevent further deliberate release of exotic bee species (such as of bumblebees in mainland Australia and South Africa). Unlike many of the other impacts that man has on the environment, introduction of exotic species is usually irreversible. Once feral bee populations are established, removal is probably impossible (Oldroyd 1998). Just because it is hard to measure potential competitive effects, and to quantify long-term impacts on plant communities, does not mean that these processes are not occurring. Given the numerous potential interactions between alien bees and their pathogens, on the one hand, and native flower visitors, native plants, and non-native weeds, on the other, it seems almost certain that introducing new bee species has serious impacts on natural ecosystems that we have not yet begun to appreciate.

References

Able, K.P. (1994) Magnetic orientation and magnetoreception in birds. *Prog. Neurobiol.* **42**, 449–473.

Abrahamovich, A.H., Telleria, M.C., and Díaz, N.B. (2001) *Bombus* species and their associated flora in Argentina. *Bee World* **82**, 76–87.

Ackerman, D.J., Mesler, M.R., Lu, K.L., and Montalvo, A.M. (1982) Food-foraging behavior of male euglossinin (Hymenoptera: Apidae): vagabonds or trapliners? *Biotropica* **14**, 241–248.

Aebischer, N.J., Blake, K.A., and Boatman, N.D. (1994) Field margins as habitats for game. In: *Field margins: Integrating agriculture and conservation. Brighton Crop Protection Conference Monograph No. 58*, edited by N.D. Boatman, pp. 95–104. British Crop Protection Council, Farnham.

Aizen, M.A. and Feinsinger, P. (1994) Forest fragmentation, pollination, and plant reproduction in a Chaco dry forest. *Argent. Ecol.* **75**, 330–351.

Alcock, J. and Alcock, J.P. (1983) Male behaviour in two bumblebees, *Bombus nevadensis auricomus* and *B. griseicollis* (Hymenoptera: Apidae). *J. Zool.* **200**, 561–570.

Alexander, H.M. (1990) Dynamics of plant–pathogen interactions in natural plant communities. In: *Pests, pathogens and plant communities*, edited by J.J. Burdon, and S.R. Leather, pp. 31–45. Blackwell Scientific Publications, Oxford.

Alexander, N. (ed.) (1996) *Australia: State of the Environment 1996*. CSIRO publishing, Melbourne.

Alfken, J.D. (1913) Die Bienenfauna von Bremen. *Abh. naturw. Ver. Bremen* **22**, 1–220.

Alford, D.V. (1968) The biology and immature stages of *Syntretus splendidus* (Marshall) (Hymenoptera: Braconidae, Euphorinae), a parasite of adult bumblebees. *Trans. R. Entomol. Soc. Lond.* **120**, 375–393.

Alford, D.V. (1969) *Sphaerularia bombi* as a parasite of bumble bees in England. *J. Apic. Res.* **8**, 49–54.

Alford, D.V. (1973) *Syntretus splendidus* attacking *Bombus pascuorum* in Hampshire. *Bee World* **54**, 176.

Alford, D.V. (1975) *Bumblebees*. Davis-Poynter, London.

Alford, D.V. (ed.) (1980) Atlas of the Bumblebees of the British Isles. Institute of Terrestrial Ecology Publications, No. 30.

Allander, K. and Schmid-Hempel, P. (2000) Immune defence reaction in bumble-bee workers after a previous challenge and parasitic coinfection. *Funct. Ecol.* **14**, 711–717.

Allen, J.A. (1989) Searching for search image. *Trend Ecol. Evol.* **4**, 361.

Allen, T., Cameron, S., McGinley, R., and Heinrich, B. (1978) The role of workers and new queens in the ergonomics of a bumblebee colony (Hymenoptera: Apoidea). *J. Kans. Entomol. Soc.* **51**, 329–342.

Anasiewicz, A. (1971) Observations on the bumble-bees in Lublin. *Ekol. Pol.* **19**, 401–417.

Anasiewicz, A. and Warakomska, Z. (1977) Pollen food of the bumblebees (*Bombus* Latr., Hymenoptera) and their associations with the plant species in the Lublin region. *Ekol. Pol.* **25**, 309–322.

Anderson, J.M.E. (1989) Honeybees in Natural Ecosystems. In: *Mediterranean landscapes in Australia: Mallee ecosystems and their management*, edited by J.C. Noble and R.A. Bradstock, pp. 300–304. CSIRO, East Melbourne.

Andersson, S. (1988) Size-dependent pollination efficiency in *Anchusa officinalis* (Boraginaceae): causes and consequences. *Oecologia* **76**, 125–130.

Andrewartha, H.G. and Birch, L.C. (1954) *The distribution and abundance of animals*. University of Chicago Press, Chicago.

Appelgren, M., Bergström, G., Svensson, B.G., and Cederberg, B. (1991) Marking pheromones of *Megabombus* bumble bee males. *Acta Chem. Scand.* **45**, 972–974.

Arizmendi, M.C., Dominguez, C.A., and Dirzo, R. (1995) The role of an avian nectar robber and of hummingbird pollinators in the reproduction of two plant species. *Funct. Ecol.* **10**, 119–127.

Armstrong, J.A. (1979) Biotic pollination mechanisms in the Australian flora—a review. *N. Z. J. Bot.* **17**, 467–508.

Arretz, P.V. and Macfarlane, R.P. (1986) The introduction of *Bombus ruderatus* to Chile for red clover pollination. *Bee World* **67**, 15–22.

Asada, S. and Ono, M. (2000) Difference in colony development of two Japanese bumblebees, *Bombus hypocrita* and *B. ignitus* (Hymenoptera: Apidae). *Appl. Entomol. Zool.* **35**, 597–603.

Augspurger, C.K. (1980) Mass-flowering of a tropical shrub (*Hybanthus prunifolius*): influence on pollinator attraction and movement. *Evolution* **34**, 475–488.

Awram, W.J. (1970) Flight Route Behaviour of Bumblebees. PhD thesis, University of London, London.

Bäckman, J.C. and Tiainen, J. (2002) Habitat quality of field margins in a Finnish farmland area for bumblebees (Hymenoptera: *Bombus* and *Psithyrus*). *Agric. Ecosys. Environ.* **89**, 53–68.

Baer, B., Maile, R., Schmid-Hempel, P., Margan, E.D., and Jones, G.R. (2000) Chemistry of a mating plug in bumblebees. *J. Chem. Ecol.* **26**, 1869–1875.

Baer, B., Morgan, E.D., and Schmid-Hempel, P. (2001) A nonspecific fatty acid within the bumblebee mating plug prevents females from remating. *Proc. Nat. Acad. Sci. USA* **98**, 3926–3928.

Baer, B. and Schmid-Hempel, P. (2000) The artificial insemination of bumblebee queens. *Insectes Soc.* **47**, 183–187.

Baer, B. and Schmid-Hempel, P. (2001) Unexpected consequences of polyandry for parasitism and fitness in the bumblebee, *Bombus terrestris*. *Evolution* **55**, 1639–1643.

Banaszak, J. (1983) Ecology of bees (Apoidea) of agricultural landscapes. *Pol. Ecol. Stud.* **9**, 421–505.

Banaszak, J. (1992) Strategy for conservation of wild bees in an agricultural landscape. *Agric., Ecosys. Environ.* **40**, 179–192.

Banaszak, J. (1996) Ecological bases of conservation of wild bees. In: *The conservation of bees*, edited by A. Matheson, S.L. Buchmann, C. O'Toole, P. Westrich, and I.H. Williams, pp. 55–62. Academic Press, London.

Banda, H.J. and Paxton, R.J. (1991) Pollination of greenhouse tomatoes by bees. *Acta Hortic.* **288**, 194–198.

Barron, M.C., Wratten, S.D., and Donovan, B.J. (2000) A four-year investigation into the efficiency of domiciles for enhancement of bumble bee populations.

Barrow, D.A. (1983) Ecological studies on bumblebees in South Wales with special reference to resource partitioning and bee size variation. PhD thesis, University of Wales, UK. 161 pp.

Barrow, D.A. and Pickard, R.S. (1984) Size-related selection of food plants by bumblebees. *Ecol. Entomol.* **9**, 369–373.

Barth, F.G. (1985) *Insects and Flowers, the biology of a partnership*. George Allen and Unwin, London.

Barthell, J.F., Randall, J.M., Thorp, R.W., and Wenner, A.M. (1994) Invader assisted invasion: pollination of yellow star-thistle by feral honey bees in island and mainland ecosystems [abstract]. *Suppl. Bull. Ecol. Soc. Am.* **75**, 10.

Bateson, P.P.G. (1983) Optimal outbreeding. In: *Mate choice*, edited by P. Bateson, pp. 257–277. Cambridge University Press, Cambridge.

Batra, S.W.T. (1979) *Osmia cornifrons* and *Pithitis smaragulda*, two asian bees introduced into the United States for crop pollination. In: Proc. IV Internat. Symposium of Pollination. MD Agric. Exp. Stn. Sp. Misc. Pub. I., edited by M.C. Dewey. University of Maryland, College Park.

Batra, S.W.T. (1995) Bees and pollination in our changing environment. *Apidologie* **26**, 361–370.

Beattie, A.J. (1976) Plant dispersion, pollination and gene flow in *Viola*. *Oecologia* **25**, 291–300.

Beekman, M. and Van Stratum, P. (1998) Bumblebee sex ratios: why do bumblebees produce so many males? *Proc. Royal. Soc. Lond. B* **265**, 1535–1543.

Beekman, M., Van Stratum, P., and Lingeman, R. (1998) Diapause survival and post-diapause performance in bumblebee queens (*Bombus terrestris*). *Entomol. Exp. Appl.* **89**, 207–214.

Bell, G. (1985) On the function of flowers. *Proc. Royal. Soc. Lond. B-Biol. Sci.* **224**, 223–265.

Bell, G., Lefebvre, L., Geraldeae, L.-A., and Weary, D. (1984) Partial preference of insects for the male flowers of an annual herb. *Oecologia* **64**, 287–294.

Bell, W.G. (1991) *Searching behaviour: the behavioral ecology of finding resources*. Chapman and Hall, London.

Bennett, A.W. (1884) On the constancy of insects in their visits to flowers. *J. Linn. Soc.* **17**, 175–185.

Bergman, P. and Bergström, G. (1997) Scent marking, scent origin, and species specificity in male premating behaviour of two Scandinavian bumblebees. *J. Chem. Ecol.* **23**, 1235–1251.

Bergman, P., Bergström, G., and Appelgren, M. (1996) Labial gland marking secretion in males of two Scandinavian cuckoo bumblebee species. *Chemoecology* **7**, 140–145.

Bergström, G., Bergman, P., Appelgren, M., and Smith, J.O. (1996) Labial gland chemistry of three species of bumblebees (Hymenoptera, Apidae) from North America. *Bioorg. Med. Chem.* **4**, 515–519.

Bergström, G., Kullenberg, B., Ställberg-Stenhagen, S., and Stenhagen, E. (1968) Studies on natural odiferous compounds. II. Identification of a 2,3-dihrdro-farnesol as the main component of the marking perfume of male bumblebees of the species *Bombus* terrestris. *Arkiv För Kemi* **28**, 453–469.

Bergström, G., Svensson, B.G., Appelgren, M., and Groth, I. (1981) Complexity of bumblebee marking pheromones: biochemical, ecological and systematical interpretations. In: *Biosystematics of social insects*, edited by P.E. Howse and J.L. Clement, pp. 175–183. Academic Press, London.

Bertin, R.I. (1982) Floral biology, hummingbird pollination and fruit production of Trumpet Creeper (*Campsis radicans*, Bignoniaceae). *Am. J. Bot.* **69**, 122–134.

Bertsch, A.H. (1983) Nectar production of *Epilobium augustifolium* L. at different air humidities; nectar sugar in individual flowers and the optimal foraging theory. *Oecologia* **59**, 40–48.

Bertsch, A.H. (1987) Flowers as food sources and the cost of outcrossing. *Ecol. Stud.* **61**, 277–293.

Best, L.S. and Bierzychudek, P. (1982) Pollinator foraging on foxglove (*Digitalis purpurea*): a test of a new model. *Evolution* **36**, 70–79.

Bierzychudek, P. (1981) Pollinator limitation of plant reproductive effort. *Am. Nat.* **117**, 838–840.

Bloch, G. (1999) Regulation of queen-worker conflict in bumble-bee (*Bombus terrestris*) colonies. *Proc. Royal. Soc. Lond. B* **266**, 2465–2469.

Bloch, G. and Hefetz, A. (1999) Regulation of reproduction by dominant workers in bumble-bee (*Bombus terrestris*) queenright colonies. *Behav. Ecol. Sociobiol.* **45**, 125–135.

Bloch, G., Borst, D.W., Huang, Z.-Y., Robinson, G.E., and Hefetz, A. (1996) Effects of social conditions on juvenile hormone mediated reproductive development in *Bombus terrestris* workers. *Physiol. Entomol.* **21**, 257–267.

Blough, D.S. (1979) Effects of the number and form of stimuli on visual search in the pigeon. *J. Exp. Psychol. Anim. Behav.* **5**, 211–223.

Blum, M.S. (1981) *Chemical defenses of arthropods*. Academic Press, New York.

Blum, M.S. (1987) Specificity of pheromonal signals: a search for its recognitive bases in terms of a unified chemisociality. In: *Chemistry and biology of social insects*, edited by J. Eder and H. Rembold, J. Peperny, pp. 401–405. München.

Boatman, N.D. (1992) Herbicides and the management of field boundary vegetation. *Pest. Outlook* **3**, 30–34.

Boatman, N.D. and Wilson, P.J. (1988) Field edge management for game and wildlife conservation. *Aspects Appl. Biol.* **16**, 53–61.

Boetius, J. (1948) Über den Verlauf der Nektarabsonderung einiger Blutenpflanzen. *Beih. Schweiz. Bienen.* **2**, 257–317.

Bogler, D.J., Neff, J.L., and Simpson, B.B. (1995) Multiple origins of the yucca–yucca moth association. *Proc. Nat. Acad. Sci. USA* **92**, 6864–6867.

Bohart, G.E. (1970) *The evolution of parasitism among bees*. Utah State University, p. 30.

Bohart, G.E. (1972) Management of wild bees for the pollination of crops. *Ann. Rev. Entomol.* **17**, 287–312.

Bohart, G.E. and Knowlton, G.F. (1953) Yearly population fluctuations of *Bombus morrisoni* at Fredonia, Arizona. *J. Econ. Entomol.* **45**, 890.

Bond, A.B. (1983) Visual-search and selection of natural stimuli in the pigeon—the attention threshold hypothesis. *J. Exp. Psychol.* **9**, 292–306.

Bond, A.B. and Riley, D.A. (1991) Searching image in the pigeon: a test of three hypothetical mechanisms. *Ethology* **87**, 203–224.

Bond, H.W. and Brown, W.L. (1979) The exploitation of floral nectar in *Eucalyptus incrassata* by honeyeaters and honeybees. *Oecologia* **44**, 105–111.

Boomsma, J.J. and Grafen, A. (1991) Colony-level sex ratio selection in the eusocial Hymenoptera. *J. Evol. Biol.* **3**, 383–407.

Boomsma, J.J. and Ratnieks, F.L.W. (1996) Paternity in eusocial Hymenoptera. *Phil. Trans. Royal. Soc. Lond. B* **351**, 947–975.

Borneck, R. and Merle, B. (1989) Essai d'une evaluation de l'incidence economique de l'abeille pollinisatrice dans l'agriculture européenne. *Apiacta* **24**, 33–38.

Bortolotti, L., Duchateau, M.J., and Sbrenna, G. (2001) Effect of juvenile hormone on caste determination and colony processes in the bumblebee *Bombus terrestris*. *Entomol. Exp. Appl.* **101**, 143–158.

Boucher, D.H., James, S., and Keeler, K. (1982) The ecology of mutualism. *Annu. Rev. Ecol. Syst.* **13**, 315–347.

Bourke, A.F.G. (1994) Worker matricide in social bees and wasps. *J. Theor. Biol.* **167**, 283–292.

Bourke, A.F.G. (1997) Sex ratios in bumble bees. *Phil. Trans. Royal. Soc. Lond. B* **352**, 1921–1933.

Bourke, A.F.G. and Ratnieks, F.L.W. (1999) Kin conflict over caste determination in social Hymenoptera. *Behav. Ecol. Sociobiol.* **46**, 287–297.

Bourke, A.F.G. and Ratnieks, F.L.W. (2001) Kin-selected conflict in the bumble-bee *Bombus terrestris* (Hymenoptera: Apidae). *Proc. Royal. Soc. Lond. B.* **268**, 347–355.

Bowers, M.A. (1985a) Bumblebee colonization, extinction, and reproduction in subalpine meadows in Northeastern Utah. *Ecology* **66**, 914–927.
Bowers, M.A. (1985b) Experimental analyses of competition between two species of bumble bees (Hymenoptera: apidae). *Oecologia* **67**, 224–230.
Boyle, R.M.D. and Philogène, B.J.R. (1983) The native pollinators of an apple orchard: variations and significance. *J. Hort. Sci.* **581**, 355–363.
Brand, C.J. and Westgate, J.M. (1909) Alfalfa in cultivated rows for seed production in semi-arid regions. US Dept Agric Bur Industry Circular No. 24: 1–23.
Braun, E., MacVicar, R.M., Gibson, D.R., and Paukiw, P. (1956) Pollinator studies on red clover. *Int. Congr. Entomol. Proc.* **10**, 1–221.
Brian, A.D. (1951) The pollen collection by bumble-bees. *J. Anim. Ecol.* **20**, 191–194.
Brian, A.D. (1952) Division of labour and foraging in *Bombus agrorum* Fabricius. *J. Anim. Ecol.* **21**, 223–240.
Brian, A.D. (1954) The foraging of bumble bees Part 1. Foraging behaviour. *Bee World* **35**, 61–67.
Brian, A.D. (1957) Differences in the flowers visited by four species of bumble-bees and their causes. *J. Anim. Ecol.* **21**, 223–240.
Brian, M.V. (1965a) *Social insect populations*. Academic Press, London.
Brian, M.V. (1965b) Caste differentiation in social insects. *Symp. Zool. Soc. Lond.* **14**, 13–38.
Brian, M.V. (1983) *Social insects*. Chapman and Hall, London.
Bringer, B. (1973) Territorial flight of bumble-bee males in the coniferous forests on the northernmost part of the island of Öland. *Zoon Suppl.* **1**, 15–22.
Brink, D. (1982) A bonanza-blank pollinator reward schedule in *Delphinium nelsonii* (Ranunculaceae). *Oecologia* **52**, 292–294.
Brink, D. and De Wet, J.M.J. (1980) Interpopulation variation in nectar production in *Aconitum columbianum* (Ranunculaceae). *Oecologia* **47**, 160–163.
Brittain, W.H. and Newton, D.E. (1933) A study in the relative constancy of hive bees and wild bees in pollen gathering. *Can. J. Res.* **9**, 334–349.
Brody, A.K. and Mitchell, R.J. (1997) Effects of experimental manipulation of inflorescence size on pollination and pre-dispersal seed predation in the hummingbird-pollinated plant *Ipomopsis aggregata*. *Oecologia* **110**, 86–93.
Bromley, S.W. (1934) The robber flies of Texas (Diptera, Asilidae). *Ann. Entomol. Soc. Am.* **27**, 74–113.
Bromley, S.W. (1936) Asilids feeding on bumblebees in New England. *Psyche* **43**, 14.
Bromley, S.W. (1949) The Missouri bee-killer, *Proctacanthus milbertii* Macq. (Asilidae: Diptera). *Bull. Brooklyn Entomol. Soc.* **44**, 21–28.
Bronstein, J.L. (1995) The plant-pollinator landscape. In: *Mosaic landscapes and ecological processes*, edited by L. Hansson, L. Fahrig, and G. Merriam, pp. 256–288. Chapman and Hall, London, UK.
Brower, J.P., Brower, J.V.Z., and Westcott, P.W. (1960) Experimental studies of mimicry. 5. The reactions of toads (*Bufo terrestris*) to bumblebees (*Bombus americanorum*) and their robberfly mimics (*Mallophora bomboides*), with a discussion of aggressive mimicry. *Am. Nat.* **94**, 343–355.
Brown, C.J.D. (1929) A morphological and systematical study of Utah Asilidae (Diptera). *Trans. Am. Entomol. Soc.* **54**, 295–320.
Brown, J.H. (1984) On the relationship between abundance and distribution of species. *Am. Nat.* **124**, 644–645.
Brown, M.J.F., Loosli, R., and Schmid-Hempel, P. (2000) Condition-dependent expression of virulence in a trypanosome infecting bumblebees. *Oikos* **91**, 421–427.
Buchmann, S.L. (1985) Bees use vibration to aid pollen collection from non-poricidal flowers. *J. Kans. Entomol. Soc.* **58**, 517–525.
Buchmann, S.L. and Nabhan, G.P. (1996) *The forgotten pollinators*. Island Press, Washington (DC).

Bulmer, M.G. (1981) Worker-queen conflict in annual social Hymenoptera. *J. Theor. Biol.* **93**, 239–251.

Bulmer, M.G. (1983) The significance of protandry in the social Hymenoptera. *J. Theor. Biol.* **93**, 239–251.

Burd, M. (1994) Bateman's principle and plant reproduction: the role of pollen limitation in fruit and seed set. *Bot. Rev.* **60**, 83–139.

Butler, M.J. and Stein, R.A. (1985) An analysis of the mechanisms governing species replacements in crayfish. *Oecologia* **66**, 168–177.

Buttermore, R.E. (1997) Observations of successful *Bombus terrestris* (L.) (Hymenoptera: Apidae) colonies in southern Tasmania. *Aust. J. Entomol.* **36**, 251–254.

Butz Huryn, V.M. (1997) Ecological impacts of introduced honey bees. *Quart. Rev. Biol.* **72**, 275–297.

Butz Huryn, V.M. and Moller, H. (1995) An assessment of the contribution of honeybees (*Apis mellifera*) to weed reproduction in New Zealand protected natural areas. *N. Z. J. Ecol.* **19**, 111–122.

Bulmer, M.G. (1983) The significance of protandry in social Hymenoptera. *Am. Nat.* **121**, 540–551.

Calam, D.H. (1969) Species and sex-specific compounds from the heads of male bumblebees (*Bombus* spp.). *Nature* **221**, 856–857.

Camargo, C.A. de (1979) Sex determination in bees. XI Production of diploid males and sex determination in *Melipona quadrifasciata*. *J. Apic. Res.* **18**, 77–84.

Cameron, S.A. (1981) Chemical signals in bumble bee foraging. *Behav. Ecol. Sociobiol.* **9**, 257–260.

Cameron, S.A. and Robinson, G.E. (1993) Juvenile hormone does not affect division of labour in bumble bee colonies (Hymenoptera: Apidae). *Ann. Entomol. Soc. Am.* **83**, 626–631.

Cane, J.H. and Payne, J.A. (1988) Foraging ecology of the bee *Habropoda laboriosa* (Hymenoptera: Anthophoridae), an oligolege of blueberries (Ericaceae: *Vaccinium*) in the southeastern United States. *Ann. Entomol. Soc. Am.* **81**, 419–427.

Cane, J.H. and Payne, J.A. (1990) Native bee pollinates rabbiteye blueberry. Alabama Agricultural Experiment Station, *Highlights Agric. Res.* **37**, 4.

Capaldi, E.A. and Dyer, F.C. (1999) The role of orientation flights on homing performance in honeybees. *J. Exp. Biol.* **202**, 1655–1666.

Cardale, J.C. (1993) Hymenoptera: Apoidea. In: *Zoological catalogue of Australia, Vol 10*, edited by W.W.K. Houston and G.V. Maynard, Australian Government Printing Service, Canberra.

Carpenter, F.L. (1976) Plant–pollinator interactions in Hawaii: pollination energetics of *Metrosideros collina* (Myrtaceae). *Ecology* **57**, 1125–1144.

Carpenter, F.L. (1978) A spectrum of nectar-eater communities. *Am. Zool.* **18**, 809–819.

Cartar, R.V. (1989) Condition-dependent foraging behavior of bumble bees. PhD thesis, Simon Fraser University, Burnaby, B.C., Canada.

Cartar, R.V. (1992a) Adjustment of foraging effort and task-switching in energy-manipulated wild bumble bee colonies. *Anim. Behav.* **44**, 75–87.

Cartar, R.V. (1992b) Morphological senescence and longevity: an experiment relating wing wear and life span in foraging wild bumble bees. *J. Anim. Ecol.* **61**, 225–231.

Cartar, R.V. and Dill, L.M. (1990) Colony energy requirements affect the foraging currency of bumble bees. *Behav. Ecol. Sociobiol.* **27**, 377–383.

Carvell, C. (2000) Studies of the distribution and habitat requirements of *Bombus sylvarum* (the shrill carder bee) and other bumblebees at Castlemartin range, Pembrokeshire and Kenfig

National Nature Reserve, Glamorgan and surrounding areas. Unpublished report for Countryside Council for Wales/UK Biodiversity Action Plan Bumblebee Working Group.

Carvell, C. (2002) Habitat use and conservation of bumblebees (*Bombus* spp.) under different grassland management regimes. *Biol. Conserv.* **103**, 33–49.

Carreck, N.L. and Williams, I.H. (1998) The economic value of bees in the UK. *Bee World* **79**, 115–123.

Cederberg, B., Svensson, B.G., Bergström, G., Appelgren, M., and Groth, I. (1984) Male marking pheromones in North European cuckoo bumble bees, *Psithyrus* (Hymenoptera, Apidae). *Nova Acta Reg. Soc. Sci. Upsal., Ser. V:C* **3**, 161–166.

Chapman, R.E. and Bourke, A.F.G.T. (2001) The influence of sociality on the conservation biology of social insects. *Ecol. Lett.* **4**, 650–662.

Charnov, E.L. (1976) Optimal foraging: the marginal value theorem. *Theor. Pop. Biol.* **9**, 129–136.

Cheverton, J., Kacelnik, A., and Krebs, J.R. (1985) Optimal foraging: constraints and currencies. In: *Experimental behavioural ecology and sociobiology*, edited by B. Holldobler, and M. Lindauer, pp. 109–126. Sinauer, Sunderland, MA.

Chen, P.S., Stumm Zollinger, E., Aigaki, T., Balmer, J., Beinz, M., and Bohlen, P. (1988) A male accessory gland peptide that regulates reproductive behaviour of female *Drosophila melanogaster*. *Cell* **54**, 291–298.

Chittka, L. (1998) Sensorimotor learning in bumblebees: Long-term retention and reversal training. *J. Exp. Biol.* **201**, 515–524.

Chittka, L., Gumbert, A., and Kunze, J. (1997) Foraging dynamics of bumblebees: correlates of movements within and between plant species. *Behav. Ecol.* **8**, 239–249.

Chittka, L., Kunze, J., and Geiger, K. (1995) The influence of landmarks on distance estimation of honeybees. *Anim. Behav.* **50**, 23–31.

Chittka, L., Shmida, A, Troje, N., and Menzel, R. (1994) Ultraviolet as a component of flower reflections, and the colour perception of Hymenoptera. *Vision Res.* **34**, 1489–1508.

Chittka, L. and Thomson, J.D. (1997) Sensori-motor learning and its relevance for task specialization in bumble bees. *Behav. Ecol. Sociobiol.* **41**, 385–398.

Chittka, L., Thomson, J.D., and Waser, N.M. (1999a) Flower constancy, insect psychology, and plant evolution. *Naturwissenschaften* **86**, 361–377.

Chittka, L., Williams, N.M., Rasmussen, H., and Thomson, J.D. (1999b) Navigation without vision: bumblebee orientation in complete darkness. *Proc. Royal. Soc. Lond. Ser. B-Biol. Sci.* **266**, 45–50.

Chmielewski, W. (1969) Obserwacje nad biologia nowego dla akarofauny polskiej gatunku *Kunizia laevis* (Dujardin 1849) (Acarina, Acaridae). *Polskie Pismo Entomol.* **39**, 603–617.

Chmurzynski, J.A., Kieruzel, M., Krzysztofiak, A., and Krzysztofiak, L. (1998) Long-distance homing ability in *Dasypoda altercator* (Hymenoptera, Melittidae). *Ethology* **104**, 421–429.

Cibula, D.A. and Zimmerman, M. (1987) Bumblebee foraging behaviour: changes in departure decisions as a function of experimental manipulations. *Am. Midland Nat.* **117**, 386–394.

Clark, M.G. (1976) Effect of 2-deoxy-D-glucose on flight and flight muscle substrate cycling by the bumble-bee. *Comp. Biochem. Physiol.* **55B**, 409–415.

Clark, M.G., Bloxham, D.P., Holland, P.C., and Lardy, H.A. (1973) Estimation of the fructuse diphosphatase-phosphofructokinase substrate cycle in the flight muscle of *Bombus affinis*. *Biochem. J.* **134**, 589–597.

Clark, T.B. (1977) *Spiroplasma* as a new pathogen in the honeybees. *J. Invertebr. Pathol.* **29**, 112–113.

Clark, T.B. (1982) Entomopoxvirus-like particles in three species of bumble-bees. *J. Invertebr. Pathol.* **39**, 119–122.

Clark, T.B., Whitcomb, R.F., Tully, J.G., Mouches, G., Saillard, C., Bové, J.M., Wroblewski, H., Carle, P., Rose, D.L., Henegar, R.B., and Williamson, D.L. (1985) *Spiroplasma melliferum*, a new species from the honeybee (*Apis mellifera*). *Intern. J. Sys. Bacteriol.* **35**, 296–308.

Clausen, C.P. (1940) *Entomophagous insects*. McGraw-Hill, New York.

Clifford, P.T.P. and Anderson, A.C. (1980) Herbage seed production. In: *Proceedings of the New Zealand Grassland Association*, edited by J.A. Lancashire, pp. 76–79. New Zealand Grassland Association, New Zealand.

Cnaani, J., Borst, D.W., Huang, Z.-Y., Robinson, G.E., and Hefetz, A. (1997) Caste determination in *Bombus terrestris*: differences in development and rates of JH biosynthesis between queen and worker larvae. *J. Insect Physiol.* **43**, 373–381.

Cnaani, J. and Hefetz, A. (1994) The effect of workers size frequency-distribution on colony development in *Bombus terrestris*. *Insectes Soc.* **41**, 301–307.

Cnaani, J. and Hefetz, A. (1996) The effect of social environment in the colony on caste determination and JH synthesis in *Bombus terrestris* larvae. In: *Proceedings of the XXth International Congresso Entomology*, p. 390. Florence, Italy.

Cnaani, J., Robinson, G.E., Bloch, G., Borst, D.W., and Hefetz, A. (2000) The effect of queen-worker conflict on caste determination in the bumblebee *Bombus terrestris*. *Behav. Ecol. Sociobiol.* **47**, 346–352.

Coffey, M.F. and Breen, J. (1997) Seasonal variation in pollen and nectar sources of honey bees in Ireland. *J. Agric. Res.* **36**, 63–76.

Cole, B.J. (1983) Multiple mating and the evolution of social behaviour in the Hymenoptera. *Behav. Ecol. Sociobiol.* **12**, 191–291.

Colville, F.V. (1890) Notes on bumble-bees. *Proc. Entomol. Soc. Wash.* **1**, 197–202.

Colwell, R.K. (1973) Competition and coexistence in a simple tropical community. *Am. Nat.* **107**, 737–760.

Comba, L., Corbet, S.A., Barron, A., Bird, A., Collinge, S., Miyazaki, N., and Powell, M. (1999) Garden flowers: Insect visits and the floral reward of horticulturally-modified variants. *Ann. B.* **83**, 73–86.

Commonwealth of Australia (1997) *The National Weeds Strategy*. Commonwealth of Australia, Canberra.

Cook, J.M. and Crozier, R.H. Sex determination and population biology in the Hymenoptera. *TREE* **10**, 281–286.

Cooper, K.W. (1984) Discovery of the first resident population of the European bee, *Megachile apicalis*, in the United States (Hymenoptera: Megachilidae). *Entomol. News* **95**, 225–226.

Corbet, S.A. (1987) More bees make better crops. *New Sci.* **115**, 40–43.

Corbet, S.A. (1995) Insects, plants and succession—advantages of long-term set-aside. *Agric. Ecosys. Environ.* **53**, 201–217.

Corbet, S.A. (2000) Conserving compartments in pollination webs. *Conserv. Biol.* **14**, 1229–1231.

Corbet, S.A., Bee, J., Dasmahapatra, K., Gale, S., Gorringe, E., La Ferla, B., Moorhouse, T., Trevail, A., Van Bergen, Y., and Vorontsova, M. (2001) Native or exotic? Double or single? Evaluating plants for pollinator-friendly gardens. *Ann. B.* **87**, 219–232.

Corbet, S.A., Cuthill, I., Fallows, M., Harrison, T., and Hartley, G. (1981) Why do nectar-foraging bees and wasps work upwards on inflorescences? *Oecologia* **51**, 79–83.

Corbet, S.A., Fussell, M., Ake, R., Fraser, A., Gunson, C., Savage, A., and Smith, K. (1993) Temperature and the pollinating activity of social bees. *Ecol. Entomol.* **18**, 17–30.

Corbet, S.A., Kerslake, C.J.C., Brown, D., and Morland, N.E. (1984) Can bees select nectar-rich flowers in a patch? *J. Apic. Res.* **23**, 234–242.

Corbet, S.A. and Morris, R.J. (1999) Mites on bumble bees and bluebells. *Entomologist's Monthly Magazine* **135**, 77–83.

Corbet, S.A., Saville, N.M., Fussell, M., Prys-Jones, O.E., and Unwin, D.M. (1995) The competition box: a graphical aid to forecasting pollinator performance. *J. Appl. Ecol.* **32**, 707–719.

Corbet, S.A., Williams, I.H., and Osborne, J.L. (1991) Bees and the pollination of crops and wild flowers in the European Community. *Bee World* **72**, 47–59.

Corbet, S.A. and Willmer, P.G. (1980) Pollination of the yellow passionfruit: nectar, pollen and carpenter bees. *J. Agric. Sci.* **95**, 655–666.

Corbet, S.A., Willmer, P.G., Beament, J.W.L., Unwin, D.M., and Prys-Jones, O.E. (1979) Post-secretory determinants of sugar concentration in nectar. *Plant Cell Environ.* **2**, 293–308.

Corbetta, M., Miezin, F.M., Dobmeyer, S., Shulman, G.L., and Peterson, G.L. (1990) Attentional modulation of neural processing of shape, colour, and velocity in humans. *Science* **248**, 1556–1559.

Cowgill, S.E., Wratten, S.D., and Sotherton, N.W. (1993) The selective use of floral resources by the hoverfly *Episyrphus balteatus* (Diptera: Syrphidae) on farmland. *Ann. Appl. Biol.* **122**, 223–231.

Crane, E. (1975) *Honey: a comprehensive survey*. Heinemann in co-operation with International Bee Research Association, London.

Crane, E. (1990a) *Bees and beekeeping: science, practice, and world resources*. Cornstock Publishing Associates, Ithaca, NY: Cornell University Press.

Crane, E. (1990b) *Bees and beekeeping*. Oxford, Heinemann Newnes.

Cresswell, J.E. (1990) How and why do nectar-foraging bumblebees initiate movements between inflorescences of wild bergamot *Monarda fistulosa* (Lamiaceae). *Oecologia* **82**, 450–460.

Cresswell, J.E. (1997) Spatial heterogeneity, pollinator behaviour and pollinator-mediated gene-flow: bumblebee movements in variously aggregated rows of oil-seed rape. *Oikos* **78**, 546–556.

Cresswell, J.E. (1999) The influence of nectar and pollen availability on pollen transfer by individual flowers of oilseed rape (*Brassica napus*) when pollinated by bumblebees (*Bombus lapidarius*). *J. Ecol.* **87**, 670–677.

Cresswell, J.E. and Galen, C. (1991) Frequency-dependent selection and adaptive surfaces for floral character combinations—the pollination of *Polemonium viscosum*. *Am. Nat.* **138**, 1342–1353.

Cresswell, J.E., Osborne, J.L., and Goulson, D. (2000) An economic model of the limits to foraging range in central place foragers with numerical solutions for bumblebees. *Ecol. Entomol.* **25**, 249–255.

Cresswell, J.E. and Robertson, A.W. (1994) Discrimination by pollen-collecting bumblebees among differentially rewarding flowers of an alpine wildflower, *Campanula rotundifolia* (Campanulaceae). *Oikos* **69**, 304–308.

Cribb, D. (1990). Pollination of tomato crops by honeybees. *Bee Craft* **72**, 228–231.

Crosswhite, F.S. and Crosswhite, C.D. (1970) Pollination of *Castilleja sessiflora* in southern Wisconsin. *Bull. Torr. Bot. Club* **97**, 100–105.

Crozier, R.H. (1979) Genetics of sociality. In: *Social Insects, Vol. I*, edited by H.R. Hermann, pp. 223–286. Academic Press, New York.

Crozier, R.H. and Page, R.E. (1985) On being the right size: male contributions and multiple mating in social hymenoptera. *Behav. Ecol. Sociobiol.* **18**, 105–116.

Crozier, R.H. and Pamilo, P. (1996) *Evolution of social insect colonies*. Oxford University Press.

Cruden, R.W., Hermanutz, L., and Shuttleworth, J. (1984) The pollination biology and breeding system of *Monarda fistulosa* (Labiatae). *Oecologia* **64**, 104–110.

Cruzan, M.B., Neal, P.R., and Willson, M.F. (1988) Floral display in *Phyla incisa*: consequences for male and female reproductive success. *Evolution* **42**, 505-515.

Cumber, R.A. (1949*a*) The biology of humblebees, with special reference to the production of the worker caste. *Trans. Royal. Entomol. Soc. Lond.* **100**, 1-45.

Cumber, R.A. (1949*b*) An overwintering nest of the bumble-bee *Bombus terrestris*. *NZ. Sci. Rev.* **7**, 76-77.

Cumber, R.A. (1949*c*) Humble-bee parasites and commensals found within a thirty mile radius of London. *Proc. Royal. Entomol. Soc. Lond. (A)* **24**, 119-127.

Cumber, R.A. (1953) Some aspects of the biology and ecology of bumble-bees bearing upon the yields of red-clover seed in New Zealand. *N. Z. J. Sci. Technol.* **11**, 227-240.

Cushman, J.H. and Beattie, A.J. (1991) Mutualisms: assessing the benefits to hosts and visitors. *Trends Ecol. Evol.* **6**, 191-195.

Dafni, A. (1998) The threat of *Bombus terrestris* spread. *Bee World* **79**, 113-114.

Dafni, A. and Shmida, A. (1996) The possible ecological implications of the invasion of *Bombus terrestris* (L.) (Apidae) at Mt Carmel, Israel. In: *The conservation of bees*, edited by A. Matheson, S.L. Buchmann, C. O'Toole, P. Westrich, and I.H. Williams, pp. 183-200. Academic Press, London.

Dalla Torre, K.W. v (1880) Unsere hummel-(*Bombus*) Arten. *Naturhistoriker* **2**, 40-41.

Dalla Torre, K.W. v (1882) Bermerkungen zur Gattung *Bombus* Latr., II. 3. Zur Synonymie und geographischen Verbreitung der Gattung Bombus Latr. *Bericht Naturwissenschaftlichmedizinischen Vereins in Innsbruck* **12**, 14-31.

Darwin, C. (1876) *On the effects of cross and self fertilization in the vegetable kingdom*. John Murray, London.

Darwin, C. (1886) Über die Wege der Hummelmännchen. In: *Ein Supplement zu seinen grösseren Werken. Vol. 2*. Krause, E. Gesammelte kleinere Schriften von Charles Darwin. Leipzig.

Davidson, A. (1894) On the parasites of wild bees in California. *Entomol. News* **5**, 170-172.

Davies, N.B. (1977) Prey selection and the search strategy of the spotted flycatcher (*Muscicapa striata*): a field study on optimal foraging. *Anim. Behav.* **25**, 1016-1033.

Davis, A.R. and Shuel, R.W. (1988) Distribution of carbofuran and dimethoate in flowers and their secretion in nectar as related to nectary vascular supply. *Can. J. Bot.* **66**, 1248-1255.

Davis, M.A. (1981) The effects of pollinators, predators, and energy constraints on the floral ecology and evolution of *Trillium erectum*. *Oecologia* **48**, 400-406.

Day, H.D. and Day, K.C. (1997) Directional preferences in the rotational play behaviors of young children. *Develop. Psychobiol.* **30**, 213-223.

Delaplane, K.S. and Mayer, D.F. (2000) *Crop pollination by bees*. CABI publishing, Wallingford, UK.

Delbrassinne, S. and Rasmont, P. (1988) Contribution à l'étude de la pollinisation du colza, *Brassica napus* L. var. *oleifera* (Moench) Delile, en Belgique. *Bull. Rech. Agronom. Gembloux* **23**, 123-152.

Delph, L.F. and Lively, C.M. (1989) The evolution of floral colour change: pollinator attraction versus physiological constraints in *Fuchsia excorticata*. *Evolution* **43**, 1252-1262.

Delph, L.F. and Lively, C.M. (1992) Pollinator visitation, floral display, and nectar production of the sexual morphs of a gynodioecious shrub. *Oikos* **63**, 161-170.

Dennis, P.D. and Fry, G.L.A. (1992) Field-margins: can they enhance natural enemy populations and general arthropod diversity on farmland. *Agric. Ecosys. Environ.* **40**, 95-116.

Descoins, C., Frerot, B., Gallois, M., Lettere, M., Bergström, G., Appelgren, M., Svensson, B.G., and Ågren, L. (1984) Identification des composé de la phéromone de marquage produite par les glandes labiales des mâles de *Megabombus pascuorum* (Scopoli) (Hymenoptera, Apidae). *Nova Acta Reg. Soc. Sci. Upsal. Ser. V:C* **3**, 149-152.

Devlin, B. and Stephenson, A.G. (1985) Sex differential floral longevity, nectar secretion, and pollinator foraging in a protandrous species. *Am. J. Bot.* **72**, 303–310.

Dias, D. (1958) Contribuição para o conhecimento da bionomia de *Bombus incarum* Franklin da Amazônia (Hymenoptera: Bombidae). *Rev. Bras. Entomol.* **8**, 1–20.

Donovan, B.J. (1975) Introduction of new bee species for pollinating lucerne. *Proc. N. Z. Grasslands Assoc.* **36**, 123–128.

Donovan, B.J. (1979) Importation, establishment and propagation of the alkali bee *Nomia melanderi* Cockerell (Hymenoptera: Halictidae) in New Zealand. Proc. IVth Inter. Symp. Pollination. Maryland Agric. Exp. Station Special Mis. Pub. **1**, 257–268.

Donovan, B.J. (1980) Interactions between native and introduced bees in New Zealand. *N. Z. J. Ecol.* **3**, 104–116.

Donovan, B.J. and Wier, S.S. (1978) Development of hives for field population increase, and studies on the life cycle of the four species of introduced bumble bees in New Zealand. *N. Z. J. Agric. Res.* **21**, 733–756.

Dornhaus, A. and Chittka, L. (1999) Insect behaviour—Evolutionary origins of bee dances. *Nature* **401**, 38.

Dornhaus, A. and Chittka, L. (2001) Food alert in bumblebees (*Bombus terrestris*): possible mechanisms and evolutionary implications. *Behav. Ecol. Sociobiol.* **50**, 570–576.

Douglas, J.M. (1973) Double generations of *Bombus jonellus subborealis* Rich. (Hym. Apidae) in an Artic summer. *Entomol. Scand.* **4**, 283–284.

Dover, J. (1992) The conservation of insects on arable farmland. In: *The conservation of insects and their habitats*, edited by N.W. Collins and J. Thomas, pp. 294–318. Academic Press, London.

Dover, J., Sotherton, N., and Gobbet, K. (1990) Reduced pesticide inputs in cereal field margins: effects on butterfly abundance. *Ecolo. Entomol.* **15**, 17–24.

Dramstad, W.E. (1996) Do bumblebees (Hymenopetra: Apidae) really forage close to their nests? *J. Insect Behav.* **2**, 163–182.

Dramstad, W. and Fry, G. (1995) Foraging activity of bumblebees (*Bombus*) in relation to flower resources in arable land. *Agric. Ecosys. Environ.* **53**, 123–135.

Dreisig, H. (1985) Movement patterns of a clear-wing hawkmoth, *Hemaris fuciformis*, foraging at red catchfly, *Viscaria vulgaris*. *Oecologia* **67**, 360–366.

Dreisig, H. (1995) Ideal free distributions of nectar foraging bumblebees. *Oikos* **72**, 161–172.

Duchateau, M.J. (1989) Agonistic behaviours in colonies of the bumblebee *Bombus terrestris*. *J. Ethol.* **7**, 141–151.

Duchateau, M.J., Hishiba, H., and Velthuis, H.H.W. (1994) Diploid males in the bumble bee *Bombus terrestris*. *Entomol. Exp. Appl.* **71**, 263–269.

Duchateau, M.J. and Mariën, J. (1995) Sexual biology of haploid and diploid males in the bumble bee *Bombus terrestris*. *Insectes Soc.* **42**, 255–266.

Duchateau, M.J. and Velthuis, H.H.W. (1988) Development and reproductive strategies in *Bombus terrestris* colonies. *Behaviour* **107**, 186–217.

Duchateau, M.J. and Velthuis, H.H.W. (1989) Ovarian development and egg laying in workers of *Bombus terrestris*. *Entomol. Exp. Appl.* **51**, 199–213.

Dudley, R. and Ellington, C.P. (1990) Mechanics of forward flight in bumblebees I. Kinematics and morphology. *J. Exp. Biol.* **148**, 19–52.

Duffield, G.E., Gibson, R.C., Gilhooly, P.M., Hesse, A.J., Inkley, C.R., Gilbert, F.S., and Barnard, C.J. (1993) Choice of flowers by foraging honey-bees (*Apis mellifera*)—possible morphological cues. *Ecol. Entomol.* **18**, 191–197.

Dukas, R. (1995) Transfer and interference in bumblebee learning. *Anim. Behav.* **49**, 1481–1490.

Dukas, R. and Edelstein-Keshet, L. (1998) The spatial distribution of colonial food provisioners. *J. Theor. Biol.* **190**, 121–134.

Dukas, R. and Ellner, S. (1993) Information processing and prey detection. *Ecology* **74**, 1337–1346.

Dukas, R. and Real, L.A. (1991) Learning foraging tasks by bees: a comparison between social and solitary species. *Anim. Behav.* **42**, 269–276.

Dukas, R. and Real, L.A. (1993a) Learning constraints and floral choice behaviour in bumblebees. *Anim. Behav.* **46**, 637–644.

Dukas, R. and Real, L.A. (1993b) Effects of recent experience on foraging decisions by bumblebees. *Oecologia* **94**, 244–246.

Dukas, R. and Real, L.A. (1993c) Effects of nectar variance on learning by bumblebees. *Anim. Behav.* **45**, 37–41.

Dukas, R. and Waser, N.M. (1994) Categorization of food types enhances foraging performance of bumblebees. *Anim. Behav.* **48**, 1001–1006.

Duncan, W. (1935) Humble-bees of S. Ronaldshay, Orkney. *Scot. Nat.* **1935**, 65–66.

Durrer, S. and Schmid-Hempel, P. (1994) Shared use of flowers leads to horizontal pathogen transmission. *Proc. Royal Soc. Lond., B* **258**, 299–302.

Durrer, S. and Schmid-Hempel, P. (1995) Parasites and the regional distribution of bumble bee species. *Ecography* **18**, 114–122.

Duvoisin, N., Baer, B., and Schmid-Hempel, P. (1999) Sperm transfer and male competition in a bumblebee. *Anim. Behav.* **58**, 743–749.

Dyer, F.C. (1996) Spatial memory and navigation by honeybees on the scale of the foraging range. *J. Exp. Biol.* **199**, 147–154.

Eaton, G.W. and Stewart, M.G. (1969) Blueberry blossom damage caused by bumblebees. *Can. Entomol.* **101**, 149–150.

Eckhart, V.M. (1991) The effects of floral display on pollinator visitation vary among populations of *Phacelia linearis* (Hydrophyllaceae). *Evol. Ecol.* **5**, 370–384.

Edwards, M. (1998) U.K. Biodiversity Action Plan Bumblebee Working Group Report 1998. Unpublished report for the UK BAP bumblebee working group, Midhurst, UK.

Edwards, M. (1999) U.K. Biodiversity Action Plan Bumblebee Working Group Report 1999. Unpublished report for the UK BAP bumblebee working group, Midhurst, UK.

Edwards, M. (2000) U.K. Biodiversity Action Plan Bumblebee Working Group Report 2000. Unpublished report for the UK BAP bumblebee working group, Midhurst, UK.

Edwards, M. (2001) U.K. Biodiversity Action Plan Bumblebee Working Group Report 2001. Unpublished report for the UK BAP bumblebee working group, Midhurst, UK.

Ehrlén, J. (1992) Proximate limits to seed production in a herbaceous perennial legume, *Lathyrus vernus*. *Ecology* **73**, 1820–1831.

Ehrlén, J. and Eriksson, O. (1995) Pollen limitation and population growth in a herbaceous perennial legume. *Ecology* **76**, 652–656.

Eickwort, G.C. (1994) Evolution and life-history patterns of mites associated with bees. In: *Mites*, edited by M.A. Houck, pp. 218–251. Chapman and Hall, New York.

Eickwort, G.C. and Ginsberg, H.S. (1980) Foraging and mating behaviour in Apoidea. *Ann Rev. Entomol.* **25**, 421–426.

Ellington, C.P., Machin, K.E., and Casey, T.M. (1990) Oxygen consumption of bumblebees in forward flight. *Nature* **347**, 472–473.

Endler, J.A. (1981) An overview of the relationships between mimicry and crypsis. *Biol. J. Linn. Soc.* **16**, 25–31.

Engels, W., Schultz, U., and Rädle, M. (1994) Use of the Tübingen mix for bee pasture in Germany. In: *Forage for bees in an agricultural landscape*, edited by A. Matheson, pp. 57–66. International Bee Research Association, Cardiff.

Ernst, W.R., Pearce, P.A., and Pollock, T.L. (1989) Environmental effects of Fenitrothion use in forestry. Conservation and Protection, Environment Canada, Atlantic Region, Canada.

Esch, H. (1967) Die Bedeutung der Lauterzeugung für die Verständigung der stachellosen Bienen. *Z. Vergl. Physiol.* **56**, 199–220.

Esch, H., Goller, F., and Heinrich, B. (1991) How do bees shiver? *Naturwissenschaften* **78**, 325–328.

Estoup, A., Scholl, A., Pouvreau, A., and Solignac, M. (1995) Monandry and polyandry in bumble bees (Hymenoptera, Bombinae) as evidenced by highly variable microsatellites. *Mol. Ecol.* **4**, 89–93.

Estoup, A., Slignac, M., Cornuet, J.-M., Goudet, J., and Scholl, A. (1996) Genetic differentiation of continental and island populations of *Bombus terrestris* (Hymenoptera: Apidae) in Europe. *Mol. Ecol.* **5**, 19–31.

Evans, D.L. and Waldbauer, G.P. (1982) Behaviour of adult and naïve birds when presented with a bumblebee and its mimic. *Zeits. Tierpsychol.* **59**, 247–259.

Evans, G.O. (1992) *Principles of Acarology*. Cambridge University Press, Cambridge.

Eysenck, M.W. and Keane, M.T. (1990) *Cognitive psychology*. Lawrence Erlbaum, London.

Fabre J.-H. (1879) Souvenirs Entomologiques Études sur l'Instinct et les Moeurs des Insectes, (1 Sér.), Delagrave, Paris 1879.

Fabre J.-H., (1882) Souvenirs Entomologiques Études sur l'Instinct et les Moeurs des Insectes, (2me Sér.), Delagrave, Paris 1882.

Farrar, C.L. and Bain, H.F. (1946) Honey bees as pollinators of cranberry. *Am. Bee J.* **86**, 503–504.

Fattig, P.W. (1933) Food of the robber fly, *Mallophora orcina* (Wied.) (Diptera). *Can. Entomol.* **65**, 119–120.

Feber, R.E. (1993) The ecology and concervation of butterflies on lowland farmland. D.Phil. Thesis, Oxford University.

Feber, R.E., Smith, H., and Macdonald, D.W. (1994) The effects of field margin restoration on the meadow brown butterfly, *Maniola jurtina*. In: *Field margins: integrating agriculture and conservation*, edited by N. Boatman, Monogr. No. 58, British Crop Protection Council, Farnham, UK, pp. 295–300.

Feber, R.E., Smith, H., and Macdonald, D.W. (1996) The effects on butterfly abundance of the management of uncropped edges of arable fields. *J. Appl. Ecol.* **33**, 1191–1205.

Ferguson, A.W. and Free, J.B. (1979) Production of forage-marking pheromone by the honeybee. *J. Apic. Res.* **18**, 128–135.

Ferton, C. (1901) Sur l'epoque du reveil des bourdons et des Psithyres à Bonifacio. *Ann. Soc. Entomol. Fr.* **70**, 84–85.

Firbank, L.G., Arnold, H.R., Eversham, B.C., Mountford, J.O., Radford, G.L., Telfer, M.G., Treweek, J.R., Webb, N.R.C., and Wells, T.C.E. (1993) *Managing set-aside for wildlife*. HMSO, London.

Firbank, L.G., Carter, N., Derbyshire, J.F., and Potts, G.R. (eds) (1991) *The ecology of temperate cereal fields*. Blackwell, Oxford, 469 pp.

Fischer, M. and Matthies, D. (1997) Mating structure and inbreeding and outbreeding depression in the rare plant *Gentianella germanica* (Genitanaceae). *Am. J. Bot.* **84**, 1685–1692.

Fisher, R.A. (1930) *The genetical theory of natural selection*. Clarendon Press, Oxford.

Fisher, R.M. (1984) Evolution and host specificity: a study of the invasion success of a specialized bumblebee social parasite. *Can. J. Zool.* **62**, 1641–1644.

Fisher, R.M. (1987) Queen-worker conflict and social parasitism in bumble bees (Hymenoptera: Apidae). *Anim. Behav.* **35**, 1026–1036.

Fisher, R.M. (1988) Observations on the behaviours of three European bumblebee species (*Psithyrus*). *Insectes Soc.* **35**, 341–354.

Fisher, R.M. (1989) Incipient colony manipulation, *Nosema* incidence and colony productivity of the bumble bee *Bombus terrestris* (Hymenoptera, Apidae). *J. Kans. Entomol. Soc.* **62**, 581–589.

Fisher, R.M. (1992) Sex ratios in bumble bee social parasites: support for queen-worker conflict theory? (Hymenoptera: Apidae). *Sociobiology* **20**, 205–217.

Fisher, R.M. and Pomeroy, N. (1989) Pollination of greenhouse muskmelons by bumble bees (Hymenoptera: Apidae). *Entomol. Soc. Am.* **82**, 1061–1066.

Forbes, W.T.M. (1923) The Lepidoptera of New York and neighboring states. New York.

Foster, R.L. (1992) Nestmate recognition as an inbreeding avoidance mechanism in bumble bees (Hymenoptera: Apidae). *J. Kans. Entomol. Soc.* **65**, 238–243.

Frank, A. (1941) Eigenartige Flugbahnen bei Hummelmännchen. *Z. vergl. Physiol.* **28**, 467–484.

Frankie, G. and Vinson, S.B. (1977) Scent-marking of passion flowers in Texas by females of *Xylocopa virginica texana* (Hym. Anthophoridae). *J. Kans. Entomol. Soc.* **50**, 613–625.

Frankie, G.W., Thorp, R.W., Newstrom-Lloyd, L.E., Rizzardi, M.A., Barthell, J.F., Griswold, T.L., Kim J.Y., and Kappagoda, S.(1998) Monitoring solitary bees in modified wildland habitats: Implications for bee ecology and conservation. *Environ. Entomol.* **27**, 1137–1148.

Free, J.B. (1955a) The division of labour within bumblebee colonies. *Insectes Soc.* **2**, 195–212.

Free, J.B. (1955b) The collection of food by bumblebees. *Insectes Soc.* **2**, 303–311.

Free, J.B. (1955c) Queen production in colonies of bumblebees. *Proc. Royal. Entomol. Soc. Lond. (A)* **30**, 19–25.

Free, J.B. (1958) The defence of bumblebee colonies. *Behaviour* **12**, 233–242.

Free, J.B. (1962) The behaviour of honeybees visiting field beans (*Vicia faba*). *J. Anim. Ecol.* **31**, 497–502.

Free, J.B. (1963) The flower constancy of honeybees. *J. Anim. Ecol.* **32**, 119–131.

Free, J.B. (1970) The flower constancy of bumblebees. *J. Anim. Ecol.* **39**, 395–402.

Free, J.B. (1971) Stumuli eliciting mating behaviour of bumblebee (*Bombus pratorum* L.) males. *Behaviour* **40**, 55–61.

Free, J.B. (1987) *Pheromones of social bees*. Chapman and Hall, London.

Free, J.B. (1993) *Insect pollination of crops*. 2nd edn. Academic Press, London.

Free, J.B. and Butler, C.G. (1959) *Bumblebees*. Collins, London

Free, J.B. and Ferguson, A.W. (1986) Foraging of bees on oil-seed rape (*Brassica napus* L.) in relation to the flowering of the crop and pest control. *J. Agric. Sci., Cambridge* **94**, 151–154.

Free, J.B., Ferguson, A.W., Pickett, J.A., and Williams, I.H. (1982b) Use of unpurified Nasonov pheromone components to attract clustering honeybees. *J. Apic. Res.* **21**, 26–29.

Free, J.B., Weinberg, I., and Whiten, A. (1969) The egg-eating behaviour of *Bombus lapidarius* L. *Behaviour* **35**, 313–317.

Free, J.B. and Williams, I.H. (1972) The role of the Nasanov gland pheromone in crop communication by honeybees (*Apis mellifera* L.). *Behaviour* **41**, 314–318.

Free, J.B. and Williams, I.H. (1973) The foraging behaviour of honeybees (*Apis mellifera*) on Brussels spouts (*Brassica oleracea* L.). *J. Appl. Ecol.* **10**, 489–499.

Free, J.B. and Williams, I.H. (1976) Pollination as a factor limiting the yield of field beans (*Vicia faba* L.). *J. Agric. Sci. UK* **87**, 395–399.

Free, J.B. and Williams, I.H. (1983) Scent-marking of flowers by honeybees. *J. Apic. Res.* **18**, 128–135.

Free, J.B., Williams, I., Pickett, J.A., Ferguson, A.W., and Martin, A.P. (1982a) Attractiveness of (Z)-11-eicosen-1-ol to foraging honeybees. *J. Apic. Res.* **21**, 151–156.

Freeman, B.A. (1966) Notes on the conopid flies including insect hostplant and phoretic relationships (Diptera: Conopidae). *J. Kans. Entomol. Soc.* **39**, 123–131.

Frehn, E. and Schwammberger, K. (2001) Social parasitism of *Psithyrus vestalis* in free-foraging colonies of *Bombus terrestris* (Hymenoptera: Apidae). *Entomol. Gen.* **25**, 103–105.

Freitas, B.M. and Paxton, R.J. (1998) A comparison of two pollinators: the introduced honey bee *Apis mellifera* and an indigenous bee *Centris tarsata* on cashew *Anacardium occidentale* in its native range of NE Brazil. *J. Appl. Ecol.* **35**, 109–121.

Fretwell, S.D. and Lucas, H.L. (1970) Territorial behaviour and other factors influencing habitat distribution in birds. *Acta Biotheor.* **19**, 16–36.

Frison, T.H. (1917) Notes on Bombidae, and on the life history of *Bombus auricomis* Robt. *Ann. Entomol. Soc. Am.* **10**, 277–286.

Frison, T.H. (1926) Contribution to the knowledge of the inter-relations of the bumblebees of Illinois with their animate environment. *Ann. Entomol. Soc. Am.* **19**, 203–235.

Frison, T.H. (1928) A contribution to the knowledge of the life history of *Bremus bimaculatus* (Cresson) (Hym.). *Entomol. Am.* **8**, 159–223.

Frison, T.H. (1929) A contribution to the knowledge of the bionomics of *Bremus impatiens* (Cresson) (Hym.). *Bull. Brooklyn Entomol. Soc.* **24**, 261–285.

Frison, T.H. (1930) Observations of the behavior of bumblebees (*Bremus*). The orientation flight. *Can. Entomol.* **62**, 49–54.

von Frisch, K. (1923) Uber die 'Sprache' der Beinen, eine tierpsychologische Untersuchung. *Zoologische Jahrbucher: Abteilung fur Allegemeine Zool. Physiol.* **40**, 1–186.

von Frisch, K. (1952) Hummeln als unfreiwillige Transportflieger. *Natur u. Volk* **82**, 171–174.

von Frisch, K. (1967) *The dance language and orientation of bees*. Harvard University Press, Cambridge, Massachusetts.

Frost, S.K. and Frost, P.G.H. (1981) Sunbird pollination of *Strelizia nicolai*. *Oecologia* **49**, 379–384.

Fuller, R.M. (1987) The changing extent and conservation interest of lowland grasslands in England and Wales: a review of grassland surveys 1930–84. *Biol. Conser.* **40**, 281–300.

Fussell, M. and Corbet, S.A. (1991) Forage for bumble bees and honey bees in farmland: a case study. *J. Apic. Res.* **30**, 87–97.

Fussell, M. and Corbet, S.A. (1992a) Flower usage by bumblebees a basis for forage plant management. *J. Appl. Ecol.* **29**, 451–465.

Fussell, M. and Corbet, S.A. (1992b) Observations on the patrolling behaviour of male bumblebees (Hym.). *Entomologist's Monthly Magazine* **128**, 229–235.

Fussell, M. and Corbet, S.A. (1992c) The nesting places of some British bumblebees. *J. Apic. Res.* **31**, 32–41.

Fussell, M., Osborne, J.L., and Corbet, S.A. (1991) Seasonal and diurnal patterns of insect visitors to winter sown field bean flowers in Cambridge. *Aspects Appl. Biol.* **27**, 95–99.

Fye, R.E. and Medler, J.T. (1954) Field domiciles for bumblebees. *J. Econom. Entomol.* **47**, 672–676.

Galen, C. (1983) The effects of nectar thieving ants on seedset in floral scent morphs of *Polemonium viscosum*. *Oikos* **41**, 245–249.

Galen, C. (1985) Regulation of seed-set in *Polemonium viscosum*: floral scents, pollination, and resources. *Ecology* **66**, 792–797.

Galen, C. (1989) Measuring pollinator-mediated selection on morphometric traits: bumblebees and the alpine sky pilot, *Polemonium viscosum*. *Evolution* **43**, 882–890.

Galen, C. and Newport, M.E.A. (1987) Bumble bee behaviour and selection on flower size in the sky pilot, *Polemonium viscosum*. *Oecologia* **74**, 20–23.

Galen, C. and Newport, M.E.A. (1988) Pollination quality, seed set, and flower traits in *Polemonium viscosum*: complementary effects of variation in flower scent and size. *Am. J. Bot.* **75**, 900–905.

Ganskopp, D. (1995) Free-ranging Angora-goats—left-handed or right-handed tendencies while grazing. *Appl. Anim. Behav. Sci.* **43**, 141–146.

Garófalo, C.A. (1974) Aspectos evolutivos da biologia da reprodução em abelhas (Hymenoptera, Apoidea). Dissertation thesis, Universidade de São Paulo, Ribeirão Prêto, Brazil.

Garófalo, C.A. (1976) Evolução do comportamento social visualizada através da ecologia de *Bombus morio* (Hymenoptera: Bombinae). PhD thesis, Universidade de São Paulo, Ribeirão Prêto, Brazil.

Gathmann, A., Greiler, J.H., and Tscharntke, T. (1994) Trap-nesting bees and wasps colonizing set-aside fields: succession and body size, management by cutting and sowing. *Oecologia* **98**, 8–14.

Geber, M.A. (1985) The relationship of plant size to self-pollination in *Mertensia ciliata*. *Ecology* **66**, 762–772.

Gegear, R.J. and Laverty, T.M. (1995) Effect of flower complexity on relearning flower-handling skills in bumble bees. *Can. J. Zool.* **73**, 2052–205.

Gegear, R.J. and Laverty, T.M. (1998) How many flower types can bumble bees work at the same time? *Can. J. Zool.* **76**, 1358–1365.

Gentry, A.H. (1978) Anti-pollinators for mass-flowering plants? *Biotropica* **10**, 68–69.

Gilbert, F.S. (1981) Foraging ecology of hoverflies: morphology of the mouthparts in relation to feeding on nectar and pollen in some common urban species. *Ecol. Entomol.* **6**, 245–262.

Gilbert, F.S., Haines, N., and Dickson, K. (1991) Empty flowers. *Funct. Ecol.* **5**, 29–39.

Gilbert, L.E. (1975) Ecological consequences of a coevolved mutualism between butterflies and plants. In: *Coevolution of animals and plants*, edited by L.E. Gilbert and P.H. Raven, pp. 210–240. University of Texas Press, Austin, Texas.

Gill, R.A. (1991) The value of honeybee pollination to society. *Acta Hortic.* **288**, 62–68.

Gilliam M (1997) Identification and roles of non-pathogenic microflora associated with honey bees. *FEMS Microbiol. Lett.* **155**, 1–10.

Gilliam, M., Lorenz, B.J., Prest, D.B., and Camazine, S. (1993) *Ascosphaera apis* from *Apis cerana* from South Korea. *J. Invertebr. Pathol.* **61**, 111–112.

Gilliam, M., Lorenz, B.J., and Buchmann, S.L. (1994) *Ascosphaera apis*, the chalkbrood pathogen of the honeybee, *Apis mellifera*, from larvae of a carpenter-bee, *Xylocopa californica arizonensis*. *J. Invert. Pathol.* **63**, 307–309.

Gilliam, M. and Taber, S. (1991) diseases, pests, and normal microflora of honeybees, *Apis mellifera*, from feral colonies. *J. Invert. Pathol.* **58**, 286–289.

Ginsberg, H.S. (1983) Foraging ecology of bees in an old field. *Ecology* **64**, 165–175.

Ginsberg, H.S. (1985) Foraging movements of *Halictus ligatus* (Hymenoptera: Halictidae) and *Ceratina calcarata* (Hymenoptera: Anthophoridae) on *Chrysanthemum leucanthemum* and *Erigeron annuus* (Asteraceae). *J. Kans. Entomol. Soc.* **58**, 19–26.

Ginsberg, H.S. (1986) Honey bee orientation behaviour and the influence of flower distribution on foraging movements. *Ecol. Entomol.* **11**, 173–179.

Giurfa, M. (1993) The repellent scent-mark of the honeybee *Apis mellifera ligustica* and its role as a communication cue during foraging. *Insectes Soc.* **40**, 59–67.

Giurfa, M., Eichmann, B., and Menzel, R. (1996) Symmetry perception in an insect. *Nature* **382**, 458–461.

Giurfa, M. and Núñez, J.A. (1992a). Foraging by honeybees on *Carduus acanthoides*—pattern and efficiency. *Ecol. Entomol.* **17**, 326–330.

Giurfa, M. and Núñez, J.A. (1992b) Honeybees mark with scent and reject recently visited flowers. *Oecologia* **89**, 113–117.

Giurfa, M., Núñez, J.A., and Backhaus, W. (1994) Odour and colour information in the foraging choice behavior of the honeybee. *J. Comp. Physiol.* **175**, 773–779.

Goerzen, D.W. (1991) Microflora associated with the alfalfa leafcutting bee, *Megachile rotundata* (Fab) (Hymenoptera, Megachilidae) in Saskatchewan, Canada. *Apidologie* **22**, 553–561.

Goerzen, D.W., Dumouchel, L., and Bissett, J. (1992) Cccurrence of chalkbrood caused by *Ascosphaera aggregata* Skou in a native leafcutting bee, *Megachile pugnata* Say (Hymenoptera, Megachilidae), in Saskatchewan. *Can. Entomol.* **124**, 557–558.

Goerzen, D.W., Erlandson, M.A., and Bissett, J. (1990) Occurrence of chalkbrood caused by *Ascosphaera aggregata* Skou in a native leafcutting bee, *Megachile relativa* Cresson (Hymenoptera, Megachilidae), in Saskatchewan. *Can. Entomol.* **122**, 1269–1270.

Goka, K. (1998) Influences of invasive species on native species—will the European bumblebee, *Bombus terrestris*, bring genetic pollution into Japanses native species? *Bull. Biogr. Soc. Jpn* **53**, 91–101.

Goka, K., Okabe, K., Yoneda, M., and Niwa, S. (2001) Bumblebee commercialization will cause worldwide migration of parasitic mites. *Mol. Ecol.* **10**, 2095–2099.

Goldblatt, J.W. (1984) Parasites and parasitization rates in bumble bee queens *Bombus* spp. (Hymenoptera, Apidae) in Southwestern Virginia. *Environ. Entomol.* **13**, 1661–1665.

Goldblatt, J.W. and Fell, R.D. (1987) Adult longevity of workers of the bumble bee *Bombus fervidus* (F.) and *Bombus pennsylvanicus* (De Geer) (Hymenoptera: Apidae). *Can. J. Zool.* **65**, 2349–2353.

Gonzalez, A., Rowe, C.L., Weeks, P.J., Whittle, D., Gilbert, F.S., and Barnard, C.J. (1995) Flower choice by honey-bees (*Apis mellifera* L)—sex-phase of flowers and preferences among nectar and pollen foragers. *Oecologia* **101**, 258–264.

Goodwin, S.G. (1995) Seasonal phenology and abundance of early-, mid- and long-season bumble bees in southern England, 1985–1989. *J. Apic. Res.* **34**, 79–87.

Gori, D.F. (1983) Post-pollination phenomena and adaptive floral changes. In: *Handbook of experimental pollination biology*, edited by C.E. Jones and R.J. Little, pp. 31–49. Van Nostrand Reinhold Co. Inc., New York.

Gori, D.F. (1989) Floral colour change in *Lupinus argenteus* (Fabaceae): why should plants advertise the location of unrewarding flowers to pollinators? *Evolution* **43**, 870–881.

Goulson, D. (1994) A model to predict the role of flower constancy in inter-specific competition between insect pollinated flowers. *J. Theor. Biol.* **168**, 309–314.

Goulson, D. (1999) Foraging strategies for gathering nectar and pollen in insects. *Perspectives Plant Ecol. Evol. System.* **2**, 185–209.

Goulson, D. (2000a) Are insects flower constant because they use search images to find flowers? *Oikos*, **88**, 547–552.

Goulson, D. (2000b) Why do pollinators visit proportionally fewer flowers in large patches? *Oikos* **91**, 485–492.

Goulson, D. and Cory, J.S. (1993) Flower constancy and learning in the foraging behaviour of the green-veined white butterfly. *Pieris napi. Ecol. Entomol.* **18**, 315–320.

Goulson, D., Hawson, S.A., and Stout, J.C. (1998b) Foraging bumblebees avoid flowers already visited by conspecifics or by other bumblebee species. *Anim. Behav.* **55**, 199–206.

Goulson, D., Hughes, W.O.H., Derwent, L.C., and Stout, J.C. (2002a) Colony growth of the bumblebee, *Bombus terrestris*, in improved and conventional agricultural and suburban habitats. *Oecologia* **130**, 267–273.

Goulson, D. and Jerrim, K. (1997) Maintenance of the species boundary between *Silene dioica* and *S. latifolia* (red and white campion). *Oikos* **78**, 254–266.

Goulson, D., Ollerton, J., and Sluman, C. (1997a) Foraging strategies in the small skipper butterfly, *Thymelicus flavus*: when to switch? *Anim. Behav.* **53**, 1009–1016.

Goulson, D. Peat, J., Stout, J.C., Tucker, J., Darvill, B., Derwent, L.C., and Hughes, W.O.H. (2002b) Can alloethism in workers of the bumblebee *Bombus terrestris* be explained in terms of foraging efficiency? *Anim. Behav.* **64**, 123–130.

Goulson, D. and Stout, J.C. (2001) Homing ability of the bumblebee *Bombus terrestris*. *Apidologie* **32**, 105–112.

Goulson, D., Stout, J.C., and Hawson, S.A. (1997b) Can flower constancy in nectaring butterflies be explained by Darwin's interference hypothesis? *Oecologia* **112**, 225–231.

Goulson, D., Stout, J.C., Hawson, S.A., and Allen, J.A. (1998a) The effects of floral display size and colour on recruitment of three bumblebee species to comfrey, *Symphytum officinale* L. (Boraginaceae), and subsequent seed set. *Oecologia* **113**, 502–508.

Goulson, D., Stout, J.C., and Kells, A.R. Effects of introduced bumblebees on native pollinator communities in Tasmania. *J. Insect Conserv.* (in press).

Goulson, D., Stout, J.C., Langley, J., and Hughes, W.O.H. (2000) The identity and function of scent marks deposited by foraging bumblebees. *J. Chem. Ecol.* **26**, 2897–2911.

Goulson, D. and Williams, P.H. (2001) *Bombus hypnorum* (Hymenoptera: Apidae), a new British bumblebee? *Br. J. Entomol. Nat. Hist.* **14**, 129–131.

Goulson, D. and Wright, N.P. (1998) Flower constancy in the hoverflies *Episyrphus balteatus* (Degeer) and *Syrphus ribesii* (L.) (Syrphidae). *Behav. Ecol.* **9**, 213–219.

Graham, L. and Jones, K.N. (1996) Resource partitioning and per-flower foraging efficiency in 2 bumble bee species. *Am. Midland Nat.* **136**, 401–406.

Grant, V. (1949) Pollination systems as isolating mechanisms in angiosperms. *Evolution* **3**, 82–97.

Grant, V. (1950) The flower constancy of bees. *Bot. Rev.* **3**, 82–97.

Grant, V. (1952) Isolation and hybridization between *Aquilegia formosa* and *A. pubescens*. *Aliso* **2**, 341–360.

Grant, V. (1992) Floral isolation between ornithophilous and sphingophilous species of *Ipomopsis* and *Aquilegia*. *Proc. Natl. Acad. Sci. USA* **89**, 11828–11831.

Grant V. (1993) Origins of floral isolation between ornithophilous and sphingophilous plant species. *Proc. Natl. Acad. Sci. USA* **90**, 7729–7733.

Grant, V. (1994) Modes and origins of mechanical and ethological isolation in angiosperms. *Proc. Natl. Acad. Sci. USA* **91**, 3–10.

Gray, H.E. (1925) Observations of tripping of alfalfa blossoms. *Can. Entomol.* **57**, 235–237.

Greaves, M.P. and Marshall, E.J.P. (1987) Field margins: definitions and statistics. In: *British Crop Protection Council Monograph No. 35: Field Margins*, edited by J.M. Way and P.W. Grieg-Smith, pp. 3–11. Thornton Heath: British Crop Protection Council.

Greco, C.F., Holland, D., and Kevan, P.G. (1996) Foraging behavior of honey-bees (*Apis mellifera* L) on staghorn sumac [*Rhus hirta* Sudworth (ex-*typhina* L)]—differences and dioecy. *Can. Entomol.* **128**, 355–366.

Greggers, U. and Menzel, R. (1993) Memory dynamics and foraging strategies of honeybees. *Behav. Ecol. Sociobiol.* **32**, 17–29.

Grönlund, S., Itämies, J., and Mikkola, H. (1970) On the food and feeding habits of the great grey shrike *Lanius excubitor* in Finland. *Ornis Fenn.* **47**, 167–171.

Gross, C.L. and Mackay D. (1998) Honeybees reduce fitness in the pioneer shrub *Melastoma affine* (Melastomataceae). *Biol. Conserv.* **86**, 169–178.

Guilford, T. and Dawkins, M.S. (1987) Search images not proven: A reappraisal of recent evidence. *Anim. Behav.* **35**, 1838–1845.

Gumbert, A. (2000) Color choices by bumble bees (*Bombus terrestris*): innate preferences and generalization after learning. *Behav. Ecol. Sociobiol.* **48**, 36–43.

Gupta, A.P. (ed.) (1986) *Hemocytic and humoral immunity in arthropods.* John Wiley & Sons, New York.

Gurr, L. (1955) A note on the efficiency of *Bombus terrestris* (L.) as a pollinator of lucerne (*Medicago sativa* L.). *N. Z. J. Sci. Technol.* **A37**, 300.

Gurr, L. (1957) Bumble bee species present in the South Island of New Zealand. *N. Z. J. Sci. Technol.* **A38**, 997–1001.

Haas, A. (1949) Arttpische Flugbahnen von Hummelmännchen. *Z. vergl. Physiol.* **31**, 281–307.

Haas, A. (1960) Vergleichende Verhaltensstudien zum Paarungsschwarm solitärer Apiden. *Z. Tierpsychol.* **17**, 402–416.

von Hagen, E. (1994) *Hummeln bestimmen. Ansiedeln, Vermehren, Schützen.* Augsburg: Naturbuch-Verlag.

Haldane, J.B.S. (1949) Disease and evolution. *La Ricerca Scientifi.* **19 (Suppl.)**, 68–76.

Hamilton, W.D. (1964) The genetical evolution of social behavior. *J. Theor. Biol.* **7**(I), 1–16; II: 17–32.

Hamilton, W.D. (1967) Extraordinary sex ratios. *Science* **156**, 477–488.

Hamilton, W.D. (1980) Sex vs. non-sex vs. parasite. *Oikos* **35**, 282–290.

Hamilton, W.D. (1987) Kinship, recognition, disease, and intelligence: Constraints of social evolution. In: *Animal societies: theory and facts*, edited by Y. Itô, J.L. Brown, and J. Kikkawa, pp. 81–102. Japanese Scientific Society Press, Tokyo.

Hanski, I. and Gilpin, M. (1991) Metapopulation dynamics: brief history and conceptual domain. *Biol. J. Linn. Soc.* **42**, 3–16.

Harder, L.D. (1985) Morphology as a predictor of flower choice by bumblebees. *Ecology* **66**, 198–210.

Harder, L.D. (1986) Effects of nectar concentration and flower depth on handling efficiency of bumblebees. *Oecologia* **69**, 309–315.

Harder, L.D. and Barrett, S.C.H. (1995) Mating cost of large floral displays in hermaphrodite plants. *Nature* **373**, 512–515.

Harder, L.D. and Real, L.A. (1987) Why are bumblebees risk averse? *Ecology* **68**, 1104–1108.

Hartfelder, K., Cnaani, J., and Hefetz, A. (2000) Caste-specific differences in ecdysteroid titers in early larval stages of the bumblebee *Bombus terrestris*. *J. Insect Physiol.* **46**, 1433–1439.

Hartling, L.K. and Plowright, R.C. (1978) Foraging by bumblebees on artificial flowers: a laboratory study. *Can. J. Bot.* **63**, 488–491.

Haydak, M.H. (1943) Larval food and development of castes in the honeybee. *J. Econ. Entomol.* **36**, 778–792.

Hasselrot, T.B. (1960) Studies on Swedish bumblebees (genus *Bombus* Latr.): their domestication and biology. *Opusc. Entomol. Suppl.* **17**, 1–192.

Hedenström, A., Ellington, C.P., and Wolf, T.J. (2001) Wing wear, earodynamics and flight energetics in bumblebees (*Bombus terrestris*): an experimental study. *Funct. Ecol.* **15**, 417–422.

Hefetz, A., Taghizadeh, T., and Francke, W. (1996) The exocrinology of the queen bumble bee *Bombus terrestris* (Hymenoptera: Apidae, Bombini). *Zeits. Für Naturforsch.* **51**, 409–422.

Heinrich, B. (1971) Temperature regulation of the sphinx moth, *Manduca sexta*. II. *J. Exp. Biol.* **54**, 141–166.

Heinrich, B. (1972a) Patterns of endothermy in bumblebee queens, drones and workers. *J. Comp. Physiol.* **77**, 65–79.

Heinrich, B. (1972b) Physiology of brood incubation in the bumblebee queen, *Bombus vosnesenskii*. *Nature* **239**, 223–225.

Heinrich, B. (1972c) Energetics of temperature regulation and foraging in a bumblebee, *Bombus terricola* Kirby. *J. Comp. Physiol.* **77**, 49–64.

Heinrich, B. (1972d) Temperature regulation in the bumblebee *Bombus vagans*: a field study. *Science* **175**, 185–187.

Heinrich, B. (1974) Thermoregulation in bumblebees. I. Brood incubation by *Bombus vosnesenskii* queens. *J. Comp. Physiol.* **88**, 129–140.

Heinrich, B. (1975a) Thermoregulation in bumblebees. II. Energetics of warmup and free flight. *J. Comp. Physiol.* **96**, 155–166.

Heinrich, B. (1975b) Energetics of pollination. *Annu. Rev. Ecol. Syst.* **6**, 139–170.

Heinrich, B. (1976a) Resource partitioning among some eusocial insects: Bumblebees. *Ecology* **57**, 874–889.

Heinrich, B. (1976b) The resource specializations of individual bumblebees. *Ecol. Monogr.* **46**, 105–128.

Heinrich, B. (1976c) Heat exchange in relation to blood flow between thorax and abdomen in bumblebees. *J. Exp. Biol.* **64**, 561–585.

Heinrich, B. (1979a) Resource heterogeneity and patterns of movement in foraging bumblebees. *Oecologia* **40**, 235–245.

Heinrich, B. (1979b) *Bumblebee economics*. Harvard University Press, Cambridge (Mass.).

Heinrich, B. (1979c) 'Majoring' and 'minoring' by foraging bumblebees, *Bombus vagans*: an experimental analysis. *Ecology* **60**, 245–255.

Heinrich, B. (1993) *The hot-blooded insects*, pp. 239–240. Springer-Verlag, New York.

Heinrich, B. (1996) *The thermal warriors*. Harvard University Press, Cambridge (Mass.).

Heinrich, B. and Heinrich, M.J.E. (1983a) Size and caste in temperature regulation by bumblebees. *Physiol. Zool.* **56**, 552–562.

Heinrich, B. and Heinrich, M.J.E. (1983b) Heterothermia in foraging workers and drones of the bumblebee *Bombus terricola*. *Physiol. Zool.* **56**, 563–567.

Heinrich, B. and Mommsen, T.P. (1985) Flight of winter moths near 0°C. *Science* **228**, 177–179.

Heinrich, B., Mudge, P.R., and Deringis, P.G. (1977) Laboratory analysis of flower constancy in foraging bumblebees: *Bombus ternarius* and *B. terricola*. *Behav. Ecol. Sociobiol.* **2**, 247–265.

Heinrich, B. and Raven, P.H. (1972) Energetics and pollination ecology. *Science* **176**, 597–602.

Helms, K.R. (1999) Colony sex ratios, conflict between queens and workers, and apparent queen control in the ant *Pheidole desertorum*. *Evolution* **53**, 1470–1478.

Henslow, G. (1867) Note on the structure of *Medicago sativa* as apparently affording facilities for the intercrossing of distinct flowers. *J. Linn. Soc. Bot.* **9**, 327–329.

Higashi, S., Ohara, M., Arai, H., and Matsuo, K. (1988) Robber-like pollinators: overwintering queen bumblebees foraging on *Corydalis ambigua*. *Ecol. Entomol.* **13**, 411–418.

Hingston, A.B. and McQuillan, P.B. (1998) Does the recently introduced bumblebee *Bombus terrestris* (Apidae) threaten Australian ecosystems? *Aust. J. Ecol.* **23**, 539–549.

Hingston, A.B. and McQuillan, P.B. (1999) Displacement of Tasmanian native megachilid bees by the recently introduced bumblebee *Bombus terrestris* (Linnaeus, 1758) (Hymenoptera: Apidae). *Aust. J. Zool.* **47**, 59–65.

Hobbs, G.A. (1962) Further studies on the food-gathering behaviour of bumble bees (Hymenoptera: Apoidea) in Alberta. *Can. Entomol.* **94**, 538–541.

Hobbs, G.A. (1964a) Phylogeny of bumble bees based on brood-rearing behaviour. *Can. Entomol.* **96**, 115–116.

Hobbs, G.A. (1964b) Ecology of species of *Bombus* Latr. (Hymenoptera: Apidae) in southern Alberta. I. Subgenus *Alpinobombus* Skor. *Can. Entomol.* **96**, 1465–1470.
Hobbs, G.A. (1965a) Ecology of species of *Bombus* (Hymenoptera: Apidae) in southern Alberta. II. Subgenus *Bombias* Robt. *Can. Entomol.* **97**, 120–123.
Hobbs, G.A. (1965b) Ecology of species of *Bombus* (Hymenoptera: Apidae) in southern Alberta. III. Subgenus *Cullumanobombus* Vogt. *Can. Entomol.* **97**, 1293–1302.
Hobbs, G.A. (1966a) Ecology of species of *Bombus* (Hymenoptera: Apidae) in southern Alberta. IV. Subgenus *Fervidobombus* Skorikov. *Can. Entomol.* **98**, 33–39.
Hobbs, G.A. (1966b) Ecology of species of *Bombus* (Hymenoptera: Apidae) in southern Alberta. V. Subgenus *Subterraneobombus* Vogt. *Can. Entomol.* **98**, 288–294.
Hobbs, G.A. (1967a) Ecology of species of *Bombus* (Hymenoptera: Apidae) in southern Alberta. VI. Subgenus *Pyrobombus*. *Can. Entomol.* **99**, 1272–1292.
Hobbs, G.A. (1967b) Obtaining and protecting red-clover pollinating species of *Bombus* (Hymenoptera: Apidae). *Can. Entomol.* **99**, 943–951.
Hobbs, G.A., Nummi, W.O., and Virostek, J.F. (1961) Food-gathering behaviour of honey, bumble, and leaf-cutter bees (*Hymenoptera: Apoidea*) in Alberta. *Can. Entomol.* **93**, 409–419.
Hobbs, G.A., Nummi, W.O., and Virostek, J.F. (1962) Managing colonies of bumble bees (Hymenoptera: Apidae) for pollination purposes. *Can. Entomol.* **94**, 1121–1132.
Hobbs, G.A., Virostek, J.F. and Nummi, W.O. (1960) Establishment of *Bombus* spp. (Hymenoptera: Apidae) in artificial domiciles in southern Alberta. *Can. Entomol.* **92**, 868–872.
Hodges, C.M. (1981) Optimal foraging in bumblebees: hunting by expectation. *Anim. Behav.* **29**, 1166–1171.
Hodges, C.M. (1985a) Bumble bees foraging: the threshold departure rule. *Ecology* **66**, 179–187.
Hodges, C.M. (1985b) Bumble bees foraging: Energetic consequences of using a threshold departure rule. *Ecology* **66**, 188–197.
Hoffer, E. (1882-3) Die Hummeln Steiermarks. Lebensgeschichte und Beschribung derselben. Graz.
Hogendoorn, K., Steen, Z., and Schwarz, M.P. (2000) Native Australian carpenter bees as a potential alternative to introducing bumble bees for tomato pollination in greenhouses. *J. Apic. Res.* **39**, 67–74.
Høiland, K. (1993) Threatened plants in the cultural landscape in Norway. 1. Weeds in arable environments. Norwegian Institute for Nature Research, Trondheim.
Hölldobler, B. and Wilson, E.O. (1990) *The ants.* Springer-Verlag, Berlin.
Holm, S.N. (1966) The utilization and management of bumblebees for red clover and alfalfa seed production. *Ann. Rev. Entomol.* **11**, 155–182.
Holm, S.N. (1972) Weight and life length of hibernating bumble bee queens (Hymenoptera: Bombidae) under controlled conditions. *Entomol. Scand.* **3**, 313–320.
Holmes, F.O. (1964) The distribution of honey bees and bumblebees on nectar secreting plants. *Am. Bee J.* 12–13.
Hopkins, I. (1914) History of the bumblebee in New Zealand: its introduction and results. *N. Z. Depart. Agric. Indus. Comm.* **46**, 1–29.
Hopper, S.D. (1987) Impact of honeybees on Western Australia's nectarivorous fauna. In: *Beekeeping and land management*, edited by J. Blyth, pp. 59–71. CALM, Western Australia.
Horskins, K. and Turner, V.B. (1999) Resource use and foraging patterns of honeybees, *Apis mellifera*, and native insects on flowers of *Eucalyptus costata*. *Aust. J. Ecol.* **24**, 221–227.

Hovorka, O., Urbanová, K., and Valterová, I. (1998) Premating behavior of *Bombus confusus* males and analysis of their labial gland secretion. *J. Chem. Ecol.* **24**, 183–193.

Huck, K., Schwarx, H.H., and Schmid-Hempel, P. (1998) Host choice in the phoretic mite *Parasitellus fucorum* (Mesostigmata: Parasitidae): which bumblebee caste is best? *Oecologia* **115**, 385–390.

Huber, P. (1802) Observations on several species of the genus *Apis*, known by the name of humble-bees, and called Bombinatrices by Linnaeus. *Trans. Linn. Soc. Lond.* **6**, 214–298.

Hughes, W.O.H., Howse, P.E., and Goulson, D. (2001) Polyethism and the importance of context in the alarm reaction of *Atta capiguara*. *Behav. Ecol. Sociobiol.* **49**, 503–508.

Hulkkonen, O. (1928) Zur Biologie der Südfinnischen Hummeln. *Ann. Univ. Aboensis (A)* **3**, 1–81

Hunter, P.E. and Husband, R.W. (1973) *Pneumolaelaps* (Acarina: Laelapidae) mites from North America and Greenland. *Florida Entomol.* **56**, 77–91.

Husband, R.W. and Sinha, R.N. (1970) A revision of the genus *Locustacarus* with a key to genera of the family Podapolipidae (Acarina). *Ann. Entomol. Soc. Am.* **63**, 1152–1162.

Imhoof, B. and Schmid-Hempel, P. (1998a) Patterns of local adaptation of a protozoan parasite to its bumblebee host. *Oikos*, **82**, 59–65.

Imhoof, B. and Schmid-Hempel, P. (1998b) Colony success of *Bombus terrestris* and microparasitic infections in the field. *Insectes Soc.* **46**, 223–238.

Imhoof, B. and Schmid-Hempel, P. (1999) Colony success of the bumble bee, *Bombus terrestris*, in relation to infections by two protozoan parasites, *Crithidia bombi* and *Nosema bombi*. *Insectes Soc.* **46**, 233–238.

Ingelög, T. (1988) Floraläget I Sverige. *Svensk. Bot. Tidskr.* **82**, 376–378.

Inglesfield, C. (1989) Pyrethroids and terrestrial non-target organisms. *Pest. Sci.* **27**, 387–428.

Inouye, D.W. (1976) Resource partitioning and community structure: a study of bumblebees in the Colorado Rocky Mountains. Doctoral dissertation. University of North Carolina, Chapel Hill, North Carolina, USA.

Inouye, D.W. (1978) Resource partitioning in bumblebees: experimental studies of foraging behavior. *Ecology* **59**, 672–678.

Inouye, D.W. (1980a) The effects of proboscis and corolla tube lengths on patterns and rates of flower visitation by bumblebees. *Oecologia* **45**, 197–201.

Inouye, D.W. (1980b) The terminology of floral larceny. *Ecology* **61**, 1251–1253.

Inouye, D.W. (1983) The ecology of nectar robbing. In: *The biology of nectaries*, edited by T.S. Elias, and B. Bentley, pp. 152–173. Columbia University Press, New York.

Irwin, R.E. and Brody, A.K. (1999) Nectar-robbing bumble bees reduce the fitness of *Ipomopsis aggregata* (Polemoniaceae). *Ecology* **80**, 1703–1712.

Ito, M. (1985) Supraspecific classification of bumblebees based on the characters of male genitalia. Contributions from the Institute of Low Temperature Science, Hokkaido University (B) **20**, 143 pp.

Ito, M. (1987) Geographic variation of an eastern Asian bumble bee *Bombus diversus* in some morphometric characters (Hymenoptera, Apidae). *Kontyu* **55**, 188–201.

Jacobs-Jessen, U.F. (1959) Zur Orientierung der Hummeln und einiger anderer Hymenopteren. *Z. vergl. Physiol.* **41**, 597–641.

Janzen, D.H. (1971) Euglossine bees as long-distance pollinators of tropical plants. *Science* **171**, 203–205.

Jennersten, O., Berg, L., and Lehman, C. (1988) Phenological differences in pollinator visitation, pollen deposition and seed set in the sticky catchfly *Viscaria vulgaris*. *J. Ecol.* **76**, 1111–1132.

Jennersten, O., Loman, J., Mæller, A.P., Robertson, J., and Widén, B. (1992) Conservation biology in agricultural habitat islands. In: *Ecological principles of nature conservation*, edited by L. Hansson, pp. 394–424. Elsevier Applied Science, London.

Johnson, R.A. (1986) Intraspecific resource partitioning in the bumble bees *Bombus ternarius* and *B. pensylvanicus*. *Ecology* **67**, 133–138.

Johnston, M.O. (1991) Pollen limitation of female reproduction in *Lobelia cardinalis* and *L. siphilitica*. *Ecology* **72**, 1500–1503.

Jones, C.E. (1978) Pollinator constancy as a pre-pollination isolating mechanism between sympatric species of *Cercidium*. *Evolution* **32**, 189–198.

Jones, C.E. and Buchmann, S.L. (1974) Ultraviolet floral patterns as functional orientation cues in hymenopterous pollination systems. *Anim. Behav.* **22**, 481–485.

Kadmon, R. (1992) Dynamics of forager arrivals and nectar renewal in flowers of *Anchusa strigosa*. *Oecologia* **92**, 552–555.

Kadmon, R. and Shmida, A. (1992) Departure rules used by bees foraging for nectar: a field test. *Evol. Ecol.* **6**, 142–151.

Kastberger, G. (1992) The ocellar control of orienting subsystems in homing honeybees, investigated under side-light-switching conditions. *Allg. Zool. Physiol. der Tiere* **96**, 459–479.

Katayama, E. (1973) Observations on the brood development in *Bombus ignitus* (Hymenoptera: Apidae): II. Brood development and feeding habits. *Kontyû* **41**, 203–216.

Katayama, E. (1975) Egg laying habits and brood development in *Bombus hypocrita* (Hymenoptera: Apidae): II. Brood development and feeding habits. *Kontyû* **43**, 478–496.

Katayama, E. (1996) Survivorship curves and longevity for workers of *Bombus ardens* Smith and *Bombus diversus* Smith (Hymenoptera, Apidae). *Jpn. J. Entomol.* **64**, 111–121.

Kato, M. (1988) Bumble bee visits to *Impatiens* spp.: pattern and efficiency. *Oecologia* **76**, 364–370.

Kato, M., Shibata, A., Yasui, T., and Nagamasu, H. (1999) Impact of introduced honeybees, *Apis mellifera*, upon native bee communities in the Bonin (Ogasawara) Islands. *Res. Popul. Ecol.* **2**, 217–228.

Kaule, G. and Krebs, S. (1989) Creating new habitats in intensively used farmland. In: *Biological habitat reconstruction*, edited by G.P. Buckley, pp. 161–170. Belhaven Press, London.

Kay, Q.O.N. (1982) Intraspecific discrimination by pollinators and its role in evolution. In: *Pollination and evolution*, edited by J.A. Armstrong, J.M. Powell, and A.J. Richards, pp. 9–28. Royal Botanical Gardens, Sydney.

Keasar, T., Motro, U., Shur, Y., and Shmida, A. (1996) Overnight memory retention of foraging skills by bumblebees is imperfect. *Anim. Behav.* **52**, 95–104.

Keller, L. and Nonacs, P. (1993) The role of queen pheromones in social insects: queen control or queen signal? *Anim. Behav.* **45**, 787–794.

Kells, A.R. and Goulson, D. (2000) Evidence for handedness in bumblebees. *J. Insect Behav.* **14**, 47–55.

Kells, A.R., Holland, J. and Goulson, D. (2001) The value of uncropped field margins for foraging bumblebees. *J. Insect Conserv.* **5**, 283–291.

Kendall, D.A. and Smith, B.D. (1976) The pollinating efficiency of honeybee and bumblebee visits to flowers of the runner bean *Phaseolus coccineus* L. *J. Appl. Ecol.* **13**, 749–752.

Kerr, W.E. (1969) Some aspects of the evolution of social bees (Apidae). *Evol. Biol.* **3**, 119–175.

Kevan, P.G. (1976) Fluorescent nectar (technical comment). *Science* **194**, 341–342.

Kevan, P.G. (1978) Floral coloration, its colormetric analysis and significance in anthecology. In: *The pollination of flowers by insects*, edited by A.J. Richards, pp. 51–78. Academic Press, London.

Kevan, P.G. (1983) Floral colours through the insect eye: what they are and what they mean. In: *Handbook of experimental pollination ecology*, edited by C.E. Jones and R.J. Little. Van Nostrand Reinhold, New York.

Kevan, P.G. (1991) Pollination: keystone process in sustainable global productivity. *Acta Hortic.* **288**, 103–110.

Kevan, P.G., Clark, E.A., and Thomas, V.G. (1990) Insect pollinators and sustainable agriculture. *Am. J. Altern. Agric.* **5**, 13–22.

Kevan, P.G. and Gadawski, S.D. (1984) Pollination of cranberries, *Vaccinium macrocarpon*, on cultivated marshes in Ontario. *Proc. Entomol. Soc. Ont.* **114**, 45–53.

Kindl, J., Oldrich, H., Urbanová, K., and Valterová, I. (1999) Scent marking in male premating behaviour of *Bombus confusus*. *J. Chem. Ecol.* **25**, 1489–1500.

King, M.J. (1993) Buzz foraging mechanism of bumble bees. *J. Apic. Res.* **32**, 41–49.

Kipp, L.R. (1987) The flight directionality of honeybees foraging on real and artificial inflorescences. *Can. J. Zool.* **65**, 587–593.

Kipp, L.R., Knight, W., and Kipp, E.R. (1989) Influence of resource topography on pollinator flight directionality of two species of bees. *J. Insect. Behav.* **2**, 453–472.

Kleijn, D., Joenje, W., Lecoeur, D., and Marshall, E.J.P. (1998) Similarities in vegetation development of newly established herbaceous strips along contrasting European field boundaries. *Agric. Ecosys. Environ.* **68**, 13–26.

Klinkhamer, P.G.L. and de Jong, T.L. (1990) Effects of plant density and sex differential reward visitation in the protandrous *Echium vulgare* (Boraginaceae). *Oikos*, **57**, 399–405.

Klinkhamer, P.G.L., de Jong, T.L., and de Bruyn, G.J. (1989) Plant size and pollinator visitation in *Cynoglossum officinale*. *Oikos* **54**, 201–204.

Klosterhalfen, S., Fischer, W., and Bitterman, M.E. (1978) Modification of attention in honey bees. *Science* **201**, 1241–1243.

Knee, W.J. and Medler, J.T. (1965) The seasonal size increase of bumblebee workers (Hymenoptera: *Bombus*). *Can. Entomol.* **97**, 1149–1155.

Knudsen, J.T. (1994) Floral scent variation in the *Pyrola rotundifolia* complex in Scandinavia and western Greenland. *Nordic J. Bot.* **14**, 277–282.

Koeman-Kwak, M. (1973) The pollination of *Pedicularis palustris* by nectar thieves (short-tongued bumblebees). *Acta Bot. Neerl.* **22**, 608–615.

Koltermann, R. (1969) Lern-und Vergessensprozesse bei der Honigbiene aufgezeigt anhand von Duftdressuren. *Z. vergl. Physiol.* **63**, 310–334.

Komai, Y. (2001) Direct measurement of oxygen partial pressure in a flying bumblebee. *J. Exp. Biol.* **204**, 2999–3007.

König, C. and Schmid-Hempel, P. (1995) Foraging activity and immunocompetence in workers of the bumble bee, *Bombus terrestris* L. *Proc. Royal. Soc. Lond. B* **260**, 225–227.

Kosior, A. (1995) Changes in the fauna of bumble-bees (*Bombus* Latr.) and cuckoo-bees (*Psithyrus* Lep.) of selected regions in southern Poland. In: *Changes in fauna of wild bees in Europe*, edited by J. Banaszak, pp. 103–111. Bydgoszcz, Pedagogical University.

Koulianos, S. (1999) Phylogenetic relationships of the bumblebee subgenus *Pyrobombus* (Hymenoptera: Apidae) inferred from mitochondrial cytochrome B and cytochrome oxidase I sequences. *Ann. Entomol. Soc. Am.* **92**, 355–358.

Koulianos, S. and Schmid-Hempel, P. (2000) Phylogenetic relationships among bumble bees (*Bombus*, Latreille) inferred from mitochondrial cytochrome b and cytochrome oxidase I sequences. *Mol. Phylogen. Evol.* **14**, 335–341.

Krüger, E. (1917) Zur Systematik der mitteleuropäischen Hummeln (Hym.). *Entomol. Mitt.* **6**, 55–66.
Krüger, E. (1920) Beiträge zur Systematik und Morphologie der mitteleuropäischen Hummeln. *Zool. Jb. Abt. Syst.* **42**, 289–464.
Krüger, E. (1951) Über die Bahnflüge der Männchen der Gattungen *Bombus* und *Psithyrus*. *Z. Tierpsychol.* **8**, 61–75.
Kugler, H. (1936) Die Ausnutzung der Saftmalsumfärbung bei den Roszkastanienblüten durch Bienen und Hummeln. *Ber. Deutc. Bot. Gesell.* **60**, 128–134.
Kugler, H. (1940) Die Bestaubung von Blumen durch Furchenbienen (*Halictus* Latr.). *Planta* **30**, 780–799.
Kugler, H. (1943) Hummeln als Blütenbesucher. *Ergeb. Biol.* **19**, 143–323.
Kugler, H. (1950) Der Blütenbesuch der Schammfliege (*Eristalomyia tenax*). *Zeitschrift für Vergleichende Physiologie*, **32**, 328–347.
Kukuk, P.F. and May, B. (1990) Diploid males in a primitely eusocial bee, *Lasioglossum (Dialictus(zephyrum* (Hymenoptera: Halictidae). *Evolution* **44**, 1522–1528.
Kullenberg, B. (1956) Field experiments with chemical sexual attractants in Aculeate Hymenoptera males I. *Zool. Bidr. Uppsala* **31**, 253–354.
Kullenberg, B. (1973) Field experiments with chemical sexual attractants in Aculeate Hymenoptera males II. *Zoon Suppl.* **1**, 31–42.
Kullenberg, B., Bergström, G., Bringer, B., Carlberg, B., and Cederberg, B. (1973) Observations on the scent marking by *Bombus* Latr. and *Psithyrus* Lep. males (Hym., Apidae) and localization of site of production of the secretion. *Zoon Suppl.* **1**, 23–30.
Kullenberg, B., Bergström, G., and Ställberg-Stenhagen, S. (1970) Volatile components of the cephalic marking secretion of male bumblebees. *Acta Chem. Scand.* **24**, 1481–1483.
Kunin, W.E. (1993) Sex and the single mustard: population density and pollinator behaviour effects on seed set. *Ecology* **74**, 2145–2160.
Kunin, W.E. (1997) Population size and density effects in pollination: Pollinator foraging and plant reproductive success in experimental arrays of *Brassica kaber*. *J. Ecol.* **85**, 225–234.
Kunin, W.E. and Iwasa, Y. (1996) Pollinator foraging strategies in mixed floral arrays—density effects and floral constancy. *Theor. Popul. Biol.* **49**, 232–263.
Kwak, M.M. (1988) Pollination ecology and seed-set in the rare annual species *Melampyrum arvense* L. (Scrophulariaceae). *Acta Bot. Neerl.* **37**, 153–163.
Kwak, M.M., Kremer, P., Boerrichter, E., and van den Brand, C. (1991a) Pollination of the rare species *Phyteuma nigrum* (Campanulaceae): Flight distances of bumblebees. *Proc. Experi. Appl. Entomol. N.E.V. Amsterdam* **2**, 131–136.
Kwak, M.M., van den Brand, C., Kremer, P., and Boerrichter, E. (1991b) Visitation, flight distances and seed set in the rare species, *Phyteuma nigrum* (Campanulaceae). *Acta Hortic.* **288**, 303–307.
Kwak, M.M., Velterop, O., and Boerrichter, E. (1996) Insect diversity and the pollination of rare plant species. In: *The conservation of bees*, edited by A. Matheson, S.L. Buchmann, C. O'Toole, P. Westrich, and I.H. Williams, pp. 115–124. Academic Press, London.
Langley, C.M. (1996) Search images: selective attention to specific visual features of prey. *J. Exp. Psychol.* **22**, 152–163.
Lanne, B.S., Bergström, G., Wassgren, A.-B., and Törnbäck, B. (1987) Biogenetic pattern of straight chain marking compounds in male bumble bees. *Comp. Biochem. Physiol. B. Biochem.* **88**, 631–636.
Laverty, T.M. (1980) Bumble bee foraging: floral complexity and learning. *Can. J. Zool.* **58**, 1324–1335.
Laverty, T.M. (1992) Plant interactions for pollinator visits: a test of the magnet species effect. *Oecologia* **89**, 502–508.

Laverty, T.M. (1994a) Costs to foraging bumblebees of switching plant species. *Can. J. Zool.* **72**, 43–47.

Laverty, T.M. (1994b) Bumblebee learning and flower morphology. *Anim. Behav.* **47**, 531–545.

Laverty, T.M. and Plowright, R.C. (1988) Flower handling by bumblebees a comparison of specialists and generalists. *Anim. Behav.* **36**, 733–740.

Lawrence, E.S. and Allen, J.A. (1983) On the term 'search image'. *Oikos* **40**, 313–314.

Leatherdale, D. (1970) The arthropod hosts of entomogenous fungi in Britain. *Entomophaga* **15**, 419–435.

Levin, D.A. (1978) Pollinator behaviour and the breeding structure of plant populations. In: *The pollination of flowers by insects*, edited by A.J. Richards, pp. 133–150. Academic Press, London.

Levin, M.D. (1983) Value of bee pollination to US agriculture. *Bull. Entomol. Soc. Am.* **29**, 50–51.

Levins, R. and MacArthur, R. (1969) An hypothesis to explain the incidence of monophagy. *Ecology* **50**, 910–911.

Lewis, A.C. (1986) Memory constraints and flower choice in *Pieris rapae*. *Science* **232**, 863–865.

Lewis, A.C. (1989) Flower visit consistency in *Pieris rapae*, the cabbage butterfly. *J. Anim. Ecol.* **58**, 1–13.

Lewis, A.C. (1993) Learning and the evolution of resources: pollinators and flower morphology. In: *Insect learning: ecology and evolutionary perspectives*, edited by D.R. Papaj and A.C. Lewis, pp. 219–242. Chapman and Hall, New York.

Lex, T. (1954) Duftmale an Bluten. *Zeitschrift für Vergleichende Physiologie* **36**, 212–234.

Lie-Pettersen, O.J. (1901) Biologische Beobachtungen an norwegischen Hummeln. *Bergens Mus. Aarb.* **6**, 3–10.

Liersch, S. and Schmid-Hempel, P. (1998) Genetic variation within social insect colonies reduces parasite load. *Proc. Royal. Soc. Lond. B* **265**, 221–225.

Lilley, J.H., Cerenius, L., and Soderhall, K. (1997) RAPD evidence for the origin of crayfish plague outbreaks in Britain. *Aquaculture* **157**, 181–185.

Linhard, E. (1912) Humlebien som Husdyr. Spredte Traek af nogle danske Humlebiarters Biologi. *Tidsskr. PlAvl* **19**, 335–352.

Lloyd, J.E. (1981) Sexual selection: individuality, identification, and recognition in a bumblebee and other insects. *Florida Entomol.* **64**, 89–111.

Lockey, K.H. (1980) Insect cuticular hydrocarbons. *Comp. Biochem. Physiol.* **65B**, 457–462.

Løken, A. (1949) Bumble bees in relation to *Aconitum septentrionale* in central Norway. *Nytt Magasin for naturvidenskapene* **87**, 1–60.

Løken, S. (1973) Studies on Scandinavian bumble bees (Hymenoptera, Apidae). *Norsk Entomologisk Tidsskrift* **20**, 1–28.

Lovell, J.H. (1918) *The flower and the bee*. Scribner's, New York.

Low, T. (1999) *Feral future*. Penguin Books Australia Ltd, Ringwood, Australia.

Ludwig, F. (1885) Die biologische Bedeutung des Farbenwechsels mancher Blumen. *Biolog. Centr.* **4**, 196.

Ludwig, F. (1887) Einige neue Falle von Farbenwechsel in verbluhenden Blüthenstanden. *Biolog. Centr.* **6**, 1–3.

Lunau, K. (1990) Colour saturation triggers innate reactions to flower signals: flower dummy experiments with bumblebees. *J. Comp. Physiol. A* **166**, 827–834.

Lunau, K., Wacht, S., and Chittka, L. (1996) Colour choice of naïve bumble bees and their implications for colour perception. *J. Comp. Physiol.* **178**, 477–489.

Lundberg, H. and Svensson, B.G. (1975) Studies on the behaviour of *Bombus* Latr. species (Hymenoptera: Apidae) parasitized by *Sphaerularia bombi* Dufour (Nematoda) in an alpine area. *Norw. J. Entomol.* **22**, 129–134.

MacDonald, M.A. (2001) The colonisation of Northern Scotland by *Bombus terrestris* (L.) and *B. lapidarius* (L.) (Hym., Apidae), and the possible role of climate change. *Entomologist's Monthly Magazine* **137**, 1–13.

Macfarlane, R.P. and Griffin, R.P. (1990) New Zealand distribution and seasonal incidence of the nematode *Sphaerularia bombi* Dufour, a parasite of bumblebees. *N. Z. J. Zool.* **17**, 191–199.

Macfarlane, R.P., Griffin, R.P., and Read, P.E.C. (1983) Bumble bee management options to improve 'grasslands pawera' red clover seed yields. *Proc. N. Z. Grasslands Assoc.* **44**, 47–53.

Macfarlane, R.P. and Gurr, L. (1995) Distribution of bumble bees in New Zealand. *New Zealand Entomol.* **18**, 29–36.

Macfarlane, R.P. and Pengelly, D.H. (1974) Conopidae and Sarcophidae (Diptera) as parasites of adult Bombinae (Hymenoptera) in Ontario. *Proc. Entomol. Soc. Ont.* **105**, 55–59.

Macior, L.W. (1966) Foraging behavior of *Bombus* (Hymenoptera: Apidae) in relation to *Aquilegia* pollination. *Am. J. Bot.* **53**, 302–309.

Macior, L.W. (1968) *Bombus* (Hym. Apid.) queen foraging in relation to vernal pollination in Wisconsin. *Ecology* **49**, 20–25.

Mackensen, O. (1951) Viability and sex determination in the honey bee (*Apis mellifera* L.). *Genetics* **36**, 500–509.

MacKenzie, K.E. (1994) The foraging behaviour of oney bees (*Apis mellifera* L) and bumble bees (*Bombus* spp) on cranberry (*Vaccinium macrocarpon* Ait). *Apidologie* **25**, 375–383.

Macuda, T., Gegear, R.J., Laverty, T.M., and Timney, B. (2001) Behavioural assessment of visual acuity in bumblebees (*Bombus impatiens*). *J. Exp. Biol.* **204**, 559–564.

MAFF (1998) *The countryside stewardship scheme*. MAFF Publications, London.

MAFF (1999) *Arable stewardship*. MAFF Publications, London.

Mal, T.K., Lovett-Doust, J., Lovett-Doust, L., and Mulligan, G.A. (1992) The biology of Canadian weeds. 100. *Lythrum salicaria*. *Can. J. Plant Sci.* **72**, 1305–1330.

Mallet, J. (1999) Causes and consequenceas of a lack of coevolution in Müllerian mimicry. *Evol. Ecol.* **13**, 777–806.

Maloof, J.E. (2000) The ecological effect of nectar robbers, with an emphasis on the reproductive biology of *Corydalis caseana*. Dissertation, University of Maryland, College Park, Maryland, USA.

Maloof, J.E. (2001) The effects of a bumble bee nectar robber on plant reproductive success and pollinator behavior. *Am. J. Bot.* **88**, 1960–1965.

Maloof, J.E. and Inouye, D.W. (2000) Are nectar robbers cheaters or mutualists? *Ecology* **81**, 2651–2661.

Mänd, M., Mänd, R., and Williams, I.H. (2002) Bumblebees in the agricultural landscape of Estonia. *Agric. Ecosys. Environ.* **89**, 69–76.

Mangel, M. (1990) Dynamic information in uncertain and changing worlds. *J. Theor. Biol.* **146**, 317–332.

Mangum, W.A. and Brooks, R.W. (1997) First records of *Megachile (Callomegachile) sculpturalis* Smith (Hymenoptera : Megachilidae) in the continental United States. *J. Kans. Entomol. Soc.* **70**, 140–142.

Manning, A. (1956) Some aspects of the foraging behaviour of bumblebees. *Behaviour* **9**, 164–201.

Marden, J.H. (1984) Intrapopulation variation in nectar secretion in *Impatiens capensis*. *Oecologia* **63**, 418–422.

Markow, T.A. (1995) Evolutionary ecology and developmental instability. *Annu. Rev. Entomol.* **40**, 105–120.

Marloth, R. (1895) The fertilisation of '*Disa uniflora*' Berg., by insects. *Trans. S. Afr. Philosoph. Soc.* **8**, 93-95.

Marshall, E.J.P. and Smith, B.D. 1987. Field margin flora and fauna; interactions with agriculture. In: *Field margins,* edited by J.M. Way, and P.W. Greig-Smith, Monograph No. 35: pp. 22-33. British Crop Protection Council, Thornton Heath, UK.

Marshall, E.J.P., Thomas, C.F.G., Joenje, W., Kleijn, D., Burel, F., and Lecoeur, D. (1994) Establishing vegetation strips in contrasted European Farm situations. In: *British Crop Protection Monograph No. 58. Field margins: integrating agriculture and conservation,* edited by N.D. Boatman pp. 335-340. Farnham: British Crop Protection Council.

Martin, E.C. (1975) *The use of bees for crop pollination in Dadant and Sons. The hive and the honeybee,* Dadant and Sons, Hamilton, Illinois.

Marucci, P.E. and Moulter, H.J. (1977) Cranberry pollination in New Jersey. *Acta Hortic.* **61**, 217-222.

Matthews, E. (1984) *To Bee or Not? Bees in National Parks—The Introduced Honeybee in Conservation Parks in South Australia,* pp. 9-14. Magazine of the South Australia National Parks Association, Adelaide.

McDade, L.A. and Kinsman, S. (1980) The impact of floral parasitism in two neotropical hummingbird-pollinated plant species. *Evolution* **34**, 944-958.

McFarlane, R.P. (1976) Bees and Pollination. In: *New Zealand insect pests,* edited by D.N. Ferro pp. 221-229. Lincoln University College of Agriculture, New Zealand.

McFarlane, R.P., Grundell, J.M., and Dugdale, J.S. (1992) Gorse on the Chatham Islands: seed formation, arthropod associates and control. Proceedings of the 45th New Zealand Plant Protection Conference: 251-255.

McGregor, S.E. (1976) Insect Pollination of Cultivated Crops. USDA Agriculture Handbook No. 496, US Government Printing Office, Washington, DC.

McGregor, S.E., Alcorn, E.B., Kuitz, E.B., Jr., and Butler, G.D., Jr. (1959) Bee visitors to Saguaro flowers. *J. Econom. Entomol.* **52**, 1002-1004.

Medler, J.T. (1962a) Morphometric studies on bumblebees. *Ann. Entomol. Soc. Am.* **55**, 212-218.

Medler, J.T. (1962b) Morphometric analyses of bumblebee mouthparts. *Trans. XI Intern. Congr. Entomol.* **2**, 517-521.

Meidell, O. (1934) Fra dagliglivet I et homlebol. *Naturen* **58**, 85-95; 108-116.

Meidell, O. (1944) Notes of the pollination of *Malampyrum pratense* and the 'honey stealing' of bumblebees. *Bergens Museum Arbok* **11**, 5-11.

Meijere, J.C.H. de (1904) Beiträge zur Kenntnis der Biologie und der systematischen Verwandtschaft der Conopiden. *Tijdschr. Entomol.* **46**, 144-224.

Meijere, J.C.H. de (1912) Neue Beiträge zur Kenntnis der Conopiden. *Tijdschr. Entomol.* **55**, 184-207.

Meineke, H. (1978) Umlernen einer Honigbiene zwischen gelbund blau-Belohnung im Dauerversuch. *J. Insect Physiol.* **24**, 155-163.

Meisels, S. and Chiasson, H. (1997) Effectiveness of *Bombus impatiens* Cr. As pollinators of greenhouse sweet peppers (*Capsicum annuum* L.). *Acta Hortic.* **437**, 425-429.

Menezes Pedro, S.R. and Camargo, J.M.F. (1991) Interactions on floral resources between the Africanized honey bee *Apis mellifera* L and the native bee community (Hymenoptera: Apoidea) in a natural 'cerrado' ecosystem in southeast Brazil. *Apidologie* **22**, 397-415.

Menke, H.F. (1954) Insect pollination in relation to alfalfa seed production in Washington. *Wash. Agric. Exp. Stat. Bull.* **555**, 24 pp.

Menzel, R. (1967) Untersuchungen zum Erlernen von Spektralfarben durch die Honigbiene (*Apis mellifica*). *Z. vgl. Physiol.* **56**, 22-62.

Menzel, R. (1969) Das Gedächtnis der Honigbiene für Spaktralfarben. II. Umlernen und Mehrfachlernen. *Z. V. Physiol.* **63**, 290-309.

Menzel, R. (1979) Behavioral access to short-term memory in bees. *Nature* **281**, 368–369.

Menzel, R. (1990) Learning, memory and 'cognition' in honey bees. In: *Neurobiology of comparative cognition*, edited by R.P. Kesner and D.S. Olten, pp. 237–292. Erlbaum Inc., Hillsdale, NJ.

Menzel, R. (1999) Memory dynamics in the honeybee. *J. Comp. Physiol. A* **185**, 323–340.

Menzel, R. and Erber, I. (1978) Learning and memory in bees. *Sci. Am.* **239**, 102–110.

Menzel, R., Geiger, K., Chittka, L., Joerges, J., Kunze, J., and Muller, U. (1996) The knowledge base of bee navigation. *J. Exp. Biol.* **199**, 141–146.

Menzel, R., Geiger, K., Joerges, J., Muller, U., and Chittka, L. (1998) Bees travel novel homeward routes by integrating separately acquired vector memories. *Anim. Behav.* **55**, 139–152.

Menzel, R., Greggers, U., and Hammer, M. (1993) Functional organization of appetitive learning and memory in a generalist pollinator, the honey bee. In: *Insect learning: ecological and evolutionary perspectives*, edited by D. Papaj and A.C. Lewis, pp. 79–125. Chapman and Hall, New York.

Menzel, R., Gumbert, A., Kunze, J., Shmida, A., and Vorobyev, M. (1997) Pollinators' strategies in finding flowers. *Israel J. Plant Sci.* **45**, 141–156.

Menzel, R., and Müller, U. (1996) Learning and memory in honeybees: from behaviour to neural substrates. *Ann. Rev. Neurosci.* **19**, 379–404.

Michener, C.D. (1965) A classification of the bees of the Australian and South Pacific regions. *Bull. Am. Mus. Nat. Hist.* **130**, 1–324.

Michener, C.D. (1974) *The Social behavior of the bees: A comparative study*, Second Edition, pp. 404 Harvard University Press (Belknap Press), Cambridge (MA).

Michener, C.D. (1979) Biogeography of bees. *Annu. Missouri Bot. Garden* **66**, 277–347.

Michener, C.D. (1990) Classification of the Apidae (Hymenoptera). *Univ. Kans. Sci. Bull.* **54**, 75–164.

Michener, C.D. and Amir, M. (1977) The seasonal cycle and habitat of a tropical bumble bee. *Pacific Insects* **17**, 75–164.

Michener, C.D. and Grimaldi, D.A. (1988) The oldest fossil bee: apoid history, evolutionary stasis, and antiquity of social behavior. *Proc. Natl. Acad. Sci. USA* **85**, 6424–6426.

Michener, C.D. and Laberge, W.E. (1954) A large *Bombus* nest from Mexico. *Psyche* **61**, 63–67.

Mikkola, K. (1978) Spring migrations of wasps and bumblebees on the southern coast of Finland (Hymenoptera, Vespidae and Apidae). *Ann. Entomol. Fenni.* **44**, 10–26.

Mikkola, K. (1984) Migration of wasp and bumble bee queens across the Gulf of Finland (Hymenoptera: Vespidae and Apidae). *Notulae Entomol.* **64**, 125–128.

Milliron, H.E. (1971) A monograph of the western hemisphere bumblebees (Hymenoptera: Apidae; Bombinae). I. The genera *Bombus* and *Megabombus* subgenus *Bombias*. *Memoirs of the Entomological Society of Canada, No. 82*.

Milliron, H.E. and Oliver, D.R. (1966) Bumblebees from northern Ellesmere Island, with observations on usurpation by *Megabombus hyperboreus* (Schöhn). *Can. Entomol.* **98**, 207–213.

Minckley, R.L., Buchmann, S.L., and Wcislo, W.T. (1991) Bioassay evidence for a sex attractant pheromone in the large carpenter bee, *Xylocopa varipuncta* (Anthophoridae, Hymenoptera). *J. Zool.* **244**, 285–291.

Miyamoto, S. (1957) Behavior study on *Bombus ardens* Smith in early stage of nesting. *Sci. Rep. Hyogo Univ. Agric. 3 (Ser. Agric. Biol.)* 1–5.

Moret, Y. and Schmid-Hempel, P. (2000) Survival for immunity: the price of immune system activation for bumblebee workers. *Science* **290**, 1166–1168.

Moritz, R.F.A. (1985) The effects of multiple mating on the worker-queen conflict in *Apis mellifera* L. *Behav. Ecol. Sociobiol.* **16**, 375–377.

Moritz, R.F.A., Kryger, P., Koeniger, G., Koeniger, N., Estoup, A., and Tingek, S. (1995) High degree of polyandry in *Apis dorsata* queens detected by DNA microsatellite variability. *Behav. Ecol. Sociobiol.* **37**, 357–363.

Morris, W.F. (1996) Mutualism denied – nectar-robbing bumble bees do not reduce female or male success of bluebells. *Ecology* **77**, 1451–1462.

Morse, D.H. (1977) Resource partitioning in bumble bees: the role of behavioral factors. *Science* **197**, 678–680.

Morse, D.H. (1978a) Foraging rate, foraging position and worker size in Bumble-Bee workers. *Proc. IV Intern. Symp. Pollination. Md. Agric. Exp. Sta. Spec. Misc. Publ.* **1**, 447–452.

Morse, D.H. (1978b) Size-related foraging differences of bumble bee workers. *Ecol. Entomol.* **3**, 189–192.

Morse, D.H. (1986a) Inflorescence choice and time allocation by insects foraging on milkweed. *Oikos* **46**, 229–236.

Morse, D.H. (1986b) Predatory risk to insects foraging at flowers. *Oikos* **46**, 223–228.

Mohr, N.A. and Kevan, P.G. (1987) Pollinators and pollination requirements of lowbush blueberry (*Vaccinium angustifolium* Ait. and *V. myrtilloides* Michx.) and cranberry (*V. macrocarpon* Ait.) in Ontario with notes on highbush blueberry (*V. corymbosum* L.) and ligonberry (*V. vitisideae* L.). *Proc. Entomol. Soc. Ont.* **118**, 149–154.

Møller, A.P. (1993) Developmental stability, sexual selection, and the evolution of secondary sexual characters. *Etologia* **3**, 199–208.

Møller, A.P. (1995) Bumblebee preference for symmetrical flowers. *Proc. Nat. Acad. Sci.* **92**, 2288–2292.

Møller, A.P. and Eriksson, M. (1995) Pollinator preference for symmetrical flowers and sexual selection in plants. *Oikos* **73**, 15–22.

Møller, A.P. and Pomiankowski, A. (1993) Fluctuating asymmetry and sexual selection. *Genetica* **89**, 267–279.

Møller, A.P. and Thornhill, R. (1998) Bilateral symmetry and sexual selection: A meta-analysis. *Am. Nat.* **151**, 174–192.

Møller, A.P. and Sorci, G. (1998) Insect preference for symmetrical artificial flowers. *Oecologia* **114**, 37–42.

Morandin, L.A., Laverty, T.M., and Kevan, P.G. (2001) Bumble bee (Hymenoptera: Apidae) activity and pollination levels on commercial tomato greenhouses. *J. Econ. Entomol.* **94**, 462–467.

Morse, D.H. (1981) Interactions among syrphid flies and bumblebees on flowers. *Ecology* **62**, 81–88.

De Los Mozos Pascual, M., and Domingo, L.M. (1991) Flower constancy in *Heliotaurus ruficollis* (Fabricius 1781, Coleoptera: Alleculidae). *Elytron (Barc)* **5**, 9–12.

Muir, R. and Muir, N. (1987) *Hedgerows: Their history and wildlife*, p. 250. Michael Joseph, London.

Müller, C.B. (1994) Parasite-induced digging behaviour in bumble bee workers. *Anim. Behav.* **48**, 961–966.

Müller, C.B. and Schmid-Hempel, P. (1992a) Variation in life history pattern in relation to worker mortality in the bumblebee *Bombus lucorum. Funct. Ecol.* **6**, 48–56.

Müller, C.B. and Schmid-Hempel, P. (1992b) Correlates of reproductive success among field colonies of *Bombus lucorum*: the importance of growth and parasites. *Ecol. Entomol.* **17**, 343–353.

Müller, C.B. and Schmid-Hempel, P. (1993a) Exploitation of cold temperature as defense against parasitoids in bumblebees. *Nature* **363**, 65–67.

Müller, C.B. and Schmid-Hempel, P. (1993b) Correlates of reproductive success among field colonies of *Bombus lucorum* L.: The importance of growth and parasites. *Ecol. Entomol.* **17**, 343–353.

Müller, C.B., Shykoff, J.A., and Sutcliffe, G.H. (1992) Life history patterns and opportunities for queen-worker conflict in bumblebees (Hymenoptera: Apidae). *Oikos* **65**, 242–248.

Müller, H. (1883) The effect of the change of colour in the flowers of *Pulmonaria officinalis* upon its fertilizers. *Nature* **28**, 81.

Neal, P.R., Dafni, A., and Giurfa, M. (1998) Floral symmetry and its role in plant-pollinator systems: Terminology, distribution, and hypotheses. *Annu. Rev. Ecol. Syst.* **29**, 345–373.

Ne'eman, G., Dafni, A., and Potts, S.G. (2000) The effect of fire on flower visitation and fruit set in four core-species in east Mediterranean scrubland. *Plant Ecol.* **146**, 157–104.

Newman, H.W. (1851) Habits of the Bombinatrices. *Proc. Entomol. Soc. Lond.* 1851; 86–92.

Newsholme, E.A., Crabtree, B., Higgins, S.J., Thornton, S.D., and Start, C. (1972) The activities of fructose diphosphatase in flight muscles from the bumble-bee and the role of this enzyme in heat generation. *Biochem. J.* **128**, 89–97.

Newton, S.D. and Hill, G.D. (1983) Robbing of field bean flowers by the short-tongued bumble bee *Bombus terrestris* L. *J. Apic. Res.* **22**, 124–129.

Nielsen, D.M., Visker, K.E., Cunningham, M.J., Keller, R.W., Glick, S.D., and Carlson, J.N. (1997) Paw preference, rotation, and dopamine function in collins HI and LO mouse strains. *Physiol. Behav.* **61**, 525–535.

Nilsson, L.A. (1992) Orchid pollination biology. *Trends. Ecol. Evol.* **7**, 255–259.

Núñez, J.A. (1967) Sammelbienen markieren versiegte Futterquellen durch Duft. *Naturwissenschaften* **54**, 322–323.

Oberrath, R., Zanke, C., and Bohninggaese, K. (1995) Triggering and ecological significance of floral color-change in lungwort (*Pulmonaria* spec). *Flora* **190**, 155–159.

O'Donnell, S. and Foster, R.L. (2001) Thresholds of response in nest thermoregulation by worker bumble bees, *Bombus bifarius nearcticus* (Hymenoptera: Apidae). *Ethology* **107**, 387–399.

O'Donnell, S. and Jeanne, R.L. (1992) Forager success increases with experience in *Polybia occidentalis* Olivier (Hymenoptera: Vespidae). *Insectes Soc.* **39**, 451–454.

O'Donnell, S. and Jeanne, R.L. (1995) Implications of senescence patterns for the evolution of age polyethism in eusocial insects. *Behav. Ecol.* **6**, 269–273.

O'Donnell, S., Reichardt, M., and Foster, R. (2000) Individual and colony factors in bumble bee division of labor (*Bombus bifarius nearcticus* Handl; Hymenoptera, Apidae). *Insectes Soc.* **47**, 164–170.

Ohara, M. and Higashi, S. (1994) Effects of inflorescence size on visits from pollinators and seed set of *Corydalis ambigua* (Papaveraceae). *Oecologia* **98**, 25–30.

Ohashi, K. and Yahara, T. (1998) Effects of variation in flower number on pollinator visits in *Cirsium purpuratum* (Asteraceae). *Am. J. Bot.* **85**, 219–224.

Ohashi, K. and Yahara, T. (1999) How long to stay on, and how often to visit a flowering plant?—a model for foraging strategy when floral displays vary in size. *Oikos* **86**, 386–392.

Oldham, N.J., Billen, J., and Morgan, E.D. (1994) On the similarity of Dufour gland secretion and their cuticular hydrocarbons of some bumblebees. *Physiol. Entomol.* **19**, 115–123.

Oldroyd, B.P. (1998) Controlling feral honey bee, *Apis mellifera* L. (Hymenoptera: Apidae), populations in Australia: Methodologies and costs. *Aust. J. Entomol.* **37**, 97–100.

Oldroyd, B.P., Smolenski, A.J., Cornuet, J.M., Wongsiri, S., Estoup, A., Rinderer, T.E., and Crozier, R.H. (1995) Levels of polyandry and intracolonial genetic relationships in *Apis florea*. *Behav. Ecol. Sociobiol.* **37**, 329–335.

Olesen, J.M. and Knudsen, J.T. (1994) Scent profiles of flower color morphs of *Corydalis cava* (Fumariaceae) in relation to foraging behavior of bumblebee queens (*Bombus terrestris*). *Biochem. System. Ecol.* **22**, 231–237.

Olmstead, A.L. and Wooten, D.B. (1987) Bee pollination and productivity growth: the case of alfalfa. *Am. J. Agric. Econ.* 56–63.

O'Neill, K.M., Evans, H.E., and Bjostad, L.B. (1991) Territorial behaviour in males of three North American species of bumblebees (Hymenoptera: Apidae, *Bombus*). *Can. J. Zool.* **69**, 604–613.

O'Neal, R.J. and Waller, G.D. (1984) On the pollen harvest by the honey bee (*Apis mellifera* L.) near Tucson, Arizona (1976–1981). *Desert Plants* **6**, 81–94.

Oostermeijer, J.G.B., den Nijs, J.C.M., Raijmann, L.E.L., and Menken, S.B.J. (1992) Population biology and management of the marsh gentian (*Gentiana pneumoanthe* L.), a rare species in The Netherlands. *Bot. J. Linn. Soc.* **108**, 117–130.

Osborne, J.L. (1994) Evaluating a pollination system: Borago officinalis and bees. PhD thesis, Cambridge University, U.K.

Osborne, J.L., Clark, S.J., Morris, R.J., Williams, I.H., Riley, J.R., Smith, A.D., Reynolds, D.R., and Edwards, A.S. (1999) A landscape study of bumble bee foraging range and constancy, using harmonic radar. *J. Appl. Ecol.* **36**, 519–533.

Osborne, J.L. and Corbet, S.A. (1994) Managing habitats for pollinators in farmland. *Aspects Appl. Biol.* **40**, 207–215.

Osborne, J.L. and Williams, I.H. (1996) Bumble bees as pollinators of crops and wild flowers. Bumble bees for pleasure and profit, edited by A. Matheson, pp. 24–32. IBRA, Cardiff.

Osborne, J.L. and Williams, I.H. (2001) Site constancy of bumble bees in an experimentally patchy habitat. *Agric. Ecosys. Environ.* **83**, 129–141.

Osborne, J.L., Williams, I.H. Carreck, N.L., Poppy, G.M., Riley, J.R., Smith, A.D., Reynolds, D.R., and Edwards, A.S. (1997) Harmonic radar: a new technique for investigating bumble bee and honey bee foraging flight. *Acta Hortic.* **437**, 159–163.

Osborne, J.L., Williams, I.H., and Corbet, S.A. (1991) Bees, pollination and habitat change in the European community. *Bee World* **72**, 99–116.

Oster, G. and Heinrich, B. (1976) Why do bumblebees major? A mathematical model. *Ecol. Monogr.* **46**, 129–133.

Oster, G.F. and Wilson, E.O. (1978) *Caste and ecology in the social insects*. Princeton University Press, Princeton, NJ.

O'Toole, C. (1994) Who cares for solitary bees? In: *Forage for Bees in an agricultural landscape*, edited by A. Matheson, pp. 47–56. IBRA, Cardiff.

O'Toole, C. and Raw, A. (1991) *Bees of the world*. Blandford, London.

Ott, J.R., Real, L.A., and Silverfine, E.M. (1985) The effect of nectar variance on bumblebee patterns of movement and potential gene dispersal. *Oikos* **45**, 333–340.

Owen, J.H. (1948) The larder of the red-backed shrike. *Brit. Birds* **41**, 200–203.

Owen, R.E. (1988) Body size variation and optimal bofy size of bumble bee queens (Hymenoptera: Apidae). *Can. Entomol.* **120**, 19–27.

Owen, R.E., Mydynski, L.J., Packer, L., and McCorquodale, D.B. (1992) Allozyme variation in bumble bees (Hymenoptera: Apidae). *Biochem. Gen.* **30**, 443–453.

Owen, R.E. and Plowright, R.C. (1982) Worker-queen conflict and male parentage in bumble bees. *Behav. Ecol. Sociobiol.* **11**, 91–99.

Owen, R.E., Rodd, F.H., and Plowright, R.C. (1980) Sex ratio in bumble bee colonies: Complications due to orphaning? *Behav. Ecol. Sociobiol.* **7**, 287–291.

Palm, N.B. (1949) The pharyngeal gland in *Bombus* Latr. And *Psithyrus* Lep. *Opusc. Entomol.* **14**, 27–47.

Palmer, A.R. (1994) Fluctuating asymmetry analyses: a primer. *Developmental instability: its origins and evolutionary implications*, edited by T.A. Markow, pp. 335–364. Kluwer, Dordrecht, The Netherlands.

Palmer, A.R. (1996) Waltzing with asymmetry. *Bioscience* **46**, 518–532.

Palmer, T.P. (1968) Establishment of bumble bees in nest boxes at Christchurch. *N. Z. J. Agric. Res.* **11**, 737–739.

Palmer-Jones, T., Forster, I.W., and Clinch, P.G. (1966) Observations of the pollination of Montgomery red clover (*Trifolium pratense* L.). *N. Z. J. Agric. Res.* **9**, 738–747.

Pamilo, P. (1991) Evolution of colony characteristics in social insects. 2. Number of reproductive individuals. *Am. Nat.* **137**, 83–107.

Pamilo, P., Pekkarinen, A., and Varvio, S.-L. (1987) Clustering of bumblebee subgenera based on interspecific genetic relationships (Hymenoptera: Apidae: *Bombus* and *Psithyrus*). *Ann. Zool. Fenn.* **24**, 19–27.

Pamilo, P., Varvio, S.-L., and Pekkarinen, A. (1984) Genetic variation in bumblebees (*Bombus*, *Psithyrus*) and putative sibling species of *Bombus lucorum*. *Hereditas* **101**, 245–251.

Parker, F.D. (1981) A candidate for red clover, *Osmia coerulescens* L. *J. Apic. Res.* **20**, 62–65.

Parker, F.D., Batra, S.W.T., and Tepedino, V.J. (1987) New pollinators for our crops. *Agric Zool. Rev.* **2**, 279–304.

Parrish, J.A.D. and Bazzaz, F.A. (1979) Difference in pollination niche relationships in early and late successional plant communities. *Ecology* **60**, 597–610.

Partridge, L. (1983) Non-random mating and offspring fitness. In: *Mate Choice*, edited by P. Bateson, pp. 227–253. Cambridge University Press, Cambridge.

Paton, D.C. (1990) Budgets for the use of floral resources in mallee heath. In: *The Malle Lands: a conservation perspective*, edited by J.C. Noble, P.J. Joss, and G.K. Jones, pp. 189–193. CSIRO, Melbourne.

Paton, D.C. (1993) Honeybees in the Australian Environment—does *Apis mellifera* disrupt or benefit the native biota? *Bioscience* **43**, 95–103.

Paton, D.C. (1995) Impact of honeybees on the flora and fauna of Banksia heathlands in Ngarkat Conservation Park. *SASTA J.* **95**, 3–11.

Paton, D.C. (1996) Overview of feral and managed honeybees in Australia: distribution, abundance, extent of interactions with native biota, evidence of impacts and future research. Australian Nature Conservation Agency, Canberra.

Patten, K.D., Shanks, C.H., and Mayer, D.F. (1993) Evaluation of herbaceous plants for attractiveness to bumble bees for use near cranberry farms. *J. Apic. Res.* **32**, 73–79.

Paxton, R.J., Thorén, P.A., Estoup, A., and Tengö, J. (2001) Queen-worker conflict over male production and the sex ratio in a facultatively polyandrous bumblebee, *Bombus hypnorum*: the consequences of nest usurpation. *Mol. Ecol.* **10**, 2489–2498.

Paxton, R.J., Thorén, P.A., Gyllenstrand, N., and Tengö, J. (2000) Microsatellite DNA analysis reveals low diploid male production in a communal bee with inbreeding. *Biol. J. Linn. Soc.* **69**, 483–502.

Pearson, W.D. and Braiden, V. (1990) Seasonal pollen collection by honeybees from grass shrub highlands in Canterbury, *N. Z. J. Apic. Res.* **29**, 206–213

Pedersen, B.V. (1996) A phylogenetic analysis of cuckoo bumblebees (*Psithyrus*, Lepeletier) and bumblebees (*Bombus*, Latreille) inferred from sequences of the mitochondrial gene cytochrome oxidase I. *Mol. Phylogen. Evol.* **5**, 289–297.

Pekkarinen, A. (1979) Morphometric, colour and enzyme variation in bumblebees (Hymenoptera, Apidae, *Bombus*) in Fennoscandia and Denmark. *Acta Zool. Fenn.* **158**, 1–60.

Pekkarinen, A., Teräs, I., Viramo, J., and Paatela, J. (1981) Distribution of bumblebees (Hymenoptera, Apidae: *Bombus* and *Psithyrus*) in eatern Fennoscandia. *Not. Entomol.* **61**, 71–89.

Pekkarinen, A., Varvio-Aho, S.-L., and Pamilo, P. (1979) Evolutionary relationships in northern European *Bombus* and *Psithyrus* species (Hymenoptera, Apidae) studied on the basis of allozymes. *Ann. Entomol. Fenn.* **45**, 77–80.

Pellet, F.C. (1976) *American honey plants*. Fifth Edition. Dadant and Sons, Hamilton (IL).

Pellmyr, O. (1986) Three pollination morphs in *Cimicifuga simplex*; incipient speciation due to inferiority in competition. *Oecologia* **68**, 304–307.

Pendrel, B.A. and Plowright, R.C. (1981) Larval feeding by adult Bumble-bee workers (Hymenoptera: Apidae). *Behav. Ecol. Sociobiol.* **8**, 71–76.

Peters, G. (1972) Ursachen fur den Rückgang der seltenen heimischen Hummelarten (Hym., *Bombus* et *Psithyrus*). *Entomol. Ber.* 85–90.

Percival, M. (1974) Floral ecology of coastal scrub in Southeast Jamaica. *Biotropica* **6**, 104–129.

Pereboom, J.J.M. (1997) 'While they banquet splendidly the future mother': the significance of trophogenic and social factors on caste determination and differentiation in the bumblebee *Bombus terrestris*. PhD thesis, University of Utrecht, The Netherlands.

Pereboom, J.J.M. (2000) The composition of larval food and the significance of exocrine secretions in the bumblebee *Bombus terrestris*. *Insectes Soc.* **47**, 11–20.

Pierce, G.J. and Ollason, J.G. (1987) 8 reasons why optimal foraging theory is a complete waste of time. *Oikos* **49**, 111–118.

Pirounakis, K., Koulianos, S., and Schmid-Hempel, P. (1998) Genetic variation among European populations of *Bombus pascuorum* (Hymenoptera, Apidae) using mitochondrial DNA sequence data. *Europ. J. Entomol.* **95**, 27–33.

Pimm, S.L., Russell, G.J., Gittleman, J.L., and Brookes, T.M. (1995) The future of biodiversity. *Science*, **269**, 347–350.

Plateau, F. (1901) Observations sur le phénomène de la constance chez quelques hyménoptères. *Ann. Soc. Entomol. Belg.* **45**, 56–83.

Plath, O.E. (1923a) Notes on the egg-eating habits of bumblebees. *Psyche* **30**, 193–202.

Plath, O.E. (1923b) The bee-eating proclivity of the skunk. *Am. Nat.* **57**, 571–574.

Plath, O.E. (1934) *Bumblebees and their ways*. New York.

Pleasants, J.M. (1981) Bumblebee response to variation in nectar availability. *Ecology* **62**, 1648–1661.

Pleasants, J.M. (1983) Nectar production patterns in *Ipomopsis aggregata* (Polemoniaceae). *Am. J. Bot.* **70**, 1468–1475.

Pleasants, J.M. (1989) Optimal foraging by nectarivores: a test of the marginal-value theorem. *Am. Nat.* **134**, 51–71.

Pleasants, J.M. and Chaplin, S.J. (1983) Nectar production rates of *Asclepias quadrifolia*: causes and consequences of individual variation. *Oecologia* **59**, 232–238.

Pleasants, J.M. and Zimmerman, M. (1979) Patchiness in the dispersion of nectar resources: evidence for hot and cold spots. *Oecologia* **41**, 283–288.

Pleasants, J.M. and Zimmerman, M. (1983) The distribution of standing crop of nectar: what does it really tell us? *Oecologia* **57**, 412–414.

Pleasants, J.M. and Zimmerman, M. (1990) The effect of inflorescence size on pollinator visitation of *Delphinium nelsonii* and *Aconitum columbianum*. *Collect. Bot.* **19**, 21–39.

Plowright, C.M.S. and Plowright, R.C. (1997) The advantage of short tongues in bumble bees (*Bombus*) — Analyses of species distributions according to flower corolla depth, and of working speeds on white clover. *Can. Entomol.* **129**, 51–59.

Plowright, R.C. and Galen, C. (1985) Landmarks or obstacles: the effects of spatial heterogeneity on bumble bee foraging behaviour. *Oikos* **44**, 459–464.

Plowright, R.C. and Jay, S.C. (1968) Caste differentiation in bumblebees (*Bombus* Latr.: Hym.) 1. The determination of female size. *Insectes Soc.* **15**, 171–192.

Plowright, R.C. and Laverty, T.M. (1987) Bumble bees and crop pollination in Ontario. *Proc. Entomol. Soc. Ont.* **118** 155–160.

Plowright, R.C. and Owen, R.E. (1980) The evolutionary significance of bumble bee colour patterns: A mimetic interpretation. *Evolution* **34**, 622–637.

Plowright, R.C. and Pallet, J.M. (1979) Worker-male conflict and inbreeding in bumble bees (Hymenoptera: Apidae). *Can. Entomol.* **111**, 289–294.

Plowright, R.C. and Pendrel, B.A. (1977) Larval growth in bumble bees (Hymenoptera, Apidae). *Can. Entomol.* **109**, 967–973.

Plowright, R.C. and Silverman, A. (2000) Nectar and pollen foraging by bumble bees (Hymenoptera: Apidae): choice and tradeoffs. *Can. Entomol.* **132**, 677–679.

Plowright, R.C. and Stephen, W.P. (1973) A numerical taxonomic analysis of the evolutionary relationships of *Bombus* and *Psithyrus* (Apidae: Hymenoptera). *Can. Entomol.* **105**, 733–743.

Posner, M.I. and Peterson, S.E. (1990) The attention system of the human brain. *Ann. Rev. Neurosci.* **13**, 25–42.

Poinar, G.O. and van del Laan, P.A. (1972) Morphology and life history of *Sphaerularia bombi*. *Nematologica* **18**, 239–252.

Pomeroy, N. (1979) Brood bionomics of *Bombus ruderatus* in New Zealand (Hymenoptera: Apidae). *Can. Entomol.* **111**, 865–874.

Pomeroy, N. (1981) Use of natural field sites and field hives by a long-tongued bumble bee *Bombus ruderatus*. *N. Z. J. Agric. Res.* **24**, 409–414.

Pomeroy, N. and Plowright, R.C. (1982) The relation between worker numbers and the production of males and queens in the bumble bee *Bombus perplexus*. *Can. J. Zool.* **60**, 954–957.

Possingham, H.P. (1989) The distribution and abundance of resources encountered by a forager. *Am. Nat.* **133**, 42–60.

Postner, M. (1952) Biologische-ökologische Untersuchungen an Hummeln und ihren Nestern. *Veröf. Übersee-Museum Bremen (A).* **2**, 45–86.

Poulsen, M.H. (1973) The frequency and foraging behaviour of honeybees and bumble bees on field beans in Denmark. *J. Apic. Res.* **12**, 75–80.

Pouvreau, A. (1967) Contribution à l'étude morphologique et biologique d'*Aphomia sociella* L. (Lepidoptera, Heteroneura, Pyraloidea, Pyralididae), parasite des nids de bourdons (Hymenoptera, Apoidea, *Bombus* Latr.). *Insectes Soc.* **14**, 57–72.

Pouvreau, A. (1974) Les enemies des bourdons. II. Organismes affectant les adultes. *Apidologie* **5**, 39–62.

Pouvreau, A. (1989) Contribution à l'étude du polyéthisme chez les bourdons, *Bombus* Latr. (Hymenoptera, Apidae). *Apidologie* **20**, 229–244.

Pouvreau, A. (1991) Morphology and histology of tarsal glands in bumble bees of the genera *Bombus*, *Pyrobombus* and *Megabombus*. *Can. J. Zool.* **69**, 866–872.

Prescott, C. and Allen, R. (1986) *The first resource; wild species in the North American economy*. Yale University Press, New Haven, Connecticut, USA.

Primack, R.B. and Hall, P. (1990) Cost of reproduction in the pink lady's slipper orchid: a four year experimental study. *Am. Nat.* **136**, 638–656.

Proctor, M. and Yeo, P. (1973) *The pollination of flowers*. William Collins Sons, Glasgow.

Proctor, M., Yeo, P., and Lack, A. (1996) *The natural history of pollination*. Harper Collins, London.

Prys-Jones, O.E. (1982) Ecological studies of foraging and life history in bumblebees. PhD thesis, University of Cambridge.

Prys-Jones, O.E. (1986) Foraging behaviour and the activity of substrate cycle enzymes in bumblebees. *Anim. Behav.* **34**, 609–611.

Prys-Jones, O.E. and Corbet, S.A. (1991) *Bumblebees*. Richmond Publishing Co. Ltd. Slough, England.

Prys-Jones, O.E., Ólafsson, E., and Kristjánsson, K. (1981) The Icelandic bumblebee fauna and its distributional ecology. *J. Apic. Res.* **20**, 189–197.

Ptácek, V. (1991) Trials to rear bumblebees. *Acta Hortic.* **288**, 144–148.

Pyke, G.H. (1978a) Optimal foraging: movement patterns of bumblebees between inflorescences. *Theor. Popul. Biol.* **13**, 72–98.

Pyke, G.H. (1978b) Optimal foraging in bumblebees and coevolution with their plants. *Oecologia* **36**, 281–293.

Pyke, G.H. (1978c) Optimal foraging in hummingbirds: testing the marginal value theorem. *Am. Zool.* **18**, 739–752.

Pyke, G.H. (1978d) Optimal body size in bumblebees. *Oecologia* **34**, 255–266.

Pyke, G.H. (1979) Optimal foraging in bumblebees: rules of movement between flowers within inflorescences. *Anim. Behav.* **27**, 1167–1181.

Pyke, G.H. (1981) Honeyeater foraging: a test of optimal foraging theory. *Anim. Behav.* **29**, 878–888.

Pyke, G.H. (1982) Local geographic distributions of bumblebees near Crested Butte, Colorado: competition and community structure. *Ecology* **63**, 555–573

Pyke, G.H. (1983) Animal movements: An optimal foraging approach. In: *The ecology of animal movement*, edited by I.R. Swingland, and P.J., Greenwood, pp. 7–31. Clarendon, Oxford.

Pyke, G.H. (1984) Optimal foraging: a critical review. *Annu. Rev. Ecol. Syst.* **15**, 523–575.

Pyke, G.H. and Cartar, R.V. (1992) The flight directionality of bumblebees—do they remember where they came from. *Oikos* **65**, 321–327.

Rackham, O. (1976) *Trees and woodland in the British landscape*. Dent, London.

Raju, B.C., Nyland, G., Meikle, T., and Purcell, A.H. (1981) Helical, motile mycoplasmas associated with flowers and honeybees in California. *Can. J. Microbiol.* **27**, 249–253.

Ramsey, M.W. (1988) Differences in pollinator effectiveness of birds and insects visiting *Banksia menziesii* (Protaceae). *Oecologia* **76**, 119–124.

Ranta, E. (1982) Species structure of North European bumblebee communities. *Oikos* **38**, 202–209.

Ranta, E. (1983) Proboscis length and the coexistence of bumblebee species. *Oikos* **43**, 189–196.

Ranta, E. and Lundberg, H. (1980) Resource partitioning in bumblebees: the significance of differences in proboscis length. *Oikos* **35**, 298–302.

Ranta, E., Lundberg, H., and Teräs, I. (1980) Patterns of resource utilization in two Fennoscandian bumblebee communities. *Oikos* **36**, 1–11.

Ranta, E. and Tiainen, M. (1982) Structure in seven bumblebee communities in eastern Finland in relation to resource availability. *Holar. Ecol.* **5**, 48–54.

Ranta, E. and Vepsäläinen K. (1981) Why are there so many species? Spatio-temporal heterogeneity and northern bumblebee communities. *Oikos* **36**, 28–34.

Rasmont, P. (1983) Catalogue of the bumble bees of the west Palearctic region. *Notes faunique de Gembloux*, **7**, 71 pp.

Rasmont, P. (1988) Monographie écologique et zoogéographique des bourdons de France et de Belgique (Hymenoptera, Apidae, Bombinae). PhD thesis, Faculté des Sciences Agronomique de l'Etat, Gembloux, Belgium.

Rasmont, P. (1995) How to restore the apoid diversity in Belgium and France? Wrong and right ways, or the end of protection paradigm! In: *Changes in fauna of wild bees in Europe*, edited by J. Banaszak, pp. 53–64. Pedagogical University, Bydgoszcz.

Rasmont, P. and Mersch, P. (1988) Première estimation de la dérive faunique chez les bourdons de la Belgique (Hymenoptera, Apidae). *Annales de la Société Royale zoologique de Belgique* **118**, 141–147.

Rathcke, B. (1983) Competition and facilitation among plants for pollination. In: *Pollination ecology*, edited by L. Real, pp. 305–325. Academic Press, New York.

Rathcke, B.J. and Jules, E.S. (1993) Habitat fragmentation and plant–pollinator interactions. *Curr. Sci.* **65**, 273–277.

Ratnieks, F. and Boomsma, J.J. (1995) Facultative sex allocation by workers and the evolution of polyandry by queens in social Hymenoptera. *Am. Nat.* **145**, 969–993.

Rau, P. (1929) Experimental studies in the homing of carpenter and mining bees. *J. Comp. Physiol.* **9**, 35–70.

Real, L. (1981) Uncertainty and pollinator-plant interactions: the foraging behaviour of bees and wasps on artificial flowers. *Ecology* **62**, 20–26.

Real, L. (1983) Microbehavior and macrostructure in pollinator plant interactions. In: *Pollination ecology*, edited by L. Real, pp. 287–302. Academic Press, New York.

Real, L., Ott, J.R., and Silverfine, E. (1982) On the tradeoff between the mean and the variance in foraging: effect of spatial distribution and colour preference. *Ecology* **63**, 1617–1623.

Real, L. and Rathcke, B.J. (1988) Patterns of individual variability of floral resources. *Ecology* **69**, 728–735.

Reid, P.J. and Shettleworth, S.J. (1992) Detection of cryptic prey: Search image or search rate? *J. Exp. Psychol.* **18**, 273–286.

Reinig, W.F. (1972) Ökologische Studien an mittel- und südosteuropäischen Hummeln (*Bombus* Latr., 1802; Hym. Apidae). *Mitt. Mue. Entomol. Ges.* **60**, 1–56.

Reuter, K. (1998) Anzucht und experimentelle Untersuchungen zur fütterungsabhängigen Kastendetermination von *Bombus pascuorum* (Scopoli), (Hymenoptera: Apidae). PhD thesis, Ruhr-Universität Bochum.

Ribbands, C.R. (1949) The foraging method of individual honeybees. *J. Anim. Ecol.* **18**, 47–66.

Ribeiro, M.F. (1994) Growth in bumble bee larvae—relation between development time, mass, and amount of pollen ingested. *Can. J. Zool.* **72**, 1978–1985.

Ribeiro, M.F. Velthuis, H.H.W., Duchateau, M.J., and van der Tweel, I. (1999) Feeding frequency and caste differentiation in *Bombus terrestris* larvae. *Insectes Soc.* **46**, 306–314.

Richards, K.W. (1973) Biology of *Bombus polaris* Curtis and *B. hyperboreus* Schönherr at Lake Hazen, Northwest Territories (Hymenoptera: Bombini). *Quaesti. Entomol.* **9**, 115–157.

Richards, K.W. (1978) Nest site selection by bumble bees (Hymenoptera: Apidae) in southern Alberta. *Can. Entomol.* **110**, 301–318.

Richards, K.W. (1993) Non-*Apis* bees as crop pollinators. *Rev. Suis. Zool.* **100**, 807–822.

Richards, L.A. and Richards, K.W. (1976) Parasatid mites associated with bumblebees in Alberta, Canada (Acarina, Parasitidae; Hymenoptera: Apidae). II. Biology. *Univ. Kans. Sci. Bull.* **51**, 1–18.

Richards, O.W. (1931) Some notes on the humblebees allied to *Bombus alpinus* L. *Tromsø Mus. Aarsh.* **50**, 1–32.

Richards, O.W. (1946) Observations on *Bombus agrorum* Fabricius. (Hymen. Bombidae) *Proc. Royal. Entomol. Soc. Lond. A* **21**, 66–71.

Rick, C.M. (1950) Pollination relations of *Lycopersicon esculentum* in native and foreign regions. *Evolution* **4**, 110–122.

Riley, J.R., Reynolds, D.R., Smith, A.D., Edwards, A.S., Osborne, J.L., and Williams, I.H. (1999) Compensation for wind drift by bumblebees. *Nature* **400**, 126.

Riley, J.R., Smith, A.D., Reynolds, D.R., Edwards, A.S., Osborne, J.L., Williams, I.H., Carreck, N.L., and Poppy, G.M. (1996) Tracking bees with harmonic radar. *Nature* **379**, 29–30.

Riley, J.R., Valeur, P., Smith, A.D., Reynolds, D.R., Poppy, G.M., and Löfstedt, C. (1998) Harmonic radar as a means of tracking the pheromone-finding and pheromone-following flight in male moths. *J. Insect Behav.* **11**, 287–296.

Roberts, R.H. and Struckmeyer, B.E. (1942) Growth and fruiting of the cranberry. *Am. Soc. Hortic. Sci. Proc.* **40**, 373–379.

Robertson, A.W. (1991) A mid-winter colony of *Bombus terrestris* L. (Hym., Apidae) in Devon. *Entomologist's. Monthly Magazine* **127**, 165–166.

Robertson, A.W. and Macnair, M.R. (1995) The effects of floral display size on pollinator service to individual flowers of *Myosotis* and *Mimulus*. *Oikos* **72**, 106–114.

Robertson, J.L. and Wyatt, R. (1990) Reproductive biology of the Yellow-Fringed Orchid, *Platanthera ciliaris*. *Am. J. Bot.* **77**, 388–398.

Robertson, P., Bennett, A.F., Lumsden, L.F., Silveira, C.E., Johnson, P.G., Yen. A.L., Milledge, G.A., Lillywhite, P.K., and Pribble, H.J. (1989) Fauna of the Mallee study area north-western Victoria. National Parks and Wildlife Division Technical Report Series, No. 87. Victoria (Australia): Department of Conservation, Forests and Lands. pp. 41–42.

Robinson, W.S. (1979) Influence of 'Delicious' apple blossom morphology on the behavior of nectar-gathering honey bees. Proceedings of the IV International Symposium on Pollination, Maryland Agricultural Experimental Station Special Miscellaneous Publications 1. pp. 393–399.

Robinson, W.S., Nowodgrodzki, R., and Morse, R.A. (1989) The value of bees as pollinators of U.S. crops. *Am. Bee. J.* **129**, 411–423; 477–487.

Rodd, F.H., Plowright, R.C., and Owen, R.E. (1980) Mortality rates of adult bumble bee workers (Hymenoptera: Apidae). *Can. J. Zool.* **58**, 1718–1721.

Röseler, P.-F. (1967a) Arbeitsteilung und Drüsenzustände in Hummelvölkern. *Naturwissenschaften* **54**, 146–147.

Röseler, P.-F. (1967b) Untersuchungen über das Auftreten der 3 Formen im Hummelstaat. *Zool. Jb. Physiol.* **74**, 178–197.

Röseler, P.-F. (1970) Unterschiede in der Kastendetermination zwischen den Hummelarten *Bombus hypnorum* und *Bombus terrestris*. *Z. Naturf.* **25**, 543–548.

Röseler, P.-F. (1973) Die Anzahl Spermien im Receptaculum seminis von Hummelköniginnen (Hymenoptera, Apidae, Bombinae). *Apidologie* **4**, 267–274.

Röseler, P.-F. (1991) Roles of morphogenetic hormones in caste polymorphism in bumblebees. In: *Morphogenetic hormones in arthropods: roles in histogenesis, organogenesis, and morphogenesis*, edited by A.P. Gupta, pp. 384–399. Rutgers University Press, New Brunswick, NJ.

Röseler, P.-F. and Van Honk, C.G.J. (1990) Castes and reproduction in bumblebees. In: *Social Insects: an evolutionary approach to castes and reproduction*, edited by W. Engels, pp. 147–166. Springer, Berlin.

Röseler, P.-F. and Röseler, I. (1974) Morphologische und physiologische Differenzierung der Kasten bei den Hummelarten *Bombus hypnorum* (L.) und *Bombus terrestris* (L.). *Zoöl. Jahrb. Abt. Allg. Zoöl. Phys. Tiere* **78**, 175–198.

Röseler, P.-F. and Röseler, I. (1977) Dominance in bumblebees. *Proc. 8th Int. Congr. IUSSI*, 232–235.

Röseler, P.-F., Röseler, I., and Van Honk, C.G.J. (1981) Evidence for inhibition of corpora allata activity in workers of *Bombus terrestris* by a pheromone from the queen's mandibular glands. *Experientia* **37**, 348–351.

Roubik, D.W. (1978) Competitive interactions between neotropical pollinators and Africanized honey bees. *Science* **201**, 1030–1032.

Roubik, D.W. (1980) Foraging behavior of commercial Africanized honeybees and stingless bees. *Ecology* **61**, 8336-8845.
Roubik, D.W. (1981) Comparative foraging behaviour of *Apis mellifera* and *Trigona corvina* (Hymenoptera: Apidae) on *Baltimora recta* (Compositae). *Rev. Biol. Trop.* **29**, 177-184.
Roubik, D.W. (1982a) Ecological impact of Africanized honeybees on native neotropical pollinators. *Social insects of the tropics*, edited by P. Jaisson, pp. 233-247. Université Paris-Nord. Roubik, D.W. (1982b) The ecological impact of nectar robbing bees and pollinating hummingbirds on a tropical shrub. *Ecology* **63**, 354-360.
Roubik, D.W. (1983) Experimental community studies: time series tests of competition between African and neotropical bees. *Ecology* **64**, 971-978.
Roubik, D.W. (1989) *Ecology and natural history of tropical bees*. Cambridge University Press, Cambridge.
Roubik, D.W. (1991) Aspects of Africanized honey bee ecology in tropical America. In: The 'African' honey bee, edited by M. Spivak, *et al.*, pp. 259-281. Westview Press, Boulder (CO).
Roubik, D.W. (1996a) Measuring the meaning of honeybees. In: *The conservation of bees*, edited by A.S.L. Matheson, C. Buchmann, P. O'Toole, Westrich, and I.H. Williams, pp. 163-172. Academic Press, London.
Roubik, D.W. (1996b) African honey bees as exotic pollinators in French Guiana. In: *The Conservation of bees*, edited by A.S.L. Matheson, C. Buchmann, P. O'Toole, Westrich, and I.H. Williams, pp. 173-182. Academic Press, London.
Roubik, D.W. and Aluja, M. (1983) Flight reanges of *Melipona* and *Trigona* in tropical forests. *J. Kans. Entomol. Soc.* **56**, 217-222.
Roubik, D.W., Holbrook, N.M., and Parrav, G. (1985) Roles of nectar robbers in reproduction of the tropical treelet *Quassia amara* (Simaroubaceae). *Oecologia* **66**, 161-167.
Roubik, D.W., Moreno, J.E., Vergara, C., and Wittman, D. (1986) Sporadic food competition with the African honey bee: projected impact on neotropical social species. *J. Tropic. Ecol.* **2**, 97-111.
Sakagami, S.F. (1976) Species differences in the bionomic characters of bumblebees. A comparative review. *J. Fac. Sci. Hokkaido Univ, Ser. 6, Zool.* **20**, 390-447.
Sakagami, S.F. and Zucchi, R. (1965) Winterverhalten einer neotropischen Hummel, *Bombus atratus*, innerhalb des Beobachtungskastens. Ein Beitrag zur Biologie der Hummeln. *J. Fac. Sci. Hokkaido Univ. Ser. 6, Zool.* **15**, 712-762.
Saunders, D., Hobbs, R., and Ehrlich, P. (eds) (1992) Reconstruction of fragmented ecosystems; local and global perspectives. Surrey Beatty, Chipping Norton, UK, 326 pp.
Saunders, E. (1909) Bombi and other aculeates collected in 1908 in the Berner Oberland by the Rev. A.E. Easton, M.A. *Entomologist's Monthly Magazine* **45**, 83-84.
Sauter, A., Brown, M.J.F., Baer, B., and Schmid-Hempel, P. (2001) Males of social insects can prevent queens from multiple mating. *Proc. Royal. Soc.* **268**, 1449-1454.
Saville, N.M. (1993) Bumblebee ecology in woodlands and arable farmland. PhD Thesis, University of Cambridge.
Saville, N.M., Dramstad, W.E., Fry, G.L.A., and Corbet, S.A. (1997) Bumblebee movement in a fragmented agricultural landscape. *Agric. Ecosys. Environ.* **61**, 145-154.
Schaal, B.A. and Leverich, W.J. (1980) Pollination and banner markings in *Lupinus texensis* (Leguminosae). *Southwest. Natur.* **25**, 280-282.
Schaffer, M.J. (1996) Spatial aspects of bumble bee (*Bombus* spp. Apidae) foraging in farm landscapes. M. Appl. Sci. Thesis. Lincoln University, Lincoln, New Zealand.
Schaffer, M. and Wratten, S.D. (1994) Bumblebee (*Bombus terrestris*) movement in an intensive farm landscape. *Proceedings 47th New Zealand Plant Protection Conference*, edited by A.J. Popay, pp. 253-256. New Zealand Plant Protection Society, Rotorua, New Zealand.

Schaffer, W.M., Jensen, D.B., Hobbs, D.E., Gurevitch, J., Todd, J.R., and Valentine Schaffer, M. (1979) Competition, foraging energetics, and the cost of sociality in three species of bees. *Ecology* **60**, 976–987.

Schaffer, W.M. and Schaffer, M.W. (1979) The adaptive significance of variations in reproductive habit in the Agavaceae II: Pollinator foraging behavior and selection for increased reproductive expenditure. *Ecology* **60**, 1051–1069.

Schaffer, W.M., Zeh, D.W., Buchmann, S.L., Kleinhans, S., Valentine Schaffer, M., and Antrim, J. (1983) Competition for nectar between introduced honey bees and native North American bees and ants. *Ecology* **64**, 564–577.

Shelly, T.E., Buchmann, S.L., Villalobos, E.M., and O'Rourke, M.K. (1991) Colony ergonomics for a desert-dwelling bumblebee species (Hymenoptera: Apidae). *Ecol. Entomol.* **16**, 361–370.

Schemske, D.W. (1983) Limits to specialization and coevolution in plant–animal mutualisms. In: *Coevolution*, edited by M.H. Nitecki, pp. 67–109. The University of Chicago Press, Chicago.

Schmid-Hempel, P. (1984) The importance of handling time for the flight directionality in bees. *Behav. Ecol. Sociobiol.* **15**, 303–309.

Schmid-Hempel, P. (1985) How do bees choose flight direction while foraging? *Physiol. Entomol.* **10**, 439–442.

Schmid-Hempel, P. (1986) The influence of reward sequence on flight directionality in bees. *Anim. Behav.* **34** 831–837.

Schmid-Hempel, P. (1998) *Parasites in social insects*. Princeton University Press, Princeton, New Jersey.

Schmid-Hempel, P. (2001) On the evolutionary ecology of host-parasite interactions: addressing the question with regard to bumblebees and their parasites. *Naturwissenschaften* **88**, 147–158.

Schmid-Hempel, P. and Loosli, R. (1998) A contribution to the knowledge of *Nosema* infections in bumble bees, *Bombus* spp. *Apidologie* **29**, 525–535.

Schmid-Hempel, P. Müller, C.B., Schmid-Hempel, R., and Shykoff, J.A. (1990) Frequency and ecological correlates of parasitism by conopid flies (Conopidae, Diptera) in populations of bumblebees. *Insectes Soc.* **37**, 14–30.

Schmid-Hempel, P. and Schmid-Hempel, R. (1986) Nectar-collecting bees use distance-sensitive movement rules. *Anim. Behav.* **34**, 605–607.

Schmid-Hempel, P. and Speiser, B. (1988) Effects of inflorescence size on pollination in *Epilobium angustifolium*. *Oikos* **53**, 98–104.

Schmid-Hempel, R. (1994) Evolutionary ecology of a host-parasitoid interaction. PhD dissertation, Basel University.

Schmid-Hempel, R. and Müller, C.B. (1991) Do parasitised bumblebees forage for their colony? *Animal Behaviour* **41**, 910–912.

Schmid-Hempel, R. and Schmid-Hempel, P. (1989) Superparasitism and larval competition in conopid flies (Dipt., Conopidae), parasitising bumblebees (Hym., Apidae). *Mitt Schweizer Entomol. Ges.* **62**, 279–289.

Schmid-Hempel, R. and Schmid-Hempel, P. (1996a) Host choice and fitness correlates for conopid flies parasitising bumblebees. *Oecologia* **107**, 71–78.

Schmid-Hempel, R. and Schmid-Hempel, P. (1996b) Larval development of two parasitic flies (Conopidae) in the common host *Bombus pascuorum*. *Ecol. Entomol.* **21**, 63–70.

Schmid-Hempel, R. and Schmid-Hempel, P. (1998) Colony performance and immunocompetence of a social insect, *Bombus terrestris*, in poor and variable environments. *Funct. Ecol.* **12**, 22–30.

Schmid-Hempel, R. and Schmid-Hempel, P. (2000) Female mating frequencies in *Bombus* spp. from Central Europe. *Insectes Soc.* **47**, 36–41.

Schmitt, D. (1980) Pollinator foraging behaviour and gene dispersal in *Senecio* (Compositae). *Evolution* **34**, 934–943.
Schmitt, J. (1983) Flowering plant density and pollinator visitation in *Senecio*. *Oecologia* **60**, 97–102.
Schmitt, U. (1990) Hydrocarbons in tarsal glands of *Bombus terrestris*. *Experientia* **46**, 1080–1082.
Schmitt, U., and Bertsch, A. (1990) Do foraging bumblebees scent-mark food sources and does it matter? *Oecologia* **82**, 137–144.
Schmitt, U., Lübke, G., and Franke, W. (1991) Tarsal secretion marks food sources in bumblebees (Hymenoptera: Apidae). *Chemoecology* **2**, 35–40.
Schneider, S.S. and McNally, L.C. (1992) Factors influencing seasonal absconding in colonies of the African honeybee, *Apis mellifera scutellata*. *Insectes Soc.* **39**, 403–423.
Schneider, S.S. and McNally, L.C. (1993) Spatial foraging patterns and colony energy status in the African honeybee, *Apis mellifera scutellata*. *J. Insect Behav.* **6**, 195–210.
Schoener, T.W. (1969) Optimal size and specialization in constant and fluctuating environments: an energy-time approach. *Brookh. Symp. Biol.* **22**, 103–114.
Scholl, A., Obrecht, E., and Owen, R.E. (1990) The genetic relationship between *Bombus moderatus* Cresson and the *Bombus lucorum* auct. species complex (Hymenoptera: Apidae). *Can. J. Zool.* **68**, 2264–2268.
Schöne, H., Harris, A.C., Schöne, H., and Mahalski, P. (1993a) Homing after displacement in open or closed containers by the diggerwasp *Argogorytes carbonarius* (Hyemnoptera: Specidae). *Ethology*, **95**, 152–156.
Schöne, H., Tengö J., Kühme, D., Schöne, H., and Kühme, L. (1993b) Homing with or without sight of surroundings and sky during displacement in the digger wasp *Bembix rostrata* (Hymenoptera: Sphecidae). *Ethol. Ecol. Evol.* **5**, 549–552.
Schousboe, C. (1986) On the biology of *Scutacarus acarorum* Goeze (Acarina: Trombidiformes). *Acarologia* **27**, 151–158.
Schousboe, C. (1987) Deutonymphs of *Parasitellus* phoretic on Danish bumble bees (Parasitidae, Mesostigmata; Apidae, Hymenoptera). *Acarologia* **28**, 37–41.
Schremmer, F. (1972) Beobachtungen zum Paarungsverhalten der Männchen von *Bombus confusus* Schenk. *Z. Tierpsychol.* **31**, 503–512.
Schwarz, H.H. and Huck, K. (1997) Phoretic mites use flowers to transfer between foraging bumblebees. *Insectes Soc.* **44**, 303–310.
Schwarz, H.H., Huck, K., and Schmid-Hempel, P. (1996) Prevalence and host preferences of mesostigmatic mites (Acari: Anactinochaeta) phoretic on Swiss bumble bees (Hymenoptera: Apidae). *J. Kans. Entomol. Soc.* **69 (Suppl.)**, 35–42.
Schwarz, M.P., Gross, C.L., and Kukuk, P.F. (1991) Assessment of competition between honeybees and native bees. July 1991 Progress Report to the World Wildlife Fund, Australia, Project P158.
Schwarz, M.P., Gross, C.L., and Kukuk, P.F. (1992a) Assessment of competition between honeybees and native bees. January 1992 Progress Report to the World Wildlife Fund, Australia, Project P158.
Schwarz, M.P., Gross, C.L., and Kukuk, P.F. (1992b) Assessment of competition between honeybees and native bees. July 1992 Progress Report to the World Wildlife Fund, Australia, Project P158.
Schwarz, M.P. and Hurst, P.S. (1997) Effects of introduced honey bees on Australia's native bee fauna. *Victor. Nat.* **114**, 7–12.
Scott, P.E., Buchmann, S.L., and O'Rourke, M.K. (1993) Evidence for mutualism between flower-piercing carpenter bee and ocotillo: use of pollen and nectar by nectaring bees. *Ecol. Entomol.* **18**, 234–240.

Seeley, T.D. (1985a) *Honeybee ecology*. Princeton University Press, Princeton, NJ.
Seeley, T.D. (1985b) The information-center strategy of honey bee foraging. In: *Experimental behavioral ecology and sociobiology*, edited by B. Hölldobler and M. Lindauer, pp. 75–90. Sinauer Associates, Sunderland, Massachusetts.
Seeley, T.D. and Heinrich, B. (1981) Regulation of temperature in nests of social insects. In: *Insect thermoregulation*, edited by B. Heinrich, pp. 159–234. John Wiley, New York.
Semmens, T.D. (1996a) Flower visitation by the bumble bee *Bombus terrestris* (L.) (Hymenoptera: Apidae) now established in Tasmania. *J. Aust. Entomol. Soc.* **32**, 346.
Semmens, T.D. (1996b) Flower visitation by the bumble bee *Bombus terrestris* (L.) (Hymenoptera: Apidae) in Tasmania. *Aust. Entomol.* **23**, 33–35.
Senft, D. (1990) Protecting endangered plants. *Agric. Res.* **May 1990**, 16–18.
Sherman, P.W., Seeley, T.D., and Reeve, H.K. (1988) Parasites, pathogens and polyandry in social Hymenoptera. *Am. Nat.* **131**, 602–610.
Shelly, T.E., Buchmann, S.L., Villalobos, E.M., and O'Rourke, M.K. (1991) Colony ergonomics for a desert-dwelling bumblebee species (Hymenoptera: Apidae). *Ecol. Entomol.* **16**, 361–370.
Shields, W.M. (1982) *Philopatry, inbreeding, and the evolution of sex*. State Univ. N.Y. Press, Albany, NY., 245 pp.
Shykoff, J.A. and Bucheli, E. (1995) Pollinator visitation patterns, floral rewards and the probability of transmission of *Microbotryum violaceum*, a venereal disease of plants. *J. Ecol.* **83**, 189–198.
Shykoff, J.A., Bucheli, E., and Kaltz, O. (1997) Anther smut disease in *Dianthus silvester* (Caryophyllaceae): natural selection on floral traits. *Evolution* **51**, 383–392.
Shykoff, J.A. and Müller, C.B. (1995) Reproductive decisions in bumble-bee colonies: the influence of worker mortality in *Bombus terrestris* (Hymenoptera, Apidae). *Funct. Ecol.* **9**, 106–112.
Shykoff, J.A. and Schmid-Hempel, P. (1991a) Genetic relatedness and eusociality: Parasite-mediated selection on the genetic composition of groups. *Behav. Ecol. Sociobiol.* **28**, 371–376.
Shykoff, J.A. and Schmid-Hempel, P. (1991b) Parasites and the advantage of genetic variability within social insect colonies. *Proc. Royal Soc. Lond., B* **243**, 55–58.
Shykoff, J.A. and Schmid-Hempel, P. (1991c) Parasites delay worker reproduction in bumblebees: consequences for eusociality. *Behav. Ecol.* **2**, 242–248.
Sih, A. and Baltus, M.S. (1987) Patch size, pollinator behaviour and pollinator limitation in catnip. *Ecology* **68**, 1679–1690.
Silander, J.A. and Primack, R.B. (1978) Pollination intensity and seed set in the Evening Primrose (*Oenothera fruticosa*). *Am. Midland Nat.* **100**, 213–216.
Silva-Matos, E.V. and Garófalo, C.A. (2000) Worker life tables, survivorship, amd longevity in colonies of *Bombus (Fervidobombus) atratus* (Hymenoptera: Apidae). *Rev. Biol. Trop.* **48**, 657–664.
Silvola, J. (1984) Respiration and energetics of the bumblebee *Bombus terrestris* queen. *Holarc. Ecol.* **7**, 177–181.
Skorikov, A.S. (1922) Palearctic bumble bees. Part I. General biology (including zoogeography). *Izvestiya Severnoi Oblastnoi Stantsii Zashchity Rastenii ot Vreditelei* **4**, 1–160.
Skou, J.P., Holm, S.N., and Haas, H. (1963) Preliminary investigations on diseases in bumble bees (*Bombus* Latr.). *Kongelige Veterinaerskole og Landbohoejskole Denmark*, 27–41.
Skovgaard, O.S. (1936) Rødkloverens Bestøvning. Humlebier og Humleboer. Det Kongelige Danske Videnskabernes Selskabs Skrifter Naturvidenskabelig og Mathematisk Afdeling. 9. Raekke **6**, 1–140.
Skovgaard, O.S. (1943) Humlebiernes bopladser og overvintringssteder. *Tidsskr. Planteavl.* **47**, 285–305.

Sladen, F.W.L. (1899) Bombi in captivity, and habits of *Psithyrus*. *Entomologist's Monthly Magazine* **10**, 230–234.

Sladen, F.W.L. (1912) *The Humble-bee, its Life History and how to Domesticate it*. Macmillan & Co. Ltd.

Smith, F. (1851) Notes on the nest of *Bombus derhamellus*. *Trans. Entomol. Soc. Lond.* **1**: Proc., 111–112.

Smith, F. (1858) Notes on Aculeate Hymenoptera, with some observations on their economy. *Entomol. Annu.* 34–46.

Smith, F. (1865) Notes on Hymenoptera. *Entomol. Annu.* 81–96.

Smith, F. (1876) *Catalogue of the British bees in the collection of the British Museum*. London.

Smith, H., Feber, R.E., Johnson, P.J, McCallum, K., Plesner Jensen, S., Younes, M., and Macdonald, D.W. (1993) *The conservation management of arable field margins*. English Nature, Peterborough, UK.

Smith, H., Feber, R.E., and Macdonald, D.W. (1994) The role of wild flower seed mixtures in field margin restoration. In: *Field margins: Integrating agriculture and conservation. Monogr. No. 58*, edited by N. Boatman, pp. 289–294. British Crop Protection Council, Farnham, UK.

Smith, H. and Macdonald, D.W. (1989) Secondary succession on extended arable field margins: its manipulation for wildlife benefit and weed control. Proc. of 1989 British Crop Protection Conference—Weeds, pp. 1063–1068. British Crop Protection Council, Farnham, UK.

Smith, H. and Macdonald, D.W. (1992) The impacts of mowing and sowing on weed populations and species richness of field margin set-aside. In: *Set-aside, BCPC Monograph No. 50*, edited by J. Clarke, pp. 117–122. BCPC Publications, Bracknell.

Smith, K.G.V. (1959) The distribution and habits of the British Conopidae (Dipt.). *Trans. Soc. Br. Entomol.* **13**, 113–136.

Smith, K.G.V. (1966) The larva of *Thecophora occidensis*, with comments upon the biology of Conopidae (Diptera). *J. Zool. Lond.* **149**, 263–276.

Smith, K.G.V. (1969) Diptera: Conopidae. *Handbook for the Identification of British Insects* **10**(3a), 1–18.

Snow, A.A. and Whigham, D.F. (1989) Cost of flower and fruit production in *Tipularia discolor*. *Ecology* **70**, 1286–1293.

Soltz, R.L. (1986) Foraging path selection in bumblebees: Hindsight or foresight? *Behaviour* **99**, 1–21.

Sotherton, N.W. (1995) Beetle Banks—helping nature to control pests. *Pest Outlook* **6**, 13–17.

Southern, H.N. and Watson, J.S. (1941) Summer food of the red fox in Great Britain: a preliminary report. *J. Anim. Ecol.* **10**, 1–11.

Southwick, A.K. and Southwick, E.E. (1983) Aging effects on nectar production in two clones of *Asclepias syriaca*. *Oecologia* **56**, 121–125.

Southwick, E.E. and Southwick, L. (1992) Estimating the economic value of honey bees (Hymenoptera: Apidae) as agricultural pollinators in the United States. *J. Econ. Entomol.* **85**, 621–633.

Southwick, E.E. and Buchmann, S.L. (1995) Effects of horizon landmarks on homing success of honeybees. *Am. Nat.* **146**, 748–764.

Sowig, P. (1989) Effects of flowering plant's patch size on species composition of pollinator communities, foraging strategies, and resource partitioning in bumble-bees (Hymenoptera: Apidae). *Oecologia* **78**, 550–558.

Spaethe, J., Tautz, J., and Chittka, L. (2001) Visual contraints in foraging bumblebees: flower size and color affect search time and flight behavior. *Proc. Nat. Acad. Sci.* **98**, 3898–3903.

Spangler, H.G. and Moffett, J.O. (1977) Honeybee visits to tomato flowers in polyethylene greenhouses. *Am. Bee J.* **117**, 580–582.

Spooner, G.M. (1937) *Hymenoptera Aculeata* from the north-west highlands. *Scot. Nat.* 15–23.

Stanghellini, M.S., Ambrose, J.T., and Schultheis, J.R. (1997) The effects of *A. mellifera* and *B. impatiens* pollination on fruit set and abortion of cucumber and watermelon. *Am. Bee J.* **137**, 386–391.

Stanghellini, M.S., Ambrose, J.T., and Schultheis, J.R. (1998) Seed production in watermelon: a comparison between two commercially available pollinators. *HortScience* **33**, 28–30.

Stanton, M.L. and Preston, R. (1988) A qualitative model for evaluating the effects of flower attractiveness on male and female fitness in plants. *Am. J. Bot.* **75**, 540–544.

Stapel, C. (1933) Undersøgelser over Humlebier (*Bombus* Latr.), deres Udbredelse, Trækplanter og Betydning for Bestøvningen of Rødkløver (*Trifolium pratense* L.). *Tidsskr. Planteavl.* **39**, 193–294.

Stapledon, R.G. (1935) *The land, now and tomorrow.* Faber and Faber, London.

Stebbing, P.D. (1965) A study of the mite *Parasitus bomborum*. *Essex Nat* **3**, 284–287.

Steffan-Dewenter, I. and Tscharntke, T. (1997) Bee diversity and seed set in fragmented habitats. *Acta Hortic.* **437**, 231–234.

Steffan-Dewenter, I. and Tscharntke, T. (1999) Effects of habitat isolation on pollinator communities and seed set. *Oecologia* **121**, 432–440.

Steen, J.J.M. van der (1994) Method development for the determination of the contact LD_{50} of pesticides for bumblebees (*Bombus terrestris* L.). *Apidologie* **25**, 463–465.

Steen, J.J.M. van der (2001) Review of the methods to determine the hazard and toxicity of pesticides to bumblebees. *Apidologie* **32**, 399–406.

Stein, G. (1963) Über den Sexuallockstoff von Hummelmännchen. *Naturwissenschaften* **50**, 305.

Stevenson, A.G. (1981) Flower and fruit abortion: proximate causes and ultimate functions. *Annu. Rev. Ecol. Syst.* **12**, 245–279

Stimec, J., ScottDupree, C.D., McAndrews, J.H. (1997) Honey Bee, *Apis mellifera*, pollen foraging in southern Ontario. *Can. Field-natur.* **111**, 454–456.

Stoddard, F.L. and Bond, D.A. (1987) The pollination requirements of the faba bean. *Bee World* **68**, 564–577.

Stone, G.N. and Willmer, P.G. (1989) Warm-up rates and body temperatures in bees: the importance of body size, thermal regime and phylogeny. *J. Exp. Biol.* **147**, 303–328.

Stout, J.C. (2000) Does size matter? Bumblebee behaviour and the pollination of *Cytisus scoparius* L. (Fabaceae). *Apidologie* **31**, 129–139.

Stout, J.C., Allen, J.A., and Goulson, D. (2000) Nectar robbing, forager efficiency and seed set: bumblebees foraging on the self incompatible plant *Linaria vulgaris* Mill. (Scrophulariaceae). *Acta Oecologia* **21**, 277–283.

Stout, J.C. and Goulson, D. (2000) Bumblebees in Tasmania: their distribution and potential impact on Australian flora and fauna. *Bee World* **81**, 80–86.

Stout, J.C. and Goulson, D. (2002) The influence of nectar secretion rates on the responses of bumblebees (*Bombus* spp.) to previously visited flowers. *Behav. Ecol. Sociobiol.* **52**, 239–246.

Stout, J.C., Goulson, D., and Allen, J.A. (1998) Repellent scent marking of flowers by a guild of foraging bumblebees (*Bombus* spp.). *Behav. Ecol. Sociobiol.* **43**, 317–326.

Stout, J.C. and Goulson, D. (2001) The use of conspecific and interspecific scent marks by foraging bumblebees and honeybees. *Anim. Behav.* **62**, 183–189.

Stout, J.C., Kells, A.R., and Goulson, D. (2002) Pollination of the invasive exotic shrub *Lupinus arboreus* (Fabaceae) by introduced bees in Tasmania. *Biol. Conserv.* **106**, 425–434.

Stubbs, C.S. and Drummond, F.A. (2001) *Bombus impatiens* (Hymenoptera: Apidae): an alternative to *Apis mellifera* (Hymenoptera: Apidae) for lowbush blueberry pollination. *J. Econ. Entomol.* **94**, 609–616.

Stubbs, C.S., Drummond, F.A., and Osgood, E.A. (1994) *Osmia ribifloris biedermannii* and *Megachile rotundata* (Hymenoptera: Megachilidae) introduced into the lowbush blueberry agroecosystem in Maine. *J. Kans. Entomol. Soc.* **67**, 173–185.

Sugden, E.A. and Pyke, G.H. (1991) Effects of honey bees on colonies of *Exoneura asimillima*, an Australian native bee. *Aust. J. Ecol.* **16**, 171–181.

Sugden, E.A., Thorp, R.W., and Buchmann, S.L. (1996) Honey bee native bee competition: Focal point for environmental change and apicultural response in Australia. *Bee World* **77**, 26–44.

Sundström, L. (1994) Sex ratio bias, relatedness asymmetry and queen mating frequency in ants. *Nature* **367**, 266–268.

Surholt, B., Greive, H., Baal, T., and Bertsch, A. (1990) Non-shivering thermogenesis in asynchronous flight muscles of bumblebees? Comparative studies on males of *Bombus terrestris*, *Xylocopa sulcatipes* and *Acherontia atropos*. *Comp. Biochem. Physiol.* **97**, 493–499.

Surholt, B., Greive, H., Baal, T., and Bertsch, A. (1991) Warm-up and substrate cycling in flight muscles of male bumblebees, *Bombus terrestris*. *Comp. Biochem. Physiol.* **98**, 299–303.

Sutcliffe, G.H. and Plowright, R.C. (1988) The effects of food supply on adult size in the bumblebee *Bombus terricola* Kirby (Hymenoptera: Apidae). *Can Entomol.* **120**, 1051–1058.

Sutcliffe, G.H. and Plowright, R.C. (1990) The effects of pollen availability on the development time in the bumble bee *Bombus terricola* K. (Hymenoptera: Apidae). *Can. J. Zool.* **68**, 1120–1123.

Svensson, B., Lagerlöf, J., and Svensson, B.G. (2000) Habitat preferences of nest-seeking bumble bees (Hymenoptera: Apidae) in an agricultural landscape. *Agric. Ecosys. Environ.* **77**, 247–255.

Svensson, B.G. (1979) *Pyrobombus lapponicus* Auct., in Europe recognised as two species: *P. lapponicus* (Fabricius, 1793) and *P. monticolla* (Smith, 1849) (Hymenoptera: Apoidea: Bombinae). *Entomol. Scand.* **10**, 275–296.

Svensson, B.G. (1980) Patrolling behaviour of bumble bee males (Hymenoptera, Apidae) in a subalpine/alpine area, Swedish Lapland. *Zoon* **7**, 67–94.

Svensson, B.G., Appelgren, M., and Bergström, G. (1984) Geranylgeranyl acetate and 2-heptadecanone as the dominant marking secretion components of the labial glands in the bumble bee *Alpigenobombus wurfleini*. *Nova Acta Regiae Soc. Sci. Ups.*, Ser. V:C **3**, 145–147.

Svensson, B.G. and Bergström, G. (1977) Volatile marking secretions from the labial gland of north European *Pyrobombus* D.T. males (Hymenoptera: Apidae). *Insectes Soc.* **24**, 213–214.

Svensson, B.G. and Bergström, G. (1979) Marking pheromones of *Alpinobombus* males. *J. Chem. Ecol.* **5**, 603–615.

Swift, J. (1733) On Poetry.

Taylor, G. and Whelan, R.J. (1988) Can honeybees pollinate *Grevillea*? *Aust. Zool.* **24**, 193–196.

Telleria, M.C. (1993) Flowering and pollen collection by the honeybee (*Apis mellifera* L Var *ligustica*) in the Pampas region of Argentina. *Apidologie* **24**, 109–120.

Tepedino, V.J. and Stanton, N.L. (1981) Diversity and competition in bee-plant communities on short-grass prairie. *Oikos* **36**, 35–44.

Tengö, J., Hefetz, A., Bertsch, A., Schmitt, U., Lübke, G., and Francke, W. (1991) Species specificity and complexity of Dufour's gland secretion of bumble bees. *Comp. Biochem. Physiol.* **99B**, 641–646.

Teräs, I. (1976) Flower visits of bumblebees, *Bombus* Latr. (Hymenoptera: Apidae) during one summer. *Ann. Zool. Fenn.* **13**, 200–232.

Teräs, I. (1985) Food plants and flower visits of bumble-bees (*Bombus*: Hymenoptera, Apidae) in southern Finland. *Acta Zool. Fenn.* **179**, 1–120.

Teuber, L.R. and Barnes, D.K. (1979) Breeding alfalfa for increased nectar production. Proceedings of the IV International Symposium on Pollination, Maryland Agricultural Experimental Station Special Miscellaneous Publications 1. pp. 109–116.

Thomas, J.A. (1983) The ecology and conservation of *Lysandra bellargus* (Lepidoptera, Lycaenidae) in Britain. *J. Appl. Ecol.* **20**, 59–83.

Thomson, J.D. (1981) Spatial and temporal components of resource assessment by flower-feeding insects. *J. Anim. Ecol.* **50**, 49–60.

Thomson, J.D. (1988) Effects of variation in inflorescence size and floral rewards on the visitation rates of trap lining pollinators of *Aralia hispida*. *Evol. Ecol.* **2**, 65–76.

Thomson, J.D. and Goodell, K. (2001) Pollen removal and deposition by honeybee and bumblebee visitors to apple and almond flowers. *J. Appl. Ecol.* **38**, 1032–1044.

Thomson, J.D., Maddison, W.P., and Plowright, R.C. (1982) Behaviour of bumblebee pollinators on *Aralia hispida* Vent. (Araliaceae). *Oecologia* **54**, 326–336.

Thomson, J.D., Peterson, S.C., and Harder, L.D. (1987) Response of traplining bumble bees to competition experiments: shifts in feeding location and efficiency. *Oecologia* **71**, 295–300.

Thomson, J.D., Slatkin, M., and Thomson, B.A. (1997) Trapline foraging by bumble bees 2. Definition and detection from sequence data. *Behav. Ecol.* **8**, 199–210.

Thompson, H.M. (2001) Assessing the exposure and toxicity of pesticides to bumblebees (*Bombus* sp.). *Apidologie* **32**, 305–321.

Thompson, H.M. and Hunt, L.V. (1999) Extrapolating from honeybees to bumblebees in pesticide risk assessment. *Ecotoxicology* **8**, 147–166.

Thompson, J.N. (1982) *Interactions and coevolution*. Wiley-Interscience, New York.

Thorp, R.N., Briggs, D.L., Estes, J.R., and Erikson, E.H. (1975) Nectar flourescence under ultaviolet irradiation. *Science* **189**, 476–478.

Thorp, R.N., Briggs, D.L., Estes, J.R., and Erikson, E.H. (1976) *Science* **194**, 342 (Reply to Kevan).

Thorp, R.W. (1979) Structural, behavioral and physiological adaptations of bees (Apoidea) for collecting pollen. *Ann. Missouri Bot. Garden* **66**, 788–812.

Thorp, R.W., Wenner, A.M., and Barthell, J.F. (1994) Flowers visited by honeybees and native bees on Santa Cruz Island. In: *The fourth Californian Islands symposium: Update on the status of resources*, edited by W.L. Halvorson and G.J. Maender, pp. 351–365. Santa Barbara Museum of Natural History, Santa Barbara (CA).

Tierney, S.M. (1994) Life cycle and social organisation of two native bees in the subgenus *Brevineura*. Unpublished BSc (Hons) Thesis, Flinders University of South Australia.

Tinbergen, L. (1960) The natural control of insects in pine woods. 1. Factors influencing the intensity of predation by songbirds. *Arch. Neerl. Zool.* **13**, 265–343.

Tollsten, L. and Bergstrom, L.G. (1993) Fragrance chemotypes of *Platanthera* (Orchidaceae)—the result of adaptation to pollinating moths. *Nordic J. Bot.* **13**, 607–613.

Tollsten, L. and Ovstedal, D.O. (1994) Differentiation in floral scent chemistry among populations of *Conopodium majus* (Apiaceae). *Nordic J. Bot.* **14**, 361–367.

Tooby, J. (1982) Pathogens, polymorphism, and the evolution of sex. *J. Theor. Biol.* **97**, 557–576.

Torchio, P.F. (1987) Use of non-honey bee species as pollinators of crops. *Proc. Entomol. Soc. Ont.* **118**, 111–124.

Torchio, P.F. (1990) Diversification of pollination strategies for U.S. crops. *Environ. Entomol.* **19**, 1649–1656.

Townsend, T.H.C. (1936) The mature larva and puparium of *Brachycoma sarcophagina* (Townsend) (Diptera: Metopiidae). *Proc. Entomol. Soc. Wash.* **38**, 92–98.

Trivers, R.L. and Hare, H. (1976) Haplodiploidy and the evolution of the social insects. *Science* **191**, 249–263.

Tuck, W.H. (1897) Note on the habits of *Bombus latreillellus*. *Entomologist's Monthly Magazine* **33**, 234–235.

Udovic, D. (1981) Determinants of fruit set in *Yucca whipplei*: reproductive expenditure and pollinator availability. *Oecologia* **48**, 389–399.

Ugolini, A. (1986) Homing ability of *Polistes gallicus* (L.) (Hymenoptera Vespidae). *Monit. Zool. Ital. (N.S.)* **20**, 1–15.

Ugolini, A., Kesller, A., and Ishay, J.S. (1987) Initial orientation and homing by oriental hornets, Vespa orientalis L (Hymenoptera: Vespidae). *Monit. Zool. Ital. (N.S.)* **21**, 157–164.

Urbanová, K., Valterová, I., Hovorka, O., and Kindl, J. (2001) Chemotaxonomical characterisation of males of *Bombus lucorum* (Hymenoptera: Apidae) collected in the Czech Republic. *Eur. J. Entomol.* **98**, 111–115.

Vallet, A., Cassier, P., and Lensky, Y. (1991) Ontogeny of the fine structure of the mandibular glands of the honeybee (*Apis mellifera* L.) workers and the pheromonal activity of 2-heptanone. *J. Insect Physiol.* **37**, 789–804.

Valterová, I., Urbanová, K., Hovorka, O., and Kindl, J. (2001) Composition of the labial gland secretion of the bumblebee males *Bombus pomorum*. *Verlad der Zeitschrift für Naturforschung* **56c**, 430–436.

Van den Eijnde, J., de Ruijter, A., and Van der Steen, J. (1991) Method for rearing *Bombus terrestris* continuously and the production of bumble bee colonies for pollination purposes. *Acta Hortic.* **288**, 154–158.

Van der Blom, J. (1986) Reproductive dominance within colonies of *Bombus terrestris* (L.). *Behaviour* **97**, 37–49.

Van Doorn, A. (1987) Investigations into the regulation of dominance behaviour and the division of labour in bumble bee colonies (*Bombus terrestris*). *Neth. J. Zool.* **37**, 255–276.

Van Doorn, A. and Heringa, J. (1986) The ontogeny of a dominance hierarchy in colonies of the bumble bee *Bombus terrestris*. *Insectes Soc.* **33**, 3–25.

Van Heemert, C., de Ruijter, A., Van den Eijnde, J., and Van der Steen, J. (1990) Bees in agriculture; Year round production of bumble bee colonies for crop production. *Bee World* **71**, 54–56.

Van Honk, C.G.J. and Hogeweg, P. (1981) The ontogeny of social structure in a captive *Bombus terrestris* colony. *Behav. Ecol. Sociobiol.* **9**, 111–119.

Van Honk, C.G.J., Röseler, P.-F., Velthuis, H.H.W., and Hoogeveen, J.C. (1981) Factors influencing the egg-laying of workers in a captive *Bombus terrestris* colony. *Behav. Ecol. Sociobiol.* **9**, 9–14.

Van Honk, C.G.J., Velthuis, H.H.W., and Röseler, P.-F. (1978) A sex pheromone from the mandibular glands in bumblebee queens. *Experientia* **34**, 838.

Vennerstrom, P., Soderhall, K., and Cerenius, L. (1998) The origin of two crayfish plague (*Aphanomyces astaci*) epizootics in Finland on noble crayfish, *Astacus astacus*. *Ann. Zool. Fenn.* **35**, 43–46.

Vieting, U.K. (1988) Untesuchungen in Hessen über Auswirkung und Bedeutung von Ackerschonstreifen. 1. Konzeption des Projektes und der Botanishee Aspekt. *Mitteilungen aus der Biologischen Bundesanstalt für Land- und Forstwirtschaft Berlin-Dahlem* **247**, 29–41.

Villalobos, E.M. and Shelly, T.E. (1987) Observations on the behavior of male *Bombus sonorus* (Hymenoptera: Apidae). *J. Kans. Entomol. Soc.* **60**, 541–548.

Visscher, P.K. and Seeley, T.D. (1982) Foraging strategy of honeybee colonies in a temperate deciduous forest. *Ecology* **63**, 1790–1801.

Vogt, D.F. (1986) Thermoregulation in bumblebee colonies. I. Thermoregulatory versus brood-maintenance behaviors during acute changes in ambient temperature. *Physiol. Zool.* **59**, 55–59.

Vogt, D.F. and Heinrich, B. (1994) Abdominal temperature regulation by Arctic bumblebees. *Physiol. Zool.* **66**, 257–269.

Vogt, D.F., Heinrich, B., Dabolt, T.O., and McBath, H.L. (1994) Ovary development and colony founding in subarctic and temperate-zone bumblebee queens. *Can. J. Zool.* **72**, 1551–1556.

Voss, R., Turner, M., Inouye, R., Fisher, M., and Cort, R. (1980) Floral biology of *Markea neurantha* Hemsley (Solanaceae), a bat-pollinated epiphyte. *Am. Midland Nat.* **103**, 262–268.

Waddington, K.D. (1980) Flight patterns of foraging bees relative to density of artificial flowers and distribution of nectar. *Oecologia* **44**, 199–204.

Waddington, K.D. (1983a) Floral-visitation-sequences by bees: models and experiments. In: *Handbook of experimental pollination ecology*, edited by C.E. Jones and R.J. Little, pp. 461–473. Van Nostrand Reinhold, New York.

Waddington, K.D. (1983b) Foraging Behaviour of pollinators. In: *Pollination biology*, edited by L.A. Real, pp. 213–239. Academic Press, New York.

Waddington, K.D., Allen, T., and Heinrich, B. (1981) Floral preferences of bumblebees (*Bombus edwardsii*) in relation to intermittent versus continuous rewards. *Anim. Behav.* **29**, 779–784.

Waddington, K.D. and Heinrich, B. (1979) The foraging movements of bumblebees on vertical 'inflorescences': an experimental analysis. *J. Comp. Physiol.* **134**, 113–117.

Waddington, K.D. and Heinrich, B. (1981) Patterns of movement and floral choice by foraging bees. In: *Foraging behavior: Ecological, ethological, and psychological approaches*, edited by A. Kamil and T. Sargent, pp. 215–230. Garland STPM Press, New York.

Waddington, K.D., Visscher, P.K., Herbert, T.J., and Richter, M.R. (1994) Comparisons of forager distributions from matched honey bee colonies in suburban environments. *Behav. Ecol. Sociobiol.* **35**, 423–429.

Waller, R. (1962) *Prophet of the New Age*. Faber and Faber, London.

Wallraff, H.G. (1980) Does pigeon homing depend on stimuli perceived during displacement? I. Experiments in Germany. *J. Comp. Physiol.* **139**, 193–201.

Wallraff, H.G. (1990) Navigation by homing pigeons. *Ethol. Ecol. Evol.* **2**, 81–115.

Wallraff, H.G., Foà A., and Ioalè P. (1980) Does pigeon homing depend on stimuli perceived during displacement? II. Experiments in Italy. *J. Comp. Physiol.* **139**, 203–208.

Walther-Hellwig, K. and Frankl, R. (2000) Foraging distances of *Bombus muscorum*, *Bombus lapidarius* and *Bombus terrestris* (Hymenoptera, Apidae). *J. Insect Behav.* **13**, 239–246.

Waser, N.M. (1982a) A comparison of distances flown by different visitors to flowers of the same species. *Oecologia* **55**, 251–257.

Waser, N.M. (1982b) Competition for pollination and floral character differences among sympatric plant species: a review of evidence. In: *Handbook of experimental pollination ecology*, edited by C.E. Jones and R.J. Little, pp. 277–293. Van Nostrand Reinhold, New York.

Waser, N.M. (1983) The adaptive significance of floral traits: ideas and evidence. In: *Pollination biology*, edited by L.A. Real, pp. 241–285. Academic Press, New York.

Waser, N.M. (1986) Flower constancy: definition, cause and measurement. *Am. Nat.* **127**, 593–603.

Waser, N.M. (1998) Pollination, angiosperm speciation, and the nature of species boundaries. *Oikos* **81**, 198–201.

Waser, N.M., Chittka, L., Price, M.V., Williams, N.M., and Ollerton, J. (1996) Generalization in pollination systems, and why it matters. *Ecology* **77**, 1043–1060.

Waser, N.M. and Mitchell, R.J. (1990) Nectar standing crops in *Delphinium nelsonii* flowers: spatial autocorrelation among plants? *Ecology* **71**, 116–123.

Watson, P.J. and Thornhill, R. (1994) Fluctuating asymmetry and sexual selection. *Trend. Ecol. Evol.* **9**, 21–25.

Webster, S. and Felton, M. (1993) *Targeting for nature conservation in agricultural policy. Land use policy,* **January,** 67–82.

Wehner, R. (1981) Spatial Vision Arthropods, In: *Handbook of sensory physiology, Vol VII/6C,* edited by H. Autrum, R. Jury, W.R. Loewenstein, D.M. Mackay, and H.-L. Teuber, pp. 98–201. Springer Verlag.

Wehner, R. (1989) Spatial Vision in Arthropods In: *Handbook of sensory physiology Vol VII/6C,* edited by H. Autrum, R. Jury, W.R. Loewenstein, D.M. Mackay, and H.-L. Teuber, pp. 98–201. Springer Verlag, 1981, Witte G.R., Seger J., and Häfner, N. Hummelschauanlagen. Veröffentl. des Schulbiologiezentrums, Hannover.

Wehner, R. (1994) The polarization-vision project: championing organismic biology, In: *Neural basis of behavioural adaptation,* edited by K. Schildberger, N. Elsner, and G. Fischer, pp. 102–207. Stuttgart, New York,

Wehner, R. (1996) Middle scale navigation: the insect case. *J. Exp. Biol.* **199**, 125–127.

Weidenmüller, A., Kleineidam, C. and Tautz, J. (2002) Collective control of nest climate parameters in bumblebee colonies. *Anim. Behav.* **63**, 1065–1071.

Weislo, W.T. (1981) The roles of seasonality, host synchrony, and behaviour in the evolutions and distributions of nest parasites in Hymenoptera (Insecta), with special reference to bees (Apoidea). *Biol. Rev.* **62**, 515–543.

Weiss, M.R. (1995a) Associative color learning in a nymphalid butterfly. *Ecol. Entomol.* **20**, 298–301.

Weiss, M.R. (1995b) Floral color-change—a widespread functional convergence. *Am. J. Bot.* **82**, 167–185.

Weiss, M.R. and Lamont, B.B. (1997) Floral color change and insect pollination: A dynamic relationship. *Israel J. Plant Sci.* **45**, 185–199.

Wells, H. and Wells, P.H. (1983) Honeybee foraging ecology: optimal diet, minimal uncertainty or individual constancy? *J. Anim. Ecol.* **52**, 829–836.

Wells, H. and Wells, P.H. (1986) Optimal diet, minimal uncertainty and individual constancy in the foraging of honeybees, *Apis mellifera. J. Anim. Ecol.* **55**, 881–891.

Wells, H., Hill, P.S., and Wells, P.H. (1992) Nectarivore foraging ecology—rewards differing in sugar types. *Ecol. Entomol.* **17**, 280–288.

Wenner, A.M. and Thorp, R.W. (1994) Removal of feral honey bee (*Apis mellifera*) colonies from Santa Cruz Island. In: *The fourth Californian islands symposium: Update on the status of resources,* edited by W.L. Halvorson and G.J. Maender, pp. 513–522. Santa Barbara Museum of Natural History, Santa Barbara (CA).

Wenner, A.M. and Wells, P.H. (1990) *Anatomy of a controversy.* Columbia University Press, New York.

West, E.L. and Laverty, T.M. (1998) Effect of floral symmetry on flower choice and foraging behaviour of bumble bees. *Can. J. Zool.* **76**, 730–739.

Westergaard, G.C. and Suomi, S.J. (1996) Lateral bias for rotational behavior in tufted capuchin monkeys (*Cebus apella*). *J. Comp. Psychol.* **110**, 199–202.

Westerkamp, C. (1991) Honeybees are poor pollinators—why? *Plant Syst. Evol.* **177**, 71–75.

Westrich, P. (1989) *Die Bienen Baden-Württembergs. Vol. 1,* Ulmer Verlag, Stuttgart, 431 pp.

Westrich, P. (1996) Habitat requirements of central European bees and the problems of partial habitats. In: *The conservation of bees,* edited by A. Matheson, S.L. Buchmann, C. O'Toole, P. Westrich, and I.H. Williams, pp. 2–16. Academic Press, London.

Westrich, P., Schwenninger, H.-R., Dathe, H., Riemann, H., Saure, C., Voith, J., and Weber, K. (1998) Rote Liste der Bienen (Hymenoptera: Apidae). In: *Rote Liste Gefährdeter Tiere*

Deutschlands, edited by Bundesamt für Naturschutz, Naturschutz 55, Bonn: Schriftenr. Landschaftspf, 119–129.

Wetherwax, P.B. (1986) Why do honeybees reject certain flowers? *Oecologia* **69**, 567–570.

Whidden, T.L. (1996) The fidelity of commercially reared colonies of *Bombus impatiens* Cresson (Hymenoptera: Apidae) to lowbush blueberry in southern New Brunswick. *Can. Entomol.* **128**, 957–958.

Whitfield, J.B. and Cameron, S.A. (1993) Comparative notes on hymenopteran parasitoids in bumble bee and honey bee colonies (Hymenoptera: Apidae) reared adjacently. *Entomol. News* **104**, 240–248.

Whitfield, J.B., Cameron, S.A., Ramírez, S.R., Roesch, K., Messinger, S., Taylor, O.M., and Cole, D. (2001) Review of the *Apanteles* species (Hymenoptera: Braconidae) attacking Lepidoptera in *Bombus (Fervidobombus)* (Hymenoptera: Apidae) colonies in the New World, with description of a new species from South America. *Ann. Entomol. Soc. Am.* **96**, 851–857.

Widmer, A. and Schmid-Hempel, P. (1999) The population genetic structure of a large temperate pollinator species, *Bombus pascuorum* (Scopoli) (Hymenoptera: Apidae). *Mol. Ecol.* **8**, 387–398.

Widmer, A., Schmid-Hempel, P., Estoup, A., and Scholl, A. (1998) Population genetic structure and colonisation history of *Bombus terrestris* s.l. (Hymenoptera: apidae) from the Canary Islands and Madeira. *Heredity* **81**, 563–572.

Wiebes, J.J. (1979) Co-evolution of figs and their insect pollinators. *Annu. Rev. Ecol. Systemat.* **10**, 1–12.

Williams, A.A., Hollands, T.A., and Tucknott, O.G. (1981) The gas chromatographic-mass spectrometric examination of the volatiles produced by the fermentation of a sucrose solution. *Zeitschrift fuer Lebensmittel- Untersuchung und Forschung*, **172**, 377–381.

Williams, C.S. (1995) Conserving Europe's bees: why all the buzz? *TREE* **10**, 309–310.

Williams, C.S. (1998) The identity of the previous visitor influences flower rejection by nectar-collecting bees. *Anim. Behav.* **56**, 673–681.

Williams, I.H. and Christian, D.G. (1991) Observations on *Phacelia tanacetifolia* Bentham (Hydrophyllaceae) as a food plant for honey bees and bumble bees. *J. Apic. Res.* **21**, 236–245.

Williams, I.H., Martin, A.P., and White, R.P. (1987) The effect of insect pollination on plant development and seed production in winter oilseed rape (*Brassica napus* L.). *J. Agric. Sci. Cambridge*, **109**, 135–139.

Williams, N.H. (1982) Floral fragrances as cues in animal behavior. In: *Handbook of experimental pollination ecology*, edited by C.E. Jones and R.J. Little, pp. 50–72. Van Nostrand Reinhold, New York.

Williams, N.M. and Thomson, J.D. (1998) Trapline foraging by bumble bees: III. Temporal patterns of visitation and foraging success at single plants. *Behav. Ecol.* **9**, 612–621.

Williams, P.A. and Timmins, S.M. (1990) Weeds in New Zealand Protected Natural Areas: a review for the Department of Conservation. Science and Research Series No. 14, Department of Conservation, Wellington, N.Z. 114 pp.

Williams, P.H. (1982) The distribution and decline of British bumble bees (*Bombus* Latr). *J. Apic. Res.* **21**, 236–245.

Williams, P.H. (1985a) A preliminary cladistic investigation of relationships among the bumble bees (Hymenoptera, Apidae). *Syst. Entomol.* **10**, 239–255.

Williams, P.H. (1985b) On the distribution of bumble bees (Hymenoptera, Apidae), with particular regard to patterns within the British Isles. PhD thesis, University of Cambridge, UK. 180 pp.

Williams, P.H. (1986) Environmental change and the distribution of British bumble bees (*Bombus* Latr.). *Bee World* **67**, 50–61.
Williams, P.H. (1988) Habitat use by bumble bees (*Bombus* spp.). *Ecol. Entomol.* **13**, 223–237.
Williams, P.H. (1989a) *Bumble bees—and their decline in Britain*, Central Association of Bee-Keepers, Ilford.
Williams, P.H. (1989b) Why are there so many species of bumble bees at Dungeness? *Bot. J. Linn. Soc.* **101**, 31–44.
Williams, P.H. (1991) The bumble bees of the Kashmir Himalaya (Hymenoptera: Apidae, Bombini). *Bull. Br. Mus. (Natu. Hist.) (Entomol.)*, **60**, 1–204.
Williams, P.H. (1994) Phylogenetic relationships among bumble bees (*Bombus* Latr.): a reappraisal of morphological evidence. *Syst. Entomol.* **19**, 327–344.
Williams, P.H. (1998) An annotated checklist of bumble bees with an analysis of patterns of description (Hymenoptera: Apidae, Bombini). *Bull. Nat. Hist. Mus. Lond. (Entomol.)* **67**, 79–152.
Willmer, P.G., Bataw, A.A.M., and Highes, J.P. (1994) The superiority of bumblebees to honeybees as pollinators: insect visits to raspberry flowers. *Ecol. Entomol.* **19**, 271–284.
Wills, R.T., Lyons, M.N., and Bell, D.T. (1990) The European honey bee in Western Australian kwongan: foraging preferences and some implications for management. *Proc. Ecol. Soc. Aust.* **16**, 167–176.
Willson, M.F. and Ågren, J. (1989) Differential floral rewards and pollination by deceit in unisexual flowers. *Oikos* **55**, 23–29.
Wilms, W., ImperatrizFonseca V.L., and Engels, W. (1996) Resource partitioning between highly eusocial bees and possible impact of the introduced Africanized honey bee on native stingless bees in the Brazilian Atlantic Rainforest. *Stud. Neotropic. Fauna Environ.* **31**, 137–151.
Wilms, W. and Wiechers, B. (1997) Floral resource partitioning between native *Melipona* bees and the introduced Africanized honey bee in the Brazilian Atlantic rain forest. *Apidologie* **28**, 339–355.
Wilson, E.O. (1971) *The insect societies*. Harvard University Press, Cambridge, Mass.
Wilson, E.O. (1990) *Success and dominance in ecosystems: The case of the social insects*. Ecology Institute, Oldendorf/Luhe, Germany, pp. 104.
Wilson, M.F. and Price, P.W. (1977) The evolution of inflorescence size in *Asclepias* (Asclepiadaceae). *Evolution* **31**, 495–511.
Wilson, P. and Stine, M. (1996) Floral constancy in bumble bees—handling efficiency or perceptual conditioning. *Oecologia* **106**, 493–499.
Wilson, P. and Thomson, J.D. (1991) Heterogeneity among floral visitors leads to discordance between removal and deposition of pollen. *Ecology* **72**, 1503–1507.
Wilson, P., Thomson, J.D., Stanton, M.L., and Rigney, L.P. (1994) Beyond floral Batemania – gender biases in selection for pollination success. *Am. Nat.* **143**, 283–296.
Winston, M.L. (1987) *The biology of the honey bee*. Harvard University Press, Cambridge, Massachusetts.
Winston, M.L. and Graf, L.H. (1982) Native bee pollinators of berry crops in the Fraser Valley of British Columbia. *J. Entomol. Soc. Br. Columbia* **79**, 14–20.
Winston, M.L. and Scott, C.D. (1984) The value of bee pollination to Canadian apiculture. *Can. Beekeeping* **11**, 134.
Witschko, W. and Witschko, R. (1988) Magnetic orientation in birds, In: *Current ornithology, Vol 5*, edited by R.F. Johnston, pp. 52–89. Plenum Press, New York, London.
Wojtowski, F. and Majewski, J. (1964) Observations and experiments on the settlement of bumble bees in nest boxes. *Roczniki Wyzszej Szkoly Rolniczej w Poznaniu*, **19**, 185–196.

Wolda, H. and Roubik, D.W. (1986) Nocturnal bee abundance and seasonal bee activity in a Panamanian forest. *Ecology* **67**, 426–433.

Wolf, T.J., Ellington, C.P., Davis, S., and Fletham, M.J. (1996) Validation of the doubly labelled water technique for bumblebees *Bombus terrestris* (L.) *J. Exp. Biol.* **199**, 959–972.

Wolf, T.J., Ellington, C.P., and Begley, I.S. (1999) Foraging costs in bumblebees: field conditions cause large individual differences. *Insectes Soc.* **46**, 291–295.

Wolf, T.J. and Schmid-Hempel, P. (1989) Extra loads and foraging life span in honeybee workers. *J. Anim. Ecol.* **58**, 943–954.

Woodward, D.R. (1990) Food demand for colony development, crop preference and food availability for *Bombus terrestris* (L.) (Hymenoptera: Apidae). PhD thesis, Massey University, Palmerston North, New Zealand.

Woodward, D.R. (1996) Monitoring for impact of the introduced leafcutting bee, *Megachile rotundata* (F) (Hymenoptera: Megachilidae), near release sites in South Australia. *Aust. J. Entomol.* **35**, 187–191.

Woodward, G.L. and Laverty, T.M. (1992) Recall of flower handling skills by bumblebees: a test of Darwin's interference hypothesis. *Anim. Behav.* **44**, 1045–1051.

Woyke, J. (1963) Drone larvae from fertilized eggs of the honeybee. *J. Apic. Res.* **2**, 19–24.

Woyke, J. (1979) Sex determination in *Apis cerana indica*. *J. Apic. Res.* **18**, 122–127.

Wratt, E.C. (1968) The pollinating activities of bumble bees and honey bees in relation to temperature, cometing forage plants, and competition from other foragers. *J. Apic. Res.* **7**, 61–66.

Yarrow, I.H.H. (1970) Is *Bombus inexspectatus* (Tkalcu) a workerless obligate parasite? (Hym. Apidae). *Insectes Soc.* **17**, 95–112.

Yazgan, M.Y., Leckman, J.F., and Wexler, B.E. (1996) A direct observational measure of whole body turning bias. *Cortex*, **32**, 173–176.

Young, A., Boyle, T., and Brown, T. (1996) The population genetic consequences of habitat fragmentation for plants. *TREE* **11**, 413–418.

Zeuner, F.E. and Manning, F.J. (1976) A monograph of fossil bees (Hymenoptera: Apoidea). *Bull. Br. Mus. (Nat. Hist.), Geol.* **27**, 151–268.

Zimmerman, J.K. and Aide, T.M. (1989) Patterns of fruit production in a neotropical orchid: pollinator vs. resource limitation. *Am. J. Bot.* **76**, 67–73.

Zimmerman, M.L. (1979) Optimal foraging: a case for random movement. *Oecologia* **43**, 261–267.

Zimmerman, M.L. (1981a) Nectar dispersion patterns in a population of *Impatiens capensis*. *Virg. J. Sci.* **32**, 150–152.

Zimmerman, M.L. (1981b) Patchiness in the dispersion of nectar resources: probable causes. *Oecologia* **49**, 154–157.

Zimmerman, M.L. (1981c) Optimal foraging, plant density and the marginal value theorem. *Oecologia* **49**, 148–153.

Zimmerman, M.L. (1982) Optimal foraging: Random movement by pollen collecting bumblebees. *Oecologia* **53**, 394–398.

Zimmerman, M.L. (1983) Plant reproduction and optimal foraging: experimental nectar manipulations in *Delphinium nelsonii*. *Oikos* **41**, 57–63.

Zimmerman, M. and Cook, S. (1985) Pollinator foraging, experimental nectar-robbing and plant fitness in *Impatiens capensis*. *Am. Midland Nat.* **113**, 84–91.

Zimmerman, M.L. and Pyke, G.H. (1986) Reproduction in *Polemonium*: patterns and implications of floral nectar production and standing crop. *Am. J. Bot.* **73**, 1405–1415.

Zollikofer, C.P.E., Wehner, R., and Fukushi, T. (1995) Optical scaling in conspecific *Cataglyphis* ants. *J. Exp. Biol.* **198**, 1637–1646.

Index

Acarina 53, 63–4, 131, 153, 169–70
Aconitum 95, 98
Aconitum columbianum 124
Aconitum napellus 98
Aconitum septentrionale 95, 152
Aconitum variegatum 98
Acrostalagmus 59
Actinidia deliciosa 130, 131, 133
Aeropetes tulbhagia 95
Africa 9, 95, 132, 136, 150, 162, 175
aggression, *see* conflict
alfalfa, *see Medicago sativa*
alloethism 13, 19, 24, 25, 29
allozymes 150
Anacardium occidentale 171
Anchusa officinalis 88
Anodontobombus 3, 5, 22, 23
Antherophagus nigricornis 64
Anthophora 58, 84
Anthophora abrupta 75
Anthophora pilipes 108
Anthophoridae 1, 116, 168
Aphanomyces astaci 170
Aphomia sociella 54, 158
Apiaceae 95
Apicystis bombi 60
Apis cerana 170
Apis mellifera 22, 24, 25, 28, 29, 31, 48, 50, 58,
    59, 73, 76, 81, 84, 93, 96, 97, 99, 101, 105,
    107, 108, 109, 110, 111, 113, 114, 115, 116,
    117, 118, 121, 130–2, 133, 134, 139, 140,
    141, 145, 147, 153, 156, 157, 162–75
    africanized 131, 164, 167, 168
Apodemus sylvaticus 54
apple, *see Pyrus malus*
Argentina 136, 162, 167
Argiope aurantia 54
artificial insemination 50
Ascosphaera aggregata 170
Ascosphaera apis 169
Asia 2, 9, 41, 162
Asilidae 54
Aspergillus candidus 59
assortative mating 43–4, 45
Astacus astacus 170
Atta 24, 25

Australia 130, 136, 161, 162, 164, 165, 166, 167,
    168, 171, 172, 174, 175
Austropotamobius pallipes 170

bacterial pathogens 58
badger, *see Meles meles*
Banksia ornate 171
barberry shrub, *see Berberis darwinii*
Beauveria 59
bee-eater, *see Merops apiaster*
behavioural changes, following parasitization
    56, 62
Berberis darwinii 173
Bettsia alvei 170
blueberry, *see Vaccinium*
Boettcharia litorosa 57
Bombus americanorum 54
Bombus appositus 124, 141
Bombus atratus 10, 28, 45
Bombus bifarius 45, 46
Bombus bimaculatus 57
Bombus bohemicus 45, 67–8
Bombus brodmannicus 95
Bombus californicus 42, 45–6
Bombus confusus 41, 46
Bombus consobrinus 95, 98, 152
Bombus crotchii 42
Bombus cullumanus 143
Bombus derstaeckeri 95
Bombus distinguendus 127, 143, 149, 152
Bombus diversus 28, 150
Bombus edwardsii 13, 29
Bombus fervidus 32, 42, 57, 58, 98
Bombus flavifrons 74, 124, 127–8
Bombus frigidus 4, 9, 45, 46
Bombus griseocollis 41
Bombus hortorum 6, 8, 12, 43, 44, 63, 64, 117,
    123, 126, 127, 131, 139, 151, 153, 155,
    156, 158, 161, 162, 169
Bombus humilis 126, 143
Bombus huntii 8
Bombus hyperboreus 66
Bombus hypnorum 8, 37, 38, 46, 48, 51, 55, 66,
    151, 162
Bombus ignitus 169
Bombus impatiens 12, 58, 99, 100, 134, 175

Bombus incarum 10
Bombus inexspectatus 66
Bombus jonellus 9, 140
Bombus lapidarius 5, 8, 12, 26, 30, 32, 43, 44,
    56, 63, 64, 68, 80, 81, 86, 87, 90, 92, 93,
    94, 116, 117, 118, 120, 124, 126, 127, 139,
    151, 152, 155, 156
Bombus lapponicus 43, 46
Bombus latreillelus 42
Bombus lucorum 3, 6, 24, 32, 38, 43, 45, 56, 57,
    63, 66, 67, 68, 74, 86, 87, 124, 126, 127,
    135, 139, 140, 149, 150, 151, 155
Bombus mastrucatus 139
Bombus melanopygus 8, 37, 38
Bombus mendax 41
Bombus monticola 43, 144
Bombus morio 6, 28
Bombus morrisoni 42
Bombus muscorum 55, 81, 143, 152
Bombus nevadensis 41
Bombus niveatus 42
Bombus occidentalis 139–40, 141
Bombus pascuorum 6, 21, 30, 45, 48, 55, 56, 57,
    63, 81, 86, 87, 103, 116, 117, 119, 126,
    127, 131, 135, 139, 140, 146, 150, 151,
    152, 153, 155, 156, 157
Bombus pennsylvanicus 30, 58, 98
Bombus perplexus 21
Bombus polaris 6, 13, 15, 66
Bombus pomorum 143
Bombus pratorum 4, 6, 9, 55, 68, 80, 115, 117,
    126, 127, 139, 151, 155
Bombus regeli 42
Bombus ruderarius 42, 66, 81, 143, 152
Bombus ruderatus 53, 123, 136, 143, 161,
    162, 169
Bombus rufocinctus 41, 46, 66, 127–8
Bombus rupestris 68
Bombus sonorus 42
Bombus soroeensis 143
Bombus subterraneus 42, 53, 127, 143, 144, 152,
    157, 161-2, 169
Bombus sylvarum 4, 81, 144, 149, 152, 157
Bombus sylvestris 43, 68, 80
Bombus ternarius 21, 30, 125
Bombus terrestris 4, 6, 9, 16, 19, 20, 21, 22, 23,
    24, 25, 26, 27, 29, 30, 31, 32, 33, 35, 36,
    37, 38, 43, 44, 48, 49, 50, 55, 56, 57, 59,
    60, 61, 62, 63, 64, 65, 66, 67, 71, 74, 75,
    77, 78, 79, 80, 81, 86, 87, 102, 103, 111,
    113, 114, 116, 117, 119, 123, 124, 126, 127,
    131, 132, 134, 135, 136, 139, 140, 146,
    147, 150, 151, 152, 155, 156, 157, 158,
    159, 161, 162, 163, 165, 167, 169, 170,
    171, 173, 174, 175
Bombus terricola 6, 21, 28, 38, 125
Bombus vagans 26, 28
Bombus vestalis 45, 57, 67
Bombus vosnesenskii 13, 16, 29
Borago officinalis 120
Brachicoma devia 57
Brachicoma sarcophagina 57
Braconidae 57
Brassica campestris 133
Brassica juncea 129
Brassica napus 129, 131, 133
Brood care 4–7, 10, 15–16, 20, 22–3, 24, 25, 66
Broom, see *Cytisus scoparius*
Buzz pollination 131

California 167
Camelina sativa 135
Canada 55, 56, 124, 130, 134, 147, 170
Candida 59
Capsicum 133
Carduus acanthoides 108
carpenter bees, see *Xylocopa*
caste determination 19–22
Centaurea 155
Centaurea solstitialis 173
Cephalosporium 59
Cerceris tuberculata 74
Cerinthe 95
Chalicodoma muraria 74
Chalkbrood, see *Ascosphaera apis*
Chile 136, 162
Cimicifuga simplex 111
Cirsium 155
Cirsium arvense 173
Citrullus lanatus 132, 133
Citrus 133
Clethrionomys 54
climate 2, 6, 9, 12, 14, 15, 29, 131, 137,
    138, 163–4
coat length 127
coexistence 123, 125, 128, 163, 168
Colorado 123, 125, 138, 165
Columba livia 76, 101
commercial rearing 31, 132, 134, 162
communication 19, 113–21, 165
compass
    magnetic 85
    sun 76
competition
    among parasitoids 56–7
    for mates 39

for nest sites 162, 169
for pollination services 96
interspecific 121, 123–8, 151, 162, 163–8
intra-colony 30, 34, 36, 80,
competition point 32–5
conflict, intra-colony 31–6, 48, 51, 61
Conopidae 29, 55–7
Conops 55
conservation 81, 143–60
conservation headlands 135, 154, 155, 157, 159
copulation 8, 34, 37, 41, 45, 46, 49, 61, 68
Cordyceps 59
corolla depth 30, 123–6, 139, 152, 156, 158, 173
Corydalis caseana 141
Cosmea maritima 135
cranberry, *see Vaccinium macrocarpon*
crayfish 170
creeping thistle, *see Cirsium arvense*
Crithidia bombii 48, 51, 59–61, 62
cuckoo bumblebees, *see Psithyrus*
cucumber, *see Cucumis sativus*
Cucumis melo 130, 133
Cucumis sativus 132, 133
Cucurbita 133
Cuphea 135
Cynoglossum officinale 88
Cytisus scoparius 26, 30, 70, 173

dandelion, *see Taraxacum officinale*
Darwin's interference hypothesis 97, 99, 102, 104
declines
 in bumblebee populations 134, 137, 143–60
 in floral diversity 138, 144–6, 158
defence, of nest 55, 57, 67
Delphinium 123
Delphinium barbeyi 124
departure rules from flower patches 86, 89, 93–4
development time of immature stages 5, 21, 22
diet breadth 152, 163, 165, 174
Digitalis purpurea 109
diploid males 31, 45, 151
dispersal
 of males 149
 of queens 149–50
distribution, geographic 2, 29, 127, 143, 151, 162, 173

Echium vulgare 88, 109
egg-eating 20, 31–2, 67

Eichhornia crassipes 161
Ellington, Charles 70
Encapsulation response 29, 65, 111
Endrosis sarcitrella 64
energetics of flight 28, 70, 78, 79, 166
Ephestia kühniella 64
Euglossinae 73, 75, 84
Euplusia surinamensis 75
Europe 2, 4, 6, 8, 16, 41, 45, 53, 54, 55, 80, 124, 125, 127, 128, 130, 132, 134, 135, 137, 139, 140, 143, 144, 145–6, 148, 150, 151, 152, 153, 154, 156, 157, 162, 169, 170, 173
eusociality 24, 47
evolution of bumblebees 2
Exoneura 168
Exoneura xanthoclypeata 166
extinction 28, 53, 125, 143–4, 146, 149, 151, 162, 175

Fabre, Jean-Henri 74
Fannia canicularis 64
fat reserves 8, 9, 19, 60
Ficus 95
fidelity to foraging sites 74, 85
field bean, *see Vicia faba*
field margins 134–5, 145, 154, 157, 160
flight
 energetics of 17, 28, 70–1, 78–9, 166
 mechanics of 70–1
 speed 28, 70–1, 75, 79
floral complexity 26, 70, 87, 99, 105, 107
floral parasites 171, 172
floral rewards, assessment of 107–8, 110, 115
floral scent 110–11, 113
floral symmetry 110
flower age 107, 108
flower constancy 96–106
flower recognition 102, 110
fluctuating asymmetry, of flowers 110
foraging, also *see* under nectar or pollen
 central place 73
 efficiency 25–7, 30, 102, 103, 115, 121, 165, 166, 172
 range 28, 73–81, 137, 152, 165
Formica neorubfibarbus 138
fossil bumblebees 2
foulbrood, *see Paenibacillus larvae*
fox, *see Vulpes vulpes*
French Guiana 167
Fructose biphosphatase 12, 126, 127

gardens 9, 55, 123, 144, 158–9, 169
gene flow 150

gland
  Dufour's 116
  hypopharyngeal 20
  labial 8, 41, 42, 43, 46, 116
  mandibular 47, 116
  tarsal 116
Glycine max 133
gorse, *see Ulex europeaus*
Gossypium 133
grazing 144, 148, 158, 159, 160
great grey shrike, *see Lanius excubitor*
Grevillea X gaudichaudii 171
guard bees 24-5
Guizotia abyssinica 135

habitat management 73, 135-6, 151, 154-5, 157, 159
Halictidae 1, 162, 168
handedness 86-7
handling, of flowers 26, 70, 77, 78, 90, 91, 92, 95-9, 102-5, 107, 110, 115, 139, 166
haplodiploidy 32, 47, 60
hay meadows 144, 160
hedgerows 134, 144-5, 154, 157, 159
Heinrich, Bernd 11, 78
Helianthus annuus 129, 133
Helicobia morionella 57
hibernation 4, 9-10, 22, 38, 59, 60, 61, 63, 65, 66, 67, 127, 128, 146, 149
Hirsutella 59
homing ability 74-6
honey stomach 19, 28, 56, 79
honeybee, *see Apis mellifera*
host range (of parasites and pathogens) 53, 57
host selection, by parasitoids 56-7
hoverflies, *see* Syrphidae
Hypoaspis 63

ideal free distribution 83, 84, 88
immune response 65-5
Impatiens capensis 171
inbreeding avoidance 45-6
inbreeding in plants 137-8, 153
incidence of infection by parasites 56, 61, 68
innate preferences 96, 110
inquilines, *see Psithyrus*
intensive farming 134-6, 146, 148, 157
introductions of bumblebees 53, 136, 161-75
Israel 162, 167
Italy 150-1

Jamaica 171
Japan 132, 150, 162, 169-70
juvenile hormone 21

kin selection 47
kiwifruit, *see Actinidia deliciosa*
Kuzinia laevis 63

Lamiastrum galeobdolon 30
Lamium album 30, 135, 155, 157
landmarks 41, 76, 77, 84, 85
Lanius excubitor 54
Lathyrus vernus 137
Lavandula stoechas 109
leaf-cutter ants, *see Atta*
learning 44, 69, 70, 96-9, 101, 105, 108-10, 113, 119-20
life cycle of bumblebees 4-10, 31, 68
Linaria vulgaris 86, 140
Locustacarus buchneri 63, 169
Lolium perenne 144
longevity 6, 28, 29, 56, 111
Lotus corniculatus 103-4, 120, 158
Lupinus 130, 133, 135
Lupinus arboreus 173, 174
Lycopersicon esculentum 129, 131, 132, 133, 134, 136, 162, 175
Lysandra bellargus 148
Lythrum salicaria 173

Mallophora bomboides 54
marginal value theorem 88-94
mark-recapture 28, 74, 75, 149
mate location 8, 10, 39, 41-6
matricide 35-6
Medicago sativa 129, 131, 133
Megachile pugnata 170
Megachile relativa 170
Megachile rotundata 162, 170
megachilidae 130, 162
Melastoma affine 171
Meles meles 6, 54
Melilotus 133
Melipona fasciata 73, 165
Meliponinae 1, 2, 164, 168
Melissococcus pluton 58
melon, *see Cucumis melo*
memory 69, 70, 76, 84, 85, 97, 99, 101, 105
Mephitis mephitis 54
Merops apiaster 54
microsatellites 35, 66, 150
Microtus 54

Mimicry
   automimicry 54
   Batesian 64
   Müllerian 3, 53, 67
Mimosa pudica 171
mink, *see Mustela vison*
minoring 105
Misumenia vatia 54
mites, *see* Acarina
mitochondrial markers 150
models
   foraging range 28, 78–9
   marginal mosaic 151
   optimality 73, 83, 84–91
mole, *see Talpa europea*
monogamy 8, 47, 51, 61
mouse, field, *see Apodemus sylvaticus*
multiple generations 9, 62
Muscicapa striata 54
musk melon, *see Cucumis melo*
Mustela nivalis 54
Mustela vison 54
mutilidae 58
Mutilla europaea 58
Myopa 55

navigation 76, 77, 84
nectar
   as food for larvae 5, 20, 23
   as fuel for incubation 7
   competition for 163–8, 172
   concentration 79
   gathering 1, 9, 24, 26, 27, 28, 29, 30, 70, 75, 78, 79–80, 83, 86, 96, 97, 107–8, 109, 111, 115–21, 123, 127, 157
   pesticides in 147
   rewards in flowers 88, 105, 107–8, 109, 115–21, 129, 155, 164
   robbing 124, 135, 138–41, 161, 171
   storage 4, 6, 7, 73, 111
nematode 58, 61–2, 169
Nepeta cataria 89
nest searching 4
nest sites 4, 65, 73, 127, 134, 135–6, 146, 157, 162, 169
   artificial 135–6, 158
nest size 6, 8, 81
Netherlands 132, 138, 169
New Zealand, bumblebees in 9, 53, 59, 62, 134–6, 140, 150, 161–2, 163, 169, 170, 173, 174

niche differentiation 123, 125, 127, 128, 163–4, 168
Niger, *see Guizotia abyssinica*
Nomia melanderi 162, 169
North America 2, 3, 11, 41, 53, 54, 55, 57, 58, 66, 127, 131, 132, 134, 139, 143, 162, 170, 171, 173, 175
Nosema apis 169
Nosema bombi 59, 60

odontobombus 3, 5, 22, 23
oilseed rape, *see Brassica napus*
Onobrychis 177
Onobrychis viciifolia 86–7
Ontario 125, 134, 172
Ophrys speculum 95
Orchidaceae 95
Organic farming 157–8, 159
Osmia cornifrons 58
Osmia rufa 153
outcrossing, of plants 137–8, 141, 153, 178
ovarian development
   of *Psithyrus* 66
   of queens 15, 57, 62
   of workers 31, 33, 35, 59, 67
overheating 14, 15, 16, 29

Paecilomyces farinosus 59
Paenibacillus larvae 58, 169
Panama 164, 168
parasite 37, 48, 51, 53, 58–64, 80, 153, 169–70
Parasitellus 63–4
Parasitellus fucorum 64
parasitoid 29, 31, 55–8, 64, 65, 111, 153
patch use when foraging 79–80, 84–94, 126
pathogens 48, 50, 58–9, 60, 64–5, 169–70, 174, 175
patrolling 8, 42–5, 46
Pedicularis palustris 140
pesticides 134, 145, 146–7, 154, 157, 171
Phacelia tanacetifolia 157
Phaseolus coccineus 140
Phaseolus lunatus 133
Phaseolus multifloris 133
phenology 4, 9, 65, 127–8
pheromones
   male-produced 8, 42–5
   queen-produced 21–2, 32–5, 46–7
   worker-produced 113
phoresis 63–4
Phylidonyris novaehollandiae 167
Physocephala 55

Physocephala rufipes 56
Physocephala texana 55
Phyteuma nigrum 137
pigeon, *see Columba livia*
Pneumolaelaps 63
pocket makers, *see Odontobombus*
Polemonium viscosum 111, 137
pollen
  as food for larvae 1, 4, 5, 6, 7, 20, 21, 23, 29, 69, 129, 157, 158
  as food for young queens 8
  gathering 1, 9, 24, 26, 27, 30, 66, 70, 79, 95, 96, 97, 107, 109, 127, 131, 132, 140, 158
pollen storers, *see Anodontobombus*
pollination 71, 73, 96, 109, 129–41, 153, 161, 162, 163, 170–5
pollination syndrome 129, 137
polyandry 8, 47, 51, 61
polyethism 24
population structure
  of bumblebees 148, 149–51
  of plants 71, 172
predation 6, 12, 25, 28–9, 47, 53–64, 80, 111,
Proctacanthus hinei 54
protandry 34, 37
Prunus avium 133
Prunus cerasus 133
Prunus communis 133
Prunus domestica 133
Psithyrus 3, 10, 32, 39, 42, 44, 45, 54, 55, 57, 61, 65, 66–8, 80
Pulmonaria 108
purple loosestrife, *see Lythrum salicaria*
Pyrobombus 3
Pyrus malus 131, 132, 133.

radar, harmonic 75, 77–9
range expansions 151, 162
recruitment of foragers 2, 87, 88, 113–14
relatedness, within nests 32, 33, 35, 47, 48, 50, 60–2, 66
resource partitioning 123, 125
Ribes 133
Ribes grossularia 133
risk aversion 105–6
robber fly, *see Asilidae*
Rubus ideaus 133
Rubus fruticosus 123

Salix 157
Salvia splendens 158
Sarcophaga 57
sarcophagidae 57

scent-marks 41, 42–5, 46, 80, 85, 115–21
Scutacarus acarorum 63
search image 99–104
search times, for flowers 28, 90, 91–2, 94, 103–4
seed set 129, 137, 138, 140–1, 153, 161, 171–5
selective attention 101
Senotainia tricuspis 57
set-aside 154, 155, 157, 159
sex determination 8, 30–1, 47
sex determining locus 31
sex ratios 35, 36–9, 48
shrew, *see Sorex*
Sicus ferrugineus 29, 56
size variation of workers 5, 22, 23, 24, 28, 29
skunks, *see Mephitis mephitis*
social behaviour, evolution of 1
social parasitism 65
Solanum melongena 133
Solidago canadensis 125
Sorex 54
South Africa 95, 136, 162, 175
South America 2, 9, 161, 171
Spain 150, 151
specialisation, behavioural 24–5, 28, 30, 105
species richness 2
sperm competition 48
sperm plugs 49, 51, 61
Sphaerularia bombi 58, 61–2
spiders 29, 54
Spiroplasma melliferum 59
Spiroplasmataceae 58
spotted flycatcher, *see Muscicapa striata*
strawberry, *see Fragaria x ananassa* 130
sunflower, *see Helianthus annuus*
swarming 1, 10
sweden 42, 57, 137, 151, 158
sweet pepper, *see Capsicum*
switching between flower species 93, 97, 99, 102, 105–6
Symphytum officinale 117
Syntretus splendidus 57
Syrphidae 64, 97, 105, 108, 154
systematic searching 76, 84–6, 88–9, 91, 94, 121

Talpa europea 54
Taraxacum officinale 155
Tasmania, bumblebees in 136, 140, 150, 162, 163, 164, 165, 167, 171, 172, 173, 174, 175
taxonomy 3
territoriality 41–2, 46
thermogenesis 11, 12, 13, 126, 127

thermoregulation
  of bumblebees 11–15, 17, 29–30, 56, 126–7, 131, 137, 165, 166
  of nest 4–6, 7, 15–16, 23, 24
Thoracobombus 4, 81
thresholds, behavioural 16–17, 24, 93, 119, 120
Tineidae 64
tomato, *see Lycopersicon esculentum*
tongue length 28, 30, 123–8, 131, 135, 136, 139, 140, 145, 152–3, 155, 156, 158, 165–6, 173
transmission, of pathogens 58, 59, 169
trap-lining 42, 84–5
tree lupin, *see Lupinus arboreus*
Trifolium 86, 129, 133, 145
Trifolium medium 158
Trifolium pratense 131, 135, 140, 157, 159, 161
Trifolium repens 90, 92
Trigona prisca 2
trigonini 73, 165
tropical bumblebees 1, 2, 9, 50

Ulex europeaus 173
unimproved grassland 144–5, 146, 148, 158, 159
United Kingdom 9, 28, 57, 59, 67, 80, 86, 123, 127, 131, 135, 136, 139, 143, 144, 145, 147, 148, 149, 150, 151, 154, 155, 156–9, 161, 162, 169
United States of America 130, 131, 134, 135, 162
Urtica dioica 156
usurpation of nests 65–7

Vaccinium 131, 147
Vaccinium angustifolium 133, 134

Vaccinium ashei 133, 171
Vaccinium corymbosum 133, 140
Vaccinium macrocarpon 133
Varroa destructor 131, 169, 170
Verticilium lecanii 59
Vicia 155
Vicia cracca 26, 103–4
Vicia faba 130, 131, 132, 133, 135, 140, 147, 153
Vicia villosa 133
virus, entomopox 58
vision 102, 104
visual acuity 28
Vitula edmandsii 64
vole 54
Volucella bombylans 64
Vulpes vulpes 54

Wahlenbergia 168
water hyacinth, *see Eichhornia crassipes*
watermelon, *see Citrullus lanatus*
wax 4–5, 7, 16, 23, 24, 63, 64, 66, 103
wax moth, *see Aphomia sociella*
weasel, *see Mustela nivalis*
weeds 146, 147, 155, 156, 162, 172–4, 175
wildflower seed mixes 154, 156–7

Xylocopa 116, 175
Xylocopa californica 170
Xylocopa varipuncta 43
Xylocopa virginica 116

yellow star thistle, *see Centaurea solstitialis*
Yucca 95